Emerging Technologies for Textile Coloration

Emerging Materials and Technologies

Series Editor
Boris I. Kharissov

2D Materials for Surface Plasmon Resonance-Based Sensors
Sanjeev Kumar Raghuwanshi, Santosh Kumar, and Yadvendra Singh

Functional Nanomaterials for Regenerative Tissue Medicines
Mariappan Rajan

Uncertainty Quantification of Stochastic Defects in Materials
Liu Chu

Recycling of Plastics, Metals, and Their Composites
R.A. Ilyas, S.M. Sapuan, and Emin Bayraktar

Viral and Antiviral Nanomaterials
Synthesis, Properties, Characterization, and Application
Devarajan Thangadurai, Saher Islam, and Charles Oluwaseun Adetunji

Drug Delivery Using Nanomaterials
Yasser Shahzad, Syed A.A. Rizvi, Abid Mehmood Yousaf, and Talib Hussain

Nanomaterials for Environmental Applications
Mohamed Abou El-Fetouh Barakat and Rajeev Kumar

Nanotechnology for Smart Concrete
Ghasan Fahim Huseien, Nur Hafizah A. Khalid, and Jahangir Mirza

Nanomaterials in the Battle against Pathogens and Disease Vectors
Kaushik Pal and Tean Zaheer

MXene-Based Photocatalysts: Fabrication and Applications
Zuzeng Qin, Tongming Su, and Hongbing Ji

Advanced Electrochemical Materials in Energy Conversion and Storage
Junbo Hou

Emerging Technologies for Textile Coloration
Mohd Yusuf and Shahid Mohammad

Emerging Pollutant Treatment in Wastewater
S.K. Nataraj

For more information about this series, please visit: https://www.routledge.com/Emerging-Materials-and-Technologies/book-series/CRCEMT

Emerging Technologies for Textile Coloration

Edited by
Mohd Yusuf and Mohammad Shahid

CRC Press is an imprint of the
Taylor & Francis Group, an **informa** business

First edition published 2022
by CRC Press
6000 Broken Sound Parkway NW, Suite 300, Boca Raton, FL 33487-2742

and by CRC Press
4 Park Square, Milton Park, Abingdon, Oxon, OX14 4RN

© 2022 Taylor & Francis Group, LLC

CRC Press is an imprint of Taylor & Francis Group, LLC

Reasonable efforts have been made to publish reliable data and information, but the author and publisher cannot assume responsibility for the validity of all materials or the consequences of their use. The authors and publishers have attempted to trace the copyright holders of all material reproduced in this publication and apologize to copyright holders if permission to publish in this form has not been obtained. If any copyright material has not been acknowledged, please write and let us know so we may rectify in any future reprint.

Except as permitted under U.S. Copyright Law, no part of this book may be reprinted, reproduced, transmitted, or utilized in any form by any electronic, mechanical, or other means, now known or hereafter invented, including photocopying, microfilming, and recording, or in any information storage or retrieval system, without written permission from the publishers.

For permission to photocopy or use material electronically from this work, access www.copyright.com or contact the Copyright Clearance Center, Inc. (CCC), 222 Rosewood Drive, Danvers, MA 01923, 978-750-8400. For works that are not available on CCC, please contact mpkbookspermissions@tandf.co.uk

Trademark notice: Product or corporate names may be trademarks or registered trademarks and are used only for identification and explanation without intent to infringe.

Library of Congress Cataloging-in-Publication Data
Names: Yusuf, Mohd, editor. | Shahid, Mohammad (Of Institute of Chemical Technology), editor.
Title: Emerging technologies for textile coloration / edited by Mohd Yusuf and Mohammad Shahid.
Description: First edition. | Boca Raton, FL : CRC Press, 2022. | Series: Emerging materials and technologies | Includes bibliographical references and index. | Summary: "This book features perspectives on advances in textile coloration technologies. It provides a comprehensive and holistic overview, supporting rapid and efficient entry of new researchers into emerging subjects within textile engineering and technology. Written for academicians, scientists, researchers, and advanced students of textile science and technology, this work aims to provide depth of understanding of both state-of-the-art and emergent topics and to spur further research leading to new opportunities and applications"— Provided by publisher.
Identifiers: LCCN 2021047607 (print) | LCCN 2021047608 (ebook) | ISBN 9780367691110 (hbk) | ISBN 9780367691127 (pbk) | ISBN 9781003140467 (ebk)
Subjects: LCSH: Dyes and dyeing—Textile fibers.
Classification: LCC TP897 .E54 2022 (print) | LCC TP897 (ebook) | DDC 667/.2—dc23/eng/20211209
LC record available at https://lccn.loc.gov/2021047607
LC ebook record available at https://lccn.loc.gov/2021047608

ISBN: 978-0-367-69111-0 (hbk)
ISBN: 978-0-367-69112-7 (pbk)
ISBN: 978-1-003-14046-7 (ebk)

DOI: 10.1201/9781003140467

Typeset in Times
by codeMantra

Contents

Preface vii
Acknowledgment ix
Editors xi
Contributors xiii

Chapter 1 Textile Coloration Technologies: Today and Tomorrow 1

    Saptarshi Maiti, Akankshya Panda, Pallavi Madiwale, and Ravindra V. Adivarekar

Chapter 2 Sustainable Approaches to Coloration of Textile Materials.............. 13

    Shanmugasundaram O. Lakshmanan, Wazed Ali, and Swagata Banerjee

Chapter 3 Aspects of Mordant Dyeing of Textile Fibers 37

    Supriyo Chakraborty and Lipika Chakraborty

Chapter 4 Insights into Electrochemical Technology for Dyeing of Textiles with Future Prospects 57

    Amit Madhu

Chapter 5 Application of Microwave Irradiation in Coloration of Textiles........ 89

    Aminoddin Haji

Chapter 6 Functional Value-Added Finishing of Textile Substrates Using Nanotechnology: A Review 101

    Ahmad Faraz, Mohammad Faizan, and Shamsul Hayat

Chapter 7 Advanced Use of Nanomaterials in Coloration and Discoloration Processes 111

    Faten Hassan Hassan Abdellatif and Hend M. Ahmed

Chapter 8 Nanotechnology-Enabled Approaches for Textile Applications: Solutions to Eliminate the Negative Impacts of Dyeing Materials and Processes 133

    Azadeh Bashari and Mina Shakeri

Chapter 9 Gamma Ray Irradiation Technology for Textile
Surface Treatment/Modification: An Overview 157

*Mohd Gulfishan, Rayees Afzal Mir, Syed Aasif Hussain,
Sajjad Khan, Fatima Shabir, and Eyram Khan*

Chapter 10 Overview on Significant Approaches to Waterless
Dyeing Technology 171

K. Murugesh Babu

Chapter 11 Nonaqueous Dyeing of Textile Materials 189

*Sushant S. Pawar, Mohammad Shahid, and
Ravindra V. Adivarekar*

Chapter 12 Plasma Technology for Textile Coloration: Solutions for Green
Dyeing of Textiles 203

Mina Shakeri and Azadeh Bashari

Chapter 13 Plasma Technology in Processing of Textile Materials 217

D. Saravanan and M. Gopalakrishnan

Chapter 14 State of the Art Colorization Techniques in Textile Industry 233

Aleem Ali, Farah Jamal Ansari, and Anushka Dhiman

Index 247

Preface

Global environmental concerns have posed significant hurdles to traditional textile coloring techniques. As a response, industrial R&D has been exploring more sustainable materials with more eco-friendly as well as compatible technologies in the textile wet processing chain in recent years, resulting in a variety of innovative strategies for sustainable perspectives in this sector.

This book is a comprehensive literature exploration of the state-of-the-art and recent practices in the field of textile coloration technologies. The book discusses the fragmented situation as well as new and emerging textile technologies, providing a comprehensive reference material for readers at all levels, including academicians, scientists, and researchers, as well as postgraduate students in the textile science and technology stream.

The book will hopefully be a useful tool for gaining an overview of the multidisciplinary dimension of textile coloration through technological perspectives in order to facilitate further exploration of alternative opportunities.

The book is divided into 14 chapters that describe new and emerging technologies such as industrial innovation in coloration technologies, principles, and applications of advanced and sustainable technologies, electrochemical technology, plasma technology, ultrasonic and microwave technology, waterless coloration technology, etc. We are delighted to introduce and dedicate this volume to the learners as well as teachers of the textile coloration technological community. We welcome our readers' insightful feedback and suggestions, which will aid in the development of future volumes.

Mohd Yusuf
Mohammad Shahid
August 2021

Acknowledgment

Foremost of all, we humble ourselves before Almighty God for giving us the strength and courage that was quintessential in undertaking and accomplishing an endeavor like this.

We owe a debt of gratitude to our esteemed authors/contributors for their valuable and irreplaceable contributions to the theme's coverage. Our deep sense of gratitude goes to our research guide, Dr. F. Mohammad, Ex-Professor, Department of Chemistry, Faculty of Natural Sciences, Jamia Millia Islamia, New Delhi, India, for being a source of inspiration through his positive strokes. We acknowledge Dr. S. A. Ahmed, Vice-Chancellor, and Dr. Rayees Mir, Research Director of Glocal University, Saharanpur, and Prof. R.V. Adivarekar, Professor & Head, Department of Fibres and Textile Processing Technology, Institute of Chemical Technology, Mumbai, for their moral support and critical remarks. Mohammad Shahid acknowledges funding through the UGC DSKPDF scheme (F.4-2/2006 (BSR)/CH/18-19/0074), University Grants Commission (UGC), Govt. of India.

In particular, we would like to express our sincere thanks to the CRC Press, especially Allison Shatkin, series editor, and Gabrielle Vernachio, editorial assistant, Taylor & Francis Group, who offered us suggestive support throughout the process of compiling this volume.

Editors

Mohd Yusuf, MSc, PhD, earned an MSc in chemistry in 2006 at CCS University, Meerut, UP, India. Currently, he is an Associate Professor and Head in the Department of Natural and Applied Sciences at Glocal University, Saharanpur, UP, India. He earned a PhD at Jamia Millia Islamia (A Central University), New Delhi, India. During his R&D career, he published more than 50 research articles, reviews, books, and chapters in edited volumes.

Mohammad Shahid, MSc, PhD, is a Dr. D.S. Kothari Fellow at the Institute of Chemical Technology, Mumbai, India. He earned a PhD in organic chemistry at Jamia Millia Islamia, New Delhi, India. He has contributed more than 40 publications and is an active reviewer for several leading journals. His publications have attracted more than 2400 citations. Dr. Shahid has received several prestigious fellowships for his doctoral (Central University PhD fellowship, BSR Junior and Senior research fellowship) and postdoctoral research (Marie-Curie Individual Fellowship, China Postdoctoral Fellowship, Dr. D.S. Kothari Fellowship).

Contributors

Faten H.H. Abdellatif
Pre-treatment and Finishing of Cellulosic Fabric Department
National Research Centre
Dokki, Giza, Egypt

Ravindra V. Adivarekar
Department of Fibres and Textile Processing Technology
Institute of Chemical Technology
Mumbai, India

Hend M. Ahmed
Pre-treatment and Finishing of Cellulosic Fabric Department
National Research Centre
Dokki, Giza, Egypt

Aleem Ali
Department of Computer Science & Engg.
Glocal University
Saharanpur, India

Wazed Ali
Department of Textile and Fibre Engineering
Indian Institute of Technology, Delhi
New Delhi, India

Farah Jamal Ansari
Section of Computer Engineering
University Polytechnic, Jamia Millia Islamia
New Delhi, India

K. Murugesh Babu
Department of Textile Technology and Research Centre
Bapuji Institute of Engineering and Technology
Davangere, India

Swagata Banerjee
Department of Textile and Fibre Engineering
Indian Institute of Technology Delhi
New Delhi, India

Azadeh Bashari
Textile Engineering Department
Amirkabir University of Technology
Industrialization Center for Applied Nanotechnology (ICAN)
Tehran, Iran

Lipika Chakraborty
Government Polytechnic
Nagpur, India

Supriyo Chakraborty
Uttar Pradesh Textile Technology Institute
Kanpur, India

Anushka Dhiman
Department of Computer Science
Delhi University
New Delhi, India

Mohammad Faizan
College of Forest Resources and Environment
Nanjing Forestry University
Nanjing, P.R. China

Ahmad Faraz
School of Life Sciences
Glocal University
Saharanpur, India

M. Gopalakrishnan
Department of Textile Technology
Bannari Amman Institute of
  Technology
Sathyamangalam, India

Mohd Gulfishan
School of Agricultural Sciences
Glocal University
Saharanpur, India

Aminoddin Haji
Textile Engineering Department
Yazd University
Yazd, Iran

Shamsul Hayat
Department of Botany
Aligarh Muslim University
Aligarh, India

Syed Aasif Hussain
School of Agricultural Sciences
Glocal University
Saharanpur, India

Eyram Khan
School of Agricultural Sciences
Glocal University
Saharanpur, India

Sajjad Khan
School of Agricultural Sciences
Glocal University
Saharanpur, India

Shanmugasundaram O. Lakshmanan
Department of Textile Technology
K. S. Rangasamy College of Technology
Tiruchengode, India

Amit Madhu
Textile Chemistry Department
The Technological Institute of Textile
  and Sciences
Bhiwani, India

Pallavi Madiwale
Department of Fibres and Textile
  Processing Technology
Institute of Chemical Technology
Mumbai, India

Saptarshi Maiti
Department of Fibres and Textile
  Processing Technology
Institute of Chemical Technology
Mumbai, India

Rayees Afzal Mir
School of Agricultural Sciences
Glocal University
Saharanpur, India

Akankshya Panda
Department of Fibres and Textile
  Processing Technology
Institute of Chemical Technology
Mumbai, India

Sushant S. Pawar
Department of Fibres and Textile
  Processing Technology
Institute of Chemical Technology
Mumbai, India

D. Saravanan
Department of Textile Technology
Kumaraguru College of Technology
Coimbatore, India

Fatima Shabir
School of Agricultural Sciences
Glocal University
Saharanpur, India

Mohammad Shahid
Department of Fibres and Textile
    Processing Technology
Institute of Chemical Technology
Mumbai, India

Mina Shakeri
Department of Materials Engineering
Tarbiat Modares University
Industrialization Center for Applied
    Nanotechnology (ICAN)
Tehran, Iran

1 Textile Coloration Technologies
Today and Tomorrow

*Saptarshi Maiti, Akankshya Panda,
Pallavi Madiwale, and Ravindra V. Adivarekar*
Institute of Chemical Technology

CONTENTS

1.1 Introduction 1
1.2 Need of the Hour: Environmental Awareness 2
1.3 Drive Toward Sustainable Technologies 2
    1.3.1 Natural Dyes 3
    1.3.2 Plasma Technology 4
    1.3.3 Spin-Dyeing 4
    1.3.4 Supercritical Fluid Dyeing 4
    1.3.5 Electrochemical Dyeing 5
    1.3.6 Microwave-Assisted Dyeing 5
    1.3.7 Ultrasonic Dyeing 6
    1.3.8 Nanotechnology 7
    1.3.9 Air-Dyeing 7
    1.3.10 Microbial Colorants 8
    1.3.11 Foam Dyeing 8
    1.3.12 Digital Printing 9
1.4 Future Perspective 9
1.5 Conclusion 10
References 10

1.1 INTRODUCTION

Textile coloration is considered to be one of the most prime sectors of the textile industry. However, it also shows a negative impact on the environment. Each year, a huge quantity of water is consumed for the coloration and after-treatments of textile goods [1]. Many times, poisonous unfixed colorants (dyes), chemicals, and auxiliaries are unable to get fully removed by wastewater treatment, thereby resulting in the generation of a large quantity of wastewater. If the wastewater doesn't undergo proper after-treatment, there remains a possibility that the ecosystem gets affected, thereby posing a risk and damage to our health. With the advent of environmental awareness of the need for conservation and building sustainability, an ecologically acceptable coloration is urgently required. To promote environmentally friendly textile coloration, we all are aiming for novel processes, process designs and treatments, colorants, chemicals and auxiliaries, and reuse of wastes.

A textile dye is one type of colorant which is primarily used to impart color to any textile material. It may be natural or synthetic or may be organic or inorganic. Textile dye is a very important component of the whole garment processing sector. Globally, the market for textile dyes is predicted to reach USD 10.13 billion by 2026 [2]. For the dyeing of textiles, numerous techniques are followed [3]. The process depends on various characteristics such as the type of material, i.e., fiber/yarn/fabric, fabric design, etc. The traditional process of textile coloration consumes a huge quantity of fresh water. Presently, the textile industry ranks third in the list of a consumer of water worldwide. On average, nearly 100–150 L of water are required to process 1 kg of textile material, out of which 20%–25% of water is used overall in the entire process of dyeing [4,5].

1.2 NEED OF THE HOUR: ENVIRONMENTAL AWARENESS

Rising environmental awareness in the textile processing sector has driven efforts in research and development to find out safety in the chemical processing of textiles. Textile processing is a combination of numerous processes for the removal of natural impurities, coloration, and suitable finishing. Due to the involvement of a huge quantity of energy as well as large use of water and subsequent rise of pollution, such methods are neither eco-friendly nor economical. Toxicological health issues have led to the advent of environmentally friendly chemical processing methods.

Textile coloration is the leading producer of wastewater as well as consumer of pollutant chemicals and water. On average worldwide, the total consumption of only textile dyes is not less than 10^7 kg/year. If a maximum of 90% of dye enters into the textiles, there would be a discharge of nearly 10^6 kg/year of dyes into the waste streams by the textile industry. It is predicted that 280,000 t/year of textile dyes get discharged into industrial effluents globally [3,6]. The auxiliaries used along with the dyes vary in chemical composition, not limited to only simple inorganic and organic compounds but also complex polymers [7]. The illustration for the route of pollution in the textile industry is expressed in Figure 1.1, which illustrates a crystal idea of the consumption and emissions that cause ecological concerns.

1.3 DRIVE TOWARD SUSTAINABLE TECHNOLOGIES

Novel dyes and improved dyeing methods of different fibers used as clothing and industrial materials have been started to get followed due to ecological concerns. Thereby, herewith, an attempt has been made to undergo a review on the environmentally friendly techniques for the coloration of textiles by biotechnology, supercritical carbon dioxide, ultrasonic, microwave, plasma, etc., to improve the definite properties without harming the environment as well as the ecological system.

Textile Coloration Technologies

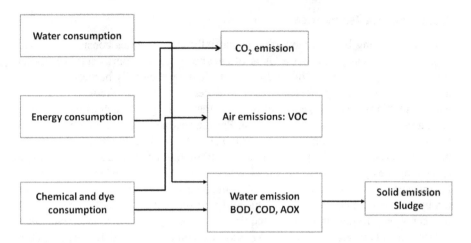

FIGURE 1.1 Pollution route in the textile processing sector.

1.3.1 Natural Dyes

Natural dyes have become popular owing to their coloration of wood, leather, food substrate as well as natural fibers such as cotton, jute, flax, wool, and silk for ages. Natural dyes are characterized by a variety of shades and can be obtained from different parts of plants such as fruit, flowers, roots, bark, and leaves (Table 1.1). Currently, there has been an upsurge in the rising interest for the use of natural dyes for natural fibers due to global environmental concerns. The use of eco-friendly, nontoxic natural dyes on textiles has become a prime concern because of the increased ecological concern to avoid some dangerous synthetic dyestuffs. Many of the natural dyes do not possess substantivity for textile fibers without the use of an external agent called mordant. They require a mordanting agent (normally complex-forming agents or metal salt) to build up substantivity between them and fiber. Natural dyes are eco-friendly as they are biodegradable and renewable, and nonhazardous to the skin, and sometimes offer some benefits to the health of the wearer [8,9].

TABLE 1.1
Common Natural Dyes of Vegetable Origin

Plant Part	Dyestuff
Seeds/Fruits	Latkan, myrobalan, beetle nut, pomegranate rind
Flowers	Kusum, marigold
Leaf	Lemongrass, coral jasmine, cardamom, tea, eucalyptus, henna, indigo
Branches/Bark	Sandalwood, khair, shillicorai, sappan wood, purple bark
Root	Beetroot, onion, madder, turmeric

1.3.2 Plasma Technology

Plasma is nothing but a combination of partially ionized gases consisting of electrons, ions, radicals, and atoms. It is often known as the fourth state of matter. Interestingly, plasma exists in different forms and can be easily harnessed to study and possibly meet the preferred results on different materials like textiles. Plasma used in textile processing generally involves the use of gases such as helium, fluorine, argon, nitrogen, oxygen, or air [10]. Plasma treatment is generally practiced in the textile processes to modify the textile surfaces. Desired functional groups can be easily introduced in the plasma-treated textiles, and they play a significant role in enhancing the dye uptake of hydrophobic fibers such as polypropylene and polyester. Postplasma treatment of such inert fibers could be taken up easily for appropriate dyeing by the dyes which are water-soluble and environmentally friendly. This technology is not only limited to hydrophobic fibers but also applicable for the treatment of hygroscopic fibers; e.g., in wool dyeing, plasma could be employed. Low-temperature plasma treatment on the wool fiber helps in lowering the dyeing temperature, thereby reducing the damage on the wool fiber [11]. Moreover, as this technology does not involve harmful chemicals, it contributes to building the steps of sustainability.

1.3.3 Spin-Dyeing

Dyeing of textiles is one such process that generally shows a negative impact on the environment through the industry. Thus, reducing actions such as the load of the environment from the process of dyeing is therefore very much important. According to a Swedish company named '*We are spin dye*', working with spin-dye technology, the spin-dyeing process is one kind of dyeing process that takes the water out of the process. The process is very simple, where the dyeing agents are mixed in the solution of spinning. On the contrary to the traditional dyeing process, spin-dye technology combines spinning and dyeing, which helps in the reduction of the number of processing steps. Such technology can be followed to achieve any color. Presently, the technology is used only for colors with high market volumes (i.e., dark blue, red, black, and brown). This technology has become successful for cellulosic and noncellulosic fibers [12].

1.3.4 Supercritical Fluid Dyeing

A supercritical fluid is a new state of compressible matter and behaves like a gas (filling up and taking any container shape), unlikely of the state of liquid (a fluid which is incompressible occupying the bottom of any container). However, a supercritical fluid possesses a density of a liquid and has high dissolving power. The supercritical state is achieved when the pressure and temperature cross the critical point. When the critical region is achieved, a substance that is usually a gas at normal conditions exhibits high solvent capacity and liquid-like density. Such behavior occurs because an increase in density reduces average intermolecular distance, thereby resulting in increased interaction between the solute and solvent. Further rise in pressure will heavily enhance the system's dielectric constant, thereby imparting dissolution power to the system. A supercritical fluid possesses the properties of both liquid and gas. Such a unique combination of liquid-like density and gas-like viscosity creates an excellent solvent. The fluid density can be easily tuned by minute changes in pressure. Supercritical fluids (SFCs) can dissolve nonpolar materials, making them very much useful for textile coloration like polyester dyeing.

Carbon dioxide (CO_2) is widely produced commercially and is mainly used as a fluid for supercritical owing to its nontoxicity, non-hazardousness, and non-corrosiveness. Supercritical fluids can easily dissolve organic hydrophobic molecules like disperse dyes. It is therefore used as a medium of dyeing for polyester dyeing [13]. It carries dual functionalities in the dyeing process, i.e., substrate heating and transportation of the dye molecules [14]. After crossing the critical point, CO_2 shows the properties of liquid which is advantageous for the easy dissolution of insoluble disperse dyes; low diffusion properties and low viscosity of gas lower the time required for dyeing. No separate drying step is required since CO_2 gets released in a gaseous state [15].

1.3.5 Electrochemical Dyeing

Sulfur and vat dyes are commonly used for cellulosic textiles. Both of them require complicated dyeing steps that involve complex reduction and oxidation. Such insoluble dyes need to get reduced to water-soluble form to get absorbed onto the fiber, followed by oxidation to the water-insoluble form for fixation [16]. The most commonly used reducing agent for vat dyes is sodium hydrosulfite (hydrose), and sulfur dyes is sodium sulfide [17]. The stability of such reducing agents is poor, which results in the decomposition of several by-products remaining in the dye bath that contains sulfur. Sulfur is of great concern for the environment owing to its water and air toxicity. It results in higher COD, thereby increasing the effluent load.

An efficient technique for the reduction and oxidation of dyes is electrochemical dyeing. This technique involves the use of electrons from the electric current [14,17]. Electrochemical dyeing is of two types: indirect or direct [17]. Vat dyes are generally reduced by indirect electrolysis technique, whereas sulfur dyes get reduced with direct electrolysis. In the case of direct electrolysis, dye reduction takes place at the cathode surface. However, in indirect electrolysis, there is a continuous regeneration of the reducing agent at the cathode. This technique can create a possibility in the reuse of dye baths along with the reducing agent.

1.3.6 Microwave-Assisted Dyeing

Microwave frequencies lie in between the radio wave and infrared radiation of the electromagnetic spectrum. The ranges of frequency and wavelength are 300 MHz to 300 GHz and 1 m to 1 mm, respectively. Heating by microwave must be considered as a dielectric-cheating form. Under the effect of an alternating field, the dipole molecules show oscillations against the changes in polarity of the field, which has a high frequency. The microwave energy of a photon is very low as compared to the energies of a chemical bond. Thereby, microwaves do not directly affect any compound's molecular structure without altering the electronic configuration of atoms. Microwave heating is better than conventional heating because it is uniform, effective, controlled, and rapid heating. This energy can easily penetrate. Microwave heating is different from traditional heating. It involves the diffusion of heat into the media from the material's surface. Microwave energy which can be directly and internally absorbed by the textile materials, gets converted into heat. Microwave heating increases the dye molecules' diffusion in the polymers, which can enhance the rate of fixation of dyes into polymeric textiles.

Microwave dyeing can consider only the thermal and dielectric properties. The dielectric property is nothing but the intrinsic electrical properties which affect the dyeing by dipolar dye rotation and influence the microwave field upon the dipoles [18]. For example, the aqueous dye solution has two polar components in the high-frequency microwave field oscillating at around 2450 MHz. It can have an impact on the vibration energy in the dye and water molecules. Here, ionic conduction takes place, which is in the form of resistance heating. The ionic acceleration through the dye solution can result in the collision of the fiber molecules with the dye molecules. The mordant enhances the penetration of the dye, thereby improving the color strength of the microwave dyed material. Microwave-promoted reactions are eco-friendly benign steps for accelerated chemical processes. This technology can reduce the energy required and the reaction time as compared to the traditional dyeing conditions. The microwave technology of heating is a volumetric phenomenon which is why it is rapid; however, the traditional heating technology is a surface phenomenon because of which it is comparatively slow, as shown in Figure 1.2.

1.3.7 Ultrasonic Dyeing

Ultrasonic wave is characterized by a frequency of about 20 kHz and a solution for the textile processing sector as it is eco-friendly, could be economical, and maintain good quality [14,16]. Ultrasonic energy is a combination of heating and cavitation [19]. Cavitation involves the formation and collapse of bubbles and ultrasound, which improves the transport of molecules toward the fiber along with the increase in the reaction rate. A lot of experiments have been carried out over the last 20 years relating to the use of ultrasonic energy for dyeing and washing. Ultrasonic technology

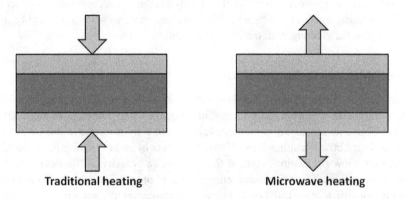

FIGURE 1.2 Traditional heating (surface) 'vs.' microwave heating (volume).

has been studied in several research works to enhance the efficiency of textile dyeing [20]. However, despite delivering good results on the laboratory scale, this technology has not yet been implemented in the industries [21]. Bi-factors responsible for this lack of commercialization are short of sufficient knowledge regarding the ultrasonic mechanism of the transfer of mass in textiles and the disadvantages found in the ultrasonic processors. However, in the near future, we all hope that the interest in ultrasonic-assisted dyeing will get increased especially in small-scale enterprises.

1.3.8 NANOTECHNOLOGY

The nanotechnology concept started more than 40 years ago. It can be best described as atomic and molecular applications in the real world within the dimensions of 1–100 nm. Globally, it is widely gaining popularity owing to its uniqueness and vast potential for different end-use applications. This technology has a real potential for the textile industry on a commercial scale. Many times, conventional processes imparting various properties to textile materials are unable to produce permanent effects. However, nanotechnology provides high durability because nanoparticles possess high surface energy and a large surface area to volume ratio, thereby increasing the affinity for textiles. Moreover, the hand feel or breathability of a fabric is unaffected by such nanoparticles. Therefore, the popularity of nanotechnological applications in the textile industry is increasing [22].

In textiles, nanotechnology was first undertaken by a US company: Nano-Tex [23]. Later, several textile companies started investing in nanotechnological developments. The coating is the most widely used technique for the application of nanoparticles onto textiles. Various coating methods can be followed that include padding, sublimation transfer printing, spraying, rinsing, and washing. In the case of padding, the nanoparticles get attached to the fabrics with the help of a padder at a particular speed and pressure, followed by drying and curing. Along with the dyeing properties, the other properties that can be achieved by nanotechnology include flame retardancy, UV-protection, antistatic, antimicrobial, wrinkle resistance, soil resistance, water and oil repellency, etc.

1.3.9 AIR-DYEING

"Colorep", a California-based company, has proposed a novel uprising airflow process of dyeing [24]. The liquor to dyeing is first atomized, followed by high airflow pressure, and finally sprayed on the fabric for dyeing in an airflow dyeing machine. In this technology, water acts as a dyeing liquor solvent. So, only a very small amount of water is consumed. It does not pollute water and uses air instead of water for the dye movement, there is no emission of hazardous waste, and most importantly, there is no waste of water [25].

In the last few years, waterless air-dyeing technology has been developed by two companies. These are Colorzenand AirDye® named American enterprises. AirDye® consumes 95% less water, 84% fewer greenhouse gases, and 86% less energy in comparison to traditional dyeing. The major benefits of air dye include less water consumption as well as less amount of hazardous waste emission [25]. However, this technology is only applicable to synthetic fibers but not natural fibers. ColorZen® is a technology that creates a revolution by making cotton dyeing more effective and eco-friendly. Their advanced technology had shown 95% fewer chemicals, 90% less water, and 75% less energy, along with zero discharge. This air-dyeing technology is comparatively novel, and the cost of the machine required for it is not economical. Therefore, the development of simple and economical approaches to air-dye coloration could be the next level of the researcher's future interest in textile dyeing.

1.3.10 Microbial Colorants

Due to low yields, seasonal production, and species extinction as a result of overexploitation, many plant and animal-derived colorants are unsustainable. Excessive use of these resources may endanger biodiversity. This has been a major issue that led to the growth of ever-increasing interest in microbial products. Many of the microbial dyes inherently possess antimicrobial properties. So, the dyeing of textiles with such microbial dyes can impart antimicrobial properties apart from the coloration effect.

Colorant-producing microorganisms are mainly found in varied habitats like industrial wastes, milk, air, fresh and marine water, soil, etc. After screening and isolation of the desired colorant-producing microorganism, the next step is the mass production of the colorant. However, since the concept of using microbial colorants for textiles is not so old and still under the research stage, the production of microbial colorants happens using the liquid fermentation technique that has been submerged. Fermentation is further followed by the colorant extraction from

the microorganism. Extraction of solvent is mostly followed because of the insolubility of dyes in water, and the colorant extract is used as a microbial dye for dyeing textile materials. The exhaust dyeing technique has been mostly reported for the dyeing of natural and synthetic fibers with microbial colorants [26].

Very few research works dealing with the dyeing of textiles have been reported. Our research group at ICT, Mumbai, in collaboration with Bhavans college Microbiology department, has reported dyeing properties of a bluish violet colorant from Chromobacterium species on cotton, wool, and silk fabrics and rosy red colorant from Micrococcus species on wool and silk [27,28].

1.3.11 Foam Dyeing

As the name suggests, the main dyeing element is foam which is obtained from an aqueous solution and then spread on a fabric. The foaming agent is used in the process. When the fabric gets covered with foam, a high temperature is applied for good dye fixation.

Foam is a dispersion of a gas in a liquid. The liquid is considered to be mostly water, whereas the gas is generally air or sometimes inert gas. This technology has an additional advantage over other dyeing methods, as foam dyeing can be done at very low wet pick-ups. The foaming agent generally used for foam dyeing must readily produce foam with a good wetting capability. Moreover, such foaming agents should be compatible with other dye bath ingredients and show rapid uniform wetting action with little or no effect on colorfastness. Most importantly, it should not be the cause of the yellowing of white materials and be well enough to form various bubble sizes. In the past few years, different forms of foam have been used for dyeing on the surface as well as producing uniform penetration or distribution of the color in the interior of the fabric or yarn [29].

Like other processes, the technology of foam dyeing is not so much critical to the environment. The benefits of the foam dyeing process are good dimensional stability, shorter dyeing times, high yield of color, enhanced dye prefixation, and migration of the dye into the fiber. This technology has been claimed to be a very attractive and economical technology to reduce energy consumption in several countries, including South America and the United States.

1.3.12 Digital Printing

Traditional printing technologies are not environmentally friendly. Concerning this, new technologies for textile printing have emerged, like digital printing. Digital printing, popularly known as inkjet printing, is similar to an office printer [30]. This technology provides high flexibility for printing. Here, the digital image is produced by drops of ink that get injected out of the nozzles of printing onto the fabric [14]. It was first applied for carpets, flags banners, and other niche products. In the last 15 years, there has been a major development of digital printing that has resulted in the rise of improvement of inks or commercial machines [31]. Today, it has a broader application in the textile industry worldwide.

Digital printing is applicable irrespective of fibers. Most of the fabrics require pretreatment before getting digitally printed. The extent of pretreatment required depends on the dye used. Plasma pretreatment has proven to be an efficient pretreatment method for pigment-printed textiles [32]. However, digital printing is mostly followed in the case of pigments, and the main advantage is that washing can be eliminated. It signifies that residues and wastewater can be well avoided, increasing the net productivity at the same time. Therefore, it can be said that when pigments are used, digital printing can be the most suitable alternative method to the traditional method.

1.4 FUTURE PERSPECTIVE

- Plasma technology has the potential to get commercialized for textile coloration at a very cheaper price as compared to the conventional methods [10]. Low-pressure plasma has been expansively investigated but not yet commercialized due to economic and technical restraints. However, atmospheric pressure plasma can overcome such limitations. Furthermore, in the case of low-pressure plasma, up-scaling is difficult for both batch and continuous processes which hinders the commercial applications [33].
- Nanotechnology is becoming popular rapidly as it offers novel materials with smart functionalities. Particularly, nanocomposite polymer, as well as nanoparticles, will produce fibers of definite properties and improve their dyeability.

- The present interest in textile process houses encourages more and more research and development for the use of supercritical CO_2 on a commercial scale for synthetic fibers using disperse-reactive and disperse dyes. However, it is a long journey toward using water-soluble dyes by this technique. It is under investigation for commercial applications [15]. A breakthrough for the technology would be for cellulosic fibers, mainly cotton, viscose, etc., for which further research needs to be carried out.
- The speed of production of digital printing is quite low, which affects and prevents digital printing from replacing conventional technology of printing. It has also been confirmed that the technology has limitations to the availability of small size of color cartridges and low viscosity dyes. Although in future digital printing is a promising technology, it is quite unsure that it will replace conventional printing technologies [34]. However, this technology will be a vital technology for the production of smaller batches and is predicted to cover at least 10% of all printed textiles [34]. On the other hand, the digital printing application is rising rapidly and can be a good replacement for traditional methods like screen and rotary printing in the near future [35].
- Furthermore, the interest in electrochemical, spin, ultrasonic, and microwave processes of dyeing has already been developed for possible applications in small-scale enterprises worldwide.

1.5 CONCLUSION

The manufacture and application of synthetic dyestuffs and auxiliaries that are often used for textile coloration have shown huge impacts on the environment and human health. However, natural dyes, microbial colorants, biodegradable auxiliaries, advanced dyeing processes, and water-free dyeing systems offer promising approaches in newer and newer techniques of coloration. The effects of dyestuffs on the environment during or after the dyeing processes should be extensively investigated and be more understandable. This will surely, in turn, offer the application of more effective existing or novel dyeing procedures and an enhancement of technology. Selection and optimization of the dyestuffs and dyeing processes in terms of stabilization, efficiency, energy, and time as well as in costs, are of utmost importance. The knowledge and up-gradation in process engineering and biomaterials are very much essential to establish processes at a commercial scale. The investigation and implementation of various novel and advanced coloration technologies should be constantly paid attention not only for the pollution reduction but also for the recycling of chemicals and auxiliaries, dyestuffs, water, and management of the final by-products for a wide range of applications.

REFERENCES

1. Bechtold, T., Mussak, R., Mahmud-Ali, A., Ganglberger, E., & Geissler, S. (2006). Extraction of natural dyes for textile dyeing from coloured plant wastes released from the food and beverage industry. *Journal of the Science of Food and Agriculture*, 86(2), 233–242.

2. Benkhaya, S., M'rabet, S., & El Harfi, A. (2020). A review on classifications, recent synthesis and applications of textile dyes. *Inorganic Chemistry Communications, 115(5)*, 1–69.
3. Chequer, F. D., De Oliveira, G. R., Ferraz, E. A., Cardoso, J. C., Zanoni, M. B., & de Oliveira, D. P. (2013). Textile dyes: dyeing process and environmental impact. *Eco-Friendly Textile Dyeing and Finishing, 6*(6), 151–176.
4. Nandhakumar, R., Kaviyarasu, R., & Kalidass, M. (2012). Dyeing of fabrics without water: a review. *The Indian Textile Journal, 2*, 37–39.
5. Kiron, M. I. (2014). Water consumption in textile industry. Textile Learners. https://textilelearner.net/water-consumption-in-textile-processing-industry/.
6. Hauser, P. J. (2000). Reducing pollution and energy requirements in cotton dyeing. *Textile Chemist & Colorist & American Dyestuff Reporter, 32*(6), 44–48.
7. Roy Choudhury, A. K. (2013). Green chemistry and the textile industry. *Textile Progress, 45*(1), 3–143.
8. Singh, H. B., & Kumar, A. B. (2014). *Handbook of natural dyes and pigments*. Woodhead Publishing India Pvt Limited, New Delhi, India.
9. Patel, B. H. (2011). Natural dyes. In: Clark, M. (Ed.), *Handbook of textile and industrial dyeing* (pp. 395–424). Woodhead Publishing Limited, Kidlington, United Kingdom.
10. Samanta, K. K., Basak, S., & Chattopadhyay, S. K. (2014). Environment-friendly textile processing using plasma and UV treatment. In: S. S. Muthu (Ed.), *Roadmap to sustainable textiles and clothing* (pp. 161–201). Springer, Singapore.
11. Thomas, H. (2007). Plasma modification of wool, In: Sishoo, R. (Ed.), *Plasma technologies for textiles* (pp. 228–246). Woodhead Publishing Limited, Kidlington, United Kingdom.
12. Terinte, N., Manda, B. K., Taylor, J., Schuster, K. C., & Patel, M. K. (2014). Environmental assessment of coloured fabrics and opportunities for value creation: spin-dyeing versus conventional dyeing of modal fabrics. *Journal of Cleaner Production, 72*, 127–138.
13. Nieminen, E., Linke, M., Tobler, M., & Vander Beke, B. (2007). EU COST Action 628: life cycle assessment (LCA) of textile products, eco-efficiency and definition of best available technology (BAT) of textile processing. *Journal of Cleaner Production, 15*(13–14), 1259–1270.
14. Hasanbeigi, A., & Price, L. (2015). A technical review of emerging technologies for energy and water efficiency and pollution reduction in the textile industry. *Journal of Cleaner Production, 95*, 30–44.
15. Bach, E., Cleve, E., & Schollmeyer, E. (2002). Past, present and future of supercritical fluid dyeing technology–an overview. *Review of Progress in Coloration and Related Topics, 32*(1), 88–102.
16. Ahmed, N. S., & El-Shishtawy, R. M. (2010). The use of new technologies in coloration of textile fibers. *Journal of Materials Science, 45*(5), 1143–1153.
17. Kulandainathan, M. A., Patil, K., Muthukumaran, A., & Chavan, R. B. (2007). Review of the process development aspects of electrochemical dyeing: its impact and commercial applications. *Coloration Technology, 123*(3), 143–151.
18. Periyasamy, A. P., & Rwahwire, S. (2019). Environmental friendly textile processing. In: Martínez, L. M. T., Kharissova, O. V., & Kharisov, B. I. (Eds.), *Handbook of ecomaterials* (pp. 1521–1558). Springer Cham, Denmark.
19. Wang, W. M., Yu, B., & Zhong, C. J. (2012). Use of ultrasonic energy in the enzymatic desizing of cotton fabric. *Journal of Cleaner Production, 33*, 179–182.
20. Vouters, M., Rumeau, P., Tierce, P., & Costes, S. (2004). Ultrasounds: an industrial solution to optimise costs, environmental requests and quality for textile finishing. *Ultrasonics Sonochemistry, 11*(1), 33–38.
21. Moholkar, V. S., & Warmoeskerken, M. M. C. G. (2004). Investigations in mass transfer enhancement in textiles with ultrasound. *Chemical Engineering Science, 59*(2), 299–311.

22. Wong, Y. W. H., Yuen, C. W. M., Leung, M. Y. S., Ku, S. K. A., & Lam, H. L. I. (2006). Selected applications of nanotechnology in textiles. *AUTEX Research Journal*, *6*(1), 1–8.
23. Russell, E. (2002). Nanotechnologies and the shrinking world of textiles. *Textile Horizons*, *9*(10), 7–9.
24. Dhanabalan, V., Sukanya, L., & Lokesh, K. V. (2015). Air dyeing technology-a review. *Textile Today*. https://www.textiletoday.com.bd/
25. Kumar, D. (2019). "Dyeing without water." *Fibre2Fashion.com*.
26. Maiti, S., Kulkarni, K., & Adivarekar, R. V. (2018). Biotechnology in textile wet processing. *Global Journal of Biomedical Science*, *2*, 7–13.
27. Nerurkar, M., Vaidyanathan, J., Adivarekar, R., & Langdana, Z. B. (2013). Use of a natural dye from *Serratia marcescens* subspecies Marcescens in dyeing of textile fabrics. *Octa Journal of Environmental Research*, *1*(2), 129–135.
28. Kulkarni, V. M., Gangawane, P. D., Patwardhan, A. V., & Adivarekar, R. V. (2014). Dyeing of silk/wool using crude pigment extract from an isolate Kocuriaflava Sp. HO-9041. *Octa Journal of Environmental Research*, *2*(4), 314–320.
29. Shang, S., Hu, E., Poon, P., Jiang, S., Kan, C. W., & Koo, R. (2011). Foam dyeing for developing the wash-out effect on cotton knitted fabrics with pigment. *Research Journal of Textile and Apparel*, *15*(1), 44–51.
30. Schönberger, H., & Schäfer, T. (2003). *Best available techniques in textile industry*. Federal Environmental Agency, Berlin.
31. Tyler, D. J. (2005). Textile digital printing technologies. *Textile Progress*, *37*(4), 1–65.
32. Zille, A., Oliveira, F. R., & Souto, A. P. (2015). Plasma treatment in textile industry. *Plasma Processes and Polymers*, *12*(2), 98–131.
33. AbdJelil, R. (2015). A review of low-temperature plasma treatment of textile materials. *Journal of Materials Science*, *50*(18), 5913–5943.
34. Gupta, S. (2001). Inkjet printing-A revolutionary ecofriendly technique for textile printing. *Indian Journal of Fibre & Textile Research*, *26*, 156–161.
35. Dehghani, A., Jahanshah, F., Borman, D., Dennis, K., & Wang, J. (2004). Design and engineering challenges for digital inkjet printing on textiles. *International Journal of Clothing Science and Technology*, *16*, 262–273.

2 Sustainable Approaches to Coloration of Textile Materials

Shanmugasundaram O. Lakshmanan
K. S. Rangasamy College of Technology

Wazed Ali and Swagata Banerjee
Indian Institute of Technology Delhi

CONTENTS

2.1 Introduction 13
2.2 Sustainable Coloration of Cellulosic Materials 14
　　2.2.1 Dyeing of Cotton Materials 14
　　2.2.2 Electrochemical Techniques 16
2.3 Sustainable Coloration of Protein Materials 17
2.4 Sustainable Coloration of Synthetic Materials 18
　　2.4.1 Dyeing of Polyester Materials 18
　　2.4.2 Dyeing of Nylon Materials 21
2.5 Other Sustainable Dyeing Techniques 22
　　2.5.1 Supercritical Carbon Dioxide Dyeing 22
　　2.5.2 Plasma-Assisted Dyeing 24
　　2.5.3 Air Dyeing 25
　　2.5.4 Ultrasound-Assisted Dyeing 26
　　2.5.5 Use of Natural Mordants in Dyeing 27
　　2.5.6 Nanotechnology in Dyeing 29
2.6 Conclusion 29
References 30

2.1 INTRODUCTION

There is an increasing demand for textile goods due to the rising population in the modern world. As a result, the need for clothes with color is always there, and the thought of continuous improvement in the coloration of textiles is an ever-green area. The coloring industry uses a huge volume of fresh water for the preparation of dresses for human beings. During the chemical processing of textile goods, these industries release pollutants such as dyes and various auxiliary chemicals (like sodium chloride) into the water stream, causing water pollution and sometimes metabolizing into mutagenic products (Chung et al., 1992; Banat et al., 1996). Langhals (2004) stated that textile industries are generating 1/5th of world industrial water pollution. Approximately 10%–15% of unreacted reactive dye and salt are released as pollutants while dyeing cotton fabric with reactive dyes. Adeel et al. (2017a) and Farizadeh et al. (2009) stated that synthetic dyes have toxic effects and are carcinogenic to living beings. The effluent generated during the dyeing process causes serious water and land pollution. Dyes extracted from plants, vegetables, fruits, flowers, animals, and minerals possess less toxicity, are nonallergic, and are eco-friendly (Adeel et al., 2017a; Kilinc et al., 2015). Many scientists have proved that natural dyes are eco-friendly and have good commercial potential for the sustainable production of textile materials (Chhipa, Srivastav, and Mehta, 2017; Datta et al., 2013; Zhang et al., 2014). 20% of the total pollution is caused by textile wet processing industries. Thus, textile chemical processing industries are really in a thirst to adopt various alternatives to overcome this issue. In recent days, air dyeing technology is gaining importance for coloring textile materials given economic and environmentally friendly aspects. Textile wet processing industries not only consume a huge volume of freshwater for processes but also generate a large volume of effluents with more than 50 toxic chemicals which are harmful to human beings. A California-based sustainable technology company, "Colorep", developed air-dyeing technology to dye textile materials. This technology uses less water/no water for the dyeing process. As a result, 90%–95% of freshwater has been saved with no effluent generation during dyeing. Air dyeing technology saves energy, freshwater, costs for effluent treatment, natural resources, reduces water pollution, and protects the environment from the discharge of solid waste. Air-dye technology is a sustainable way of dyeing textile materials without the use of water.

2.2 SUSTAINABLE COLORATION OF CELLULOSIC MATERIALS

2.2.1 Dyeing of Cotton Materials

Anuradhi Liyanapathiranage et al. (2020) demonstrated a nanofibrillated cellulose dyeing method for cotton. The results show a 30% increase in reactive dye fixation on cotton fabric and a 60% reduction in dye discharge in the washing process. Further, this technology reduces the use of freshwater by six times in the dyeing process. The authors also observed that there was no change in the physical properties of fabrics. Kim et al. (2017) and Minko et al. (2016) introduced (nanofibrillated cellulose) a novel dyeing technology for dyeing cotton material. Nanofibrillated cellulose hydrogel is a nontoxic material and acts as an efficient carrier for dyes. Nanofibrillated cellulose is a green technology based on life cycle assessment.

Many researchers (Kang et al., 1998; Yang et al., 2010, 2003, 1997, 2000; Ghosh et al., 2012) have found an improvement in reactive dye fixation on cotton material and a reduction in dye contaminants in effluent by combining nanofibrillated cellulose dyeing and polycarboxylic acid posttreatment. Furthermore, a notable wrinkle-resistant property imparted to cotton was also observed by Beisson et al. (2012). Maamoun (2014) observed an increase in color depth and a good burnout effect on polyester/cotton material using nanodisperse dye refluxed in sulfuric acid. The Burnout style of printing on textile materials was achieved by dissolving or destroying agents such as fibers, and this was popular in the fashion industry during 2003 (Kinnersly-Taylor, 2003). Natural dyes are gaining importance due to an increase in awareness of eco-friendly sources. The radiation tools such as ultraviolet (Adeel et al., 2012), ultrasound (Baaka et al., 2017), gamma rays (Adeel et al., 2017b), and plasma (Haji, Shoushtari, and Mirafshar, 2014) technology were used to extract dyes and assist the dyeing process to improve the extraction yield and fastness properties. In the same way, Adeel et al. (2020) used microwave radiation for the extraction of dyes from Arjun bark and dyeing of silk material. Adeel et al. (2018) used Arjun bark to dye textile materials to obtain a brown color. Both silk and bark powder were irradiated for 1–6 minutes before the dyeing process. Bio-mordants such as turmeric and pomegranate may be used for dyeing. The researchers obtained in-depth shades and improved fastness characteristics by adopting this approach.

Moon et al. (2011) and Klemm et al. (2011) described the production method and analyzed the size of various nanocelluloses (nanocrystalline cellulose, bacterial cellulose, and nanofibrillated cellulose). Berglund and Peijs (2010) and Dufresne (2013) elaborated on the application of nanocellulose in numerous areas, such as bio-composites, food packaging, transparent films, and optics. Kim et al. (2017) produced nanofibrillated cellulose and used it to dye cotton fibers with reactive dyes. In this technique, very small amounts of salt, water, and alkali are utilized for the dyeing of cotton material. The performance of dyeings such as shade, exhaustion percentage, degree of fixation, and fastness properties was compared with the conventional exhaust dyeing technique. A reduction in the usage of water and dyeing auxiliaries was obtained in nanofibrillated cellulose-based dyeing. The paste containing NFC hydrogel, reactive dye, salt, and alkali is printed on cotton material as pigment printing. This is an eco-friendly dyeing technique that reduces the pollutant load in the effluent.

A novel approach to reactive dyeing of cotton material was introduced by Chen et al. (2015). Nonnucleophilic solvents have been used as processing media in dyeing instead of water. As a result, there is a saving of a huge volume of water. Besides, a lower quantity of dyes, approx. 2.5% of alkali, and no electrolytes are necessary to dye cotton material. Also, 97% of waste disposal is reduced compared to conventional dyeing. This method is feasible for 100% recyclable and hydrolysis-free reactive dyeing. Many researchers have claimed that reactive dye effluents are not only difficult to treat but also problematic effluents (Allegre et al., 2006; Rosa et al., 2015). During the dyeing operation, the dyes make covalent bonds with the fiber, and at the same time, a partial amount of dye reacts with water and becomes hydrolyzed dye; it won't further react with the material. Roughly about 50% of hydrolyzed dyes are in the nonrecoverable form (Khatri et al., 2015).

Amin and Blackburn (2015) and Dasgupta et al. (2015) demonstrated the use of freshwater for various wet processing of textile materials and also discussed the effluent generation during processing. As prevention is better than cure, it is better to prevent waste generation during dyeing of cotton material with reactive dyes (Tang et al., 2008).

A few problems are associated with the use of nonpolar media in the dyeing of cotton with reactive dyes, such as poor shade, and levelling due to insufficient swelling of cotton and poor solubility of dyes. (Chavan, 1976; Shipman, 1971). Scientists used a low quantity of alcohol as a co-solvent (Cid et al., 2007; Van der Kraan et al., 2007) or low volume of water (Jun et al., 2004) to achieve deep shades in supercritical carbon dioxide dyeing techniques.

2.2.2 Electrochemical Techniques

Electrochemical techniques are being imposed in the textile industry, especially in dyeing, bleaching (color removal from jeans), and effluent treatment (Sala and Guitierrez-Bouzin, 2012). Chong and Chu (1998) demonstrated the application of electrochemical techniques to the bleaching of cotton fabrics. Furthermore, this technique is used by many researchers (Bechtold et al., 2006; Amano and Tanaka, 1990; Bechtold et al., 2005; Maier et al., 2004) to remove color present in finished jeans to produce faded effects. Appreciable progress has been made by electrochemical techniques in sulfur and vat dyeing processes, and strong reducing agents have replaced this technique in the above dyeing method. As a result, cleaner processes of dyeing without the use of salt (Teli et al., 2009; Sawada and Ueda, 2007; Bozic and Kokol, 2008; Thetford and Chorlton, 2004; Kulandainathan et al., 2007; Roessler and Jin, 2003; Roessler et al., 2002) have been achieved. Another interesting use of electrochemical techniques is the removal of reactive dyes from wastewater. These techniques are cleaner and do not produce any solid waste (Carneiro et al., 2010; Raghu and Ahmed Basha, 2007; Martinez-Huitle and Ferro, 2006; Naim and El Abd, 2002; Daneshvar et al., 2007). The researchers achieved 99% direct dye degradation efficiency while using iron and conducting polymer doped with chromium/boron-doped diamond electrode on textile effluent. Moreover, the boron-doped diamond electrode obtained 87% COD removal in the effluent (Lopes et al., 2004). Bechtold and his team did not obtain effective bleaching on denim with boron-doped diamond electrodes. The removal of color from denim is a combination of mechanical agitation and an oxidizing agent. Maier et al. (2004) achieved a good bleaching effect on jeans materials as compared to conventional methods using electrochemical bleaching techniques. Moreover, Bechtold et al. (2006) also obtained an intensive bleaching effect (de-colorization) on denim by using an electrochemical oxidation technique with the generation of in-situ hypochlorite during these processes. The authors have also claimed that this process is an efficient way to achieve consistent effects and results in lower process costs and effluent generation. Many researchers have attempted to reduce pollutants, solid waste, use of strong reducing agents and effluent load by electrochemical reduction processes in both direct and indirect reduction methods of dyeing of cotton with sulfur and vat dyes (Thomas and Aurora, 2006; Kulandainathan et al., 2007; Roessler et al., 2002; Roessler and Jin, 2003). The dye contains chromophore (azo-color giving group) and auxochrome ("-OH" – hue to color giving group). Most dyes have an azo group (Carneiro et al., 2010); Guaratini and Zanoni, 2000) and are toxic to human beings. Nowadays, color removal in wastewater is carried out by an electrochemical technique. Several methods have been tried by researchers for the removal of color from wastewater. These methods can remove 50%–60% of dye in the effluent. Moreover, chemical oxidation methods are inexpensive (Naim and El Abd, 2002). In general, electrochemical techniques have been proved to be an effective method for reduction and oxidation processes in wet processing. These techniques possess remarkable applications in bleaching of denim, reduction of vat and sulfur dyes, and the production of smart conductive textiles. Cui-Juan Ning et al. (2016) used natural dye to color nanofibers. The dye was extracted from roselle calyx and incorporated during the Bubbfil electrospinning process. (Chen et al., 2014, 2015).

2.3 SUSTAINABLE COLORATION OF PROTEIN MATERIALS

Tokino et al. (1993), Wakida et al. (1993,1996), (1996), and Sun & Stylios (2006) found that plasma treatment can enhance dye uptake, improve electronegativity on the wool fiber surface, and result in less felting properties on wool fabric. Furthermore, Wakida et al. (1993) found that argon plasma treatment can enhance the dyeability characteristics of wool material with acid dye. The combined treatment of enzyme and low-temperature plasma treatment improved the dyeing rate of cotton (Yoon et al., 1996). Kan et al. (1998) achieved good dyeing characteristics on nitrogen plasma treated wool material. Ghoranneviss et al. (2011) found that plasma sputtering treatment can improve the dyeing properties of wool and also impart antibacterial activity. Wool is another protein fabric that is used in the textile industry. An attempt was made to dye wool using different types of disperse dyes in a $scCO_2$ medium. The presence of thiazole ring and ethyl at the terminal of the dye molecule increases the hydrophobic nature of the molecule and hence improves the solubility of the dye in $scCO_2$. The increased molecular weight of the dye resulted in enhanced affinity toward the fiber. The dye-fiber interaction was mainly based on hydrogen bonding and van der Waals' forces. On increasing the dyeing temperature, the density of the $scCO_2$ increased, facilitating dye solubility. The kinetic energy of the dye molecules also increased with the increase in dyeing temperature. This resulted in more availability of the dye molecules for adsorption and hence better K/S values as achieved for the samples dyed at higher temperatures. Adequate dyeing time should be maintained to allow the transfer of the dye from the surface to the interior, followed by its fixation inside the fiber structure. The dyed samples showed good color fastness properties with acceptable mechanical properties (Zheng et al., 2017). The level dyeing properties of $scCO_2$ medium dyed samples were also studied. It was found that an increase in system temperature aided in the level of dyeing of the samples. Owing to the increased temperature, the interactions between the dye, $scCO_2$ medium, and the polymer chains increase, which helps to achieve level dyeing. In a $scCO_2$ dyeing medium, pressure also plays an important role in the level of dyeing of the samples. An increase in pressure at a constant temperature increases the solvating power of the medium. This helps to dissolve more dyes in the medium. Moreover, the increase in pressure increases the swelling of the fibers that help to obtain level-dyeing of the fabrics. Adequate time for dyeing assists in better penetration and diffusion of the dye molecules inside the fiber, which helps to obtain levelness in the dyed samples. The levelness of dyeing also depends on the dye used for dyeing. Dyes with small molecular structures help to give more level dyeing. Levelness in dyeing was better for dyes with low to moderate application temperature compared to high-temperature application dyes (Long, Ma, and Zhao, 2011).

2.4 SUSTAINABLE COLORATION OF SYNTHETIC MATERIALS

2.4.1 Dyeing of Polyester Materials

Polyester fiber is known for its superior mechanical and chemical properties. It is the most widely used synthetic fiber in the textile industry owing to its versatility. Polyester (PET) is hydrophobic with a compact structure. The dyeing of polyester is associated with high temperature and pressure conditions, which makes the process energy-intensive. The use of chemical compounds called carriers helps to reduce the dyeing temperature of polyester. However, they are sometimes carcinogenic in nature and not recommended for use. Therefore, modifications in the form of chemically treated polyester or the replacement of chemicals and auxiliaries have been introduced in the dyeing process to make it more sustainable.

In an attempt toward eco-friendly dyeing of polyester fabrics, polyethylene glycol (PEG) aqueous solution was used as the dyeing medium. The dye bath consisted of PEG aqueous solution with the disperse dye in the absence of auxiliaries. The pad-dry-cure technique was used for dyeing with 2dip-2nip padding at room temperature. The PEG performs two main functions, one from the dye perspective and the other from the fiber. PEG acts as a good dispersing and solubilizing agent for the disperse dye while acting as a plasticizer and swelling agent for the fiber. Hence, it helped to improve the dye diffusion and dyeability of the PET fabrics. An optimum PEG:water of 5:15 provided the best color depth results which indicate that beyond the aforementioned ratio, the dye retention in the PEG aqueous medium was higher. The as-dyed PET fabrics had improved levelling and fastness properties due to the multiple roles of PEG as a solubilizer, plasticizer, dispersant, and swelling agent (Ahmed et al., 2020). To reduce the water consumption in the dyeing of polyester, solvent dyeing techniques have also been used. Chlorinated solvents are banned due to their toxic nature, and hence, the search for ecofriendly solvents for polyester dyeing is an essential task. A glycine-based eutectic solvent (GES) has been used by Pawar et al. as a green medium for polyester dyeing. GES was prepared from choline chloride, urea, and glycerine. The dyeing of polyester was carried out in a high-temperature, high-pressure beaker dyeing machine, followed by reduction clearing and rinsing. The color strength of the dyed fabrics increased with the increase in temperature up to equilibrium and then decreased due to the shift of dye equilibrium to dye bath. A similar trend is observed with the increase in dyeing time. On comparing the color coordinates of 1% aqueous dyed and GES solution-dyed samples, it was found that the GES samples were darker in the shade. The wash fastness and light fastness of the GES dyed samples were comparable to aqueous dyed samples, while the sublimation fastness of GES dyed samples were superior. Moreover, the XRD results reveal the unchanged physical structure of polyester after the use of GES solvent dyeing (Pawar et al., 2019). Another solvent dyeing process of polyester using liquid paraffin has been reported by Xu et al. The usual process of exhaust dyeing was carried out followed by washing of the dyed fabrics using n-hexane and acetone. At the end of a dyeing cycle, the dye concentration in the residual dye bath was checked and restored to the required dye concentration, if required. In this way, reuse of the spent dye liquor took place. Dyeing of polyester in liquid paraffin required the optimization of temperature without having to worry about the pH and other auxiliaries.

This method of dyeing did not need any other chemicals, except for the dye, in the dyeing of polyester. The fastness properties of the dyed samples were comparable to those of the aqueous dyed ones. A significant result of the dyeing method was the removal of surface oligomers. Use of liquid paraffin as a dyeing medium led to a reduction in surface oligomers due to the high solubility of the hydrophobic oligomers in the lipophilic solvent (Xu et al., 2016).

In a study, the effect of microwave heating on the dyeing of polyester fabrics was carried out. In this study, both the polyester fabric and the dye solution were heated using a microwave. The microwave heating of polyester before dyeing helps to tune the polyester surface evenly which helps in better interaction of the fiber and the dye, leading to the uniform adsorption of the dye on the surface. Microwave heating of the dye bath leads to the disintegration of larger molecules to smaller ones that ease the penetration of the dye molecules into smaller voids of the polymer structure. This overall leads to the high K/S values of the dyed fabrics. The dyeing temperature is also reduced as compared to the conventional process, making the process energy efficient. The time of dyeing also needs to be optimum. Too short time for dyeing may not give good results as the time would not be enough for the molecules to accelerate. However, long heating hours may result in stripping of dye due to a shift of equilibrium toward dye bath. The fastness properties of the dyed samples were also good (Adeel et al. 2018). Moving further from water-intensive to waterless dyeing technology, super critical carbon dioxide (scCO$_2$) dyeing is an important technology. This process eliminates the use of water, chemicals and drying steps in the dyeing process. In a recent study, curcumin, a natural compound obtained from turmeric has been used to dye polyester fabrics using scCO$_2$, in the absence of any other chemicals or fabric pretreatment. Curcumin dyes polyester in a similar mechanism as that of the disperse dyes, owing to its nonpolar nature. It is also soluble in scCO$_2$ and hence can be used to dye polyester through this technique. It was seen that on increasing the system pressure, the dye uptake of the PET fabrics increased. Increasing the pressure increased the density of the scCO$_2$ and hence improved the interactions between the dye and solvent, thereby increasing the solubility of the dye in solvent. The increase in pressure also increased the swelling of the polymer that further helped in the adsorption and diffusion of dyes into the fiber interior. Increasing the temperature of the system helped to improve the flexibility and mobility of the polymer chains that resulted in higher dye uptake and superior color yield. The K/S values of the dyed samples increased with the increase in dye concentration due to more availability of dye molecules for adsorption and diffusion. The levelness was hardly affected by increase in dye concentration (Tadesse et al., 2019).

Apart from curcumin, natural dyes extracted from other natural sources have also been used to dye polyester. Dye extracted from watermelon rind and flesh has been used as a natural alternative to disperse dyes for coloration of polyester. They are rich in carotenoids and lycopene content and have a variety of chromophores. Such chromophores are also known to have UV protective characteristics. The exhaustion and the mechanism of fixation of the different dyes vary as per their chromophore structure and type of chromophore. The diffusion of the dye depends on the shape of the chromophore and on the type of interaction between the dye and the fiber. It has been found that the chromophore of lower molecular weight and suitable shape as the intermolecular voids in polyester, will have higher access and thus easy penetration into the polyester structure. An increase of dyeing temperature increased the depth of shade of the dyed sample owing to the enhanced kinetic energy of the dye molecules and hence better penetration of the dye into polyester fiber. The wash fastness, rubbing fastness, and light fastness of the dyed samples ranged from good to excellent. An additional property of UV shielding has also been imparted to the dyed samples. The UV shielding property was improved with the increase in dyeing temperature due to the enhanced dye adsorption at elevated temperatures (Liman et al., 2020). Disperse dyes have been synthesized from natural colorants to make the dyeing of polyester more sustainable. Natural colorants like alizarin and purpurin have been derivatized at the 2-hydroxy position to yield derivatives that are used for sustainable dyeing of polyester. The as-synthesized dyes were not pH sensitive and did not dissolve in alkali. However, the exhaustion percentage and the color depth was higher for the parent compound in comparison to the derivatives, owing to the larger size and bulky nature of the derivatives (Drivas, Blackburn, and Rayner, 2011).

The techniques such as $scCO_2$ dyeing, solvent dyeing, have certain drawbacks. The high-pressure equipment of $scCO_2$ dyeing is expensive which affects the economy of the process. The solvent dyeing process is a waterless method but requires the recovery of solvent for reuse. Hence, the field of nanotechnology is being explored for the sustainable dyeing of polyester. In a study, silicone nanomicelles containing highly dispersed dyes were used for dyeing PET fabrics. It was observed that the K/S values of the dyed PET fabrics improved with the increase in dyeing time. This implies that the dye gets released slowly from the micelle with an increase in dyeing time and provides level dyeing of the PET fabrics. The mechanism of dyeing is shown in Figure 2.1. The dyed samples acquired an additional handle property of softness imparted by amino silicone oil (Gao et al., 2018).

Carriers used in the dyeing of polyester help to reduce the harsh temperature conditions of the process. However, they have serious ecological concerns associated with them and hence natural products are being investigated as an alternative. One such natural compound is vanilla that comes from the *Vanilla planifolia* flowers. The structure of the compound is similar to the synthetic carriers and hence was used in the dyeing of polyester fabric as reported by Jalali et al. Vanillin is the main ingredient of vanilla and has antimicrobial and antioxidant properties. The K/S values of the dyed polyester samples increased with the increase in vanilla concentration as this compound helped to lower the Tg of the fiber, allowing easy penetration of the disperse dye. However, the K/S values of PET samples dyed in the presence of vanilla were lower compared to those dyed in the presence of commercial carriers. Mixture

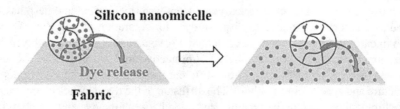

FIGURE 2.1 Mechanism of silicon nanomicelle dyeing on PET fabrics (Gao et al., 2018).

dyeing using vanilla and commercial carriers gave better results in terms of K/S values. The color fastness properties in terms of light, washing, and rubbing fastness of the PET dyed samples in the presence of vanilla was also good, which proves vanilla to be a promising alternative to synthetic carriers (Jalali et al., 2019). A study on the toxicity of o-vanillin and p-vanillin was carried out to assess their suitability for use as carriers. The two vanillins are nontoxic to humans and possess antimutagenic properties. However, inhalation of o-vanillin may cause problems like eye irritation and respiratory problems (Pasquet et al., 2013).

2.4.2 Dyeing of Nylon Materials

Nylon belongs to the family of polyamide fibers. These fibers find their applications in a number of ways in the textile industry due to their good mechanical and chemical properties. Dyeing of nylon fabrics has also been researched to make the process more sustainable and eco-friendly. Accordingly, the nylon-6 fabrics have been dyed using certain synthesized dyes through the super critical carbon dioxide ($scCO_2$) dyeing process. The dyes synthesized were antimicrobial disperse dyes, that would add functionality along with dyeing. It was found that as the dye concentration increased, the K/S of the dyed samples increased for some of the dyes while decreased for others. This was due to the lower saturation values of some of the dyes. The K/S values of the $scCO_2$ dyed samples were superior to that of the aqueous dyed samples which proved better absorbance of the dye in $scCO_2$ medium compared to water. The increase in dyeing temperature increased the kinetic energy of the dye molecules which could penetrate better into the nylon structure, which was now more flexible owing to the elevated temperatures. Thus, higher temperatures of dyeing provided samples with higher K/S values. The increased pressure of the dyeing system increased the solvating power of $scCO_2$ and hence helped to obtain darker shades. The wash fastness, light fastness and rubbing fastness of the dyed samples were also noteworthy. Some of the functional groups like cyano, chloro, in the synthesized dyes, imparted antimicrobial functionality to the dyed samples (Elmaaty et al., 2015).

The dyeing of nylon fibers is associated with the use of the large quantity of water. Recently, due to the pressure of the environmental norms, this water-intensive dyeing system of nylon is being tried to be replaced by other waterless dyeing techniques. One such method is to use decamethylcyclopentasiloxane (D5) as the dyeing medium. This process overcomes the drawbacks of supercritical dyeing and solvent dyeing procedures. D5 is a chemical that is compatible with other chemicals, is odorless, nonflammable, and stable under various chemical conditions. Disperse dyes are soluble in D5, and hence, the use of auxiliaries is not necessary. The D5 solvent is able to swell the fiber appropriately so that the dyes could penetrate and diffuse deep into the nylon structure. The high solubility of disperse dye in the solvent results in poor exhaustion of dye on the substrate. Hence, accelerants are added to reduce the solubility of the dye in D5 solvent which helps to increase their exhaustion onto the substrate. Consequently, the addition of accelerants resulted in greater color depth compared to those in the absence of accelerants. With the increase of dyeing temperature, the solubility of the dye in the solvent increased, thus decreasing the exhaustion of the dye onto the substrate. This reduced the color depth of the samples dyed at higher temperatures. However, the decreased dye exhaustion promoted the uniform dyeing of the nylon samples. The color fastness of the samples dyed at higher temperatures got improved due to better penetration of the dye into the fiber and less deposition on the surface. Since the dye is highly soluble in D5, there is no need of a separate washing process to remove the unfixed dyes. Moreover, the recovery of D5 using a pressure drying machine makes this process more sustainable (Saleem et al., 2020).

Not just for disperse dyes, D5 has been used with acid dyes as well for dyeing nylon six fabrics. The acid dyes were first dissolved in a small amount of water and then added to D5 solvent for the sake of uniformity. Through XPS analysis, it was found that D5 had the power to penetrate the interior of the fiber and cause effective dyeing. The FTIR analysis of the treated samples confirms that there was no change in the chemical structure of the fiber after dyeing in D5 medium. The treatment also does not affect the thermal stability of the fabric. The exhaustion of the acid dyes was much higher in the D5 medium compared to the aqueous medium. This was because the affinity of the dye for the fiber was higher than that for the solvent. Consequently, the fabrics dyed in D5 medium showed higher color strength than those dyed in an aqueous medium. The wash fastness, rubbing and perspiration fastness of the dyed samples in D5 medium was also good. Since most of the dye molecules were exhausted onto the substrate, the residual bath had only the D5 solvent that was further used for subsequent dyeing cycles. The solution however became turbid on repeated usage due to the impurities from the fabric and the dye. The original D5 solvent can be recovered by different filtration techniques (Saleem et al., 2021).

2.5 OTHER SUSTAINABLE DYEING TECHNIQUES

2.5.1 Supercritical Carbon Dioxide Dyeing

The critical point of any substance denotes the coexistence of its liquid and vapour phase in equilibrium. The respective temperature and pressure are known as the critical temperature and critical pressure. Beyond the critical points, the substance exists in its super critical phase. The critical temperature and critical pressure for carbon dioxide are 304.25 K and 7.38 MPa, respectively (Abou Elmaaty and Abd El-Aziz, 2018). The phase diagram for pure carbon dioxide is shown in Figure 2.2 below.

The super critical fluids have low viscosity and high solubilizing power. This makes them suitable for use as a dyeing medium. They can dissolve dyes and penetrate the remote regions inside the fiber structure without any agitation (Saus, Knittel, and Schollmeyer, 1993). Among other super critical fluids, CO_2 is considered for dyeing applications due to its nontoxic and noninflammable properties. The apparatus used for supercritical carbon dioxide (scCO_2) dyeing is shown in Figure 2.3.

The dyeing in scCO_2 is characterized by low mass transfer resistance and a high rate of dissolution of dyes. This helps in the penetration of the dye into the deep core of the fiber structure. Moreover, this is a waterless process eliminating the energy-intensive drying process. There are also negligible chances of the dye getting hydrolyzed, and hence, most of the dye is available for reaction with the fiber. The scCO_2 used in the process can be recovered. These are some of the advantages of using

Coloration of Textile Materials

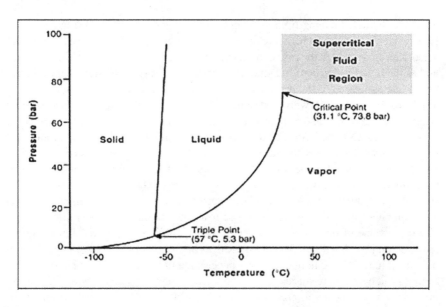

FIGURE 2.2 Phase diagram depicting the critical temperature and pressure of carbon dioxide (Abou Elmaaty and Abd El-Aziz, 2018).

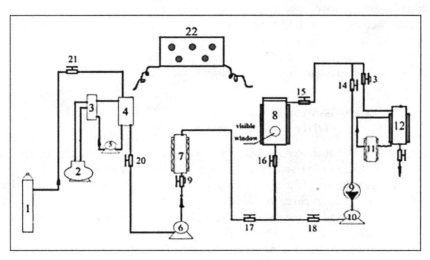

FIGURE 2.3 Apparatus for scCO$_2$ dyeing. (1) CO$_2$ cylinder; (2) refrigerating machine; (3) heat exchanger; (4) cooling unit; (5) liquid pump; (6) high-pressure syringe pump; (7, 11) preheaters; (8) dyeing autoclave/dyestuff vessel; (9) flow gauge; (10) CO$_2$ circulating pump; (12) cold trap; (13–21) valves; (22) main controller.

scCO$_2$ as a dyeing medium over aqueous dyeing. Below are discussed some dyeing processes of textiles using scCO$_2$ as dyeing medium.

Super critical carbon dioxide has been used to dye natural as well as man-made fibers. Cotton has been dyed using disperse reactive dyes in scCO$_2$ medium. It has been found that the increase in dyeing temperature helps in dye-fiber interaction in the scCO$_2$

medium. Increased dyeing time helps in deeper penetration of the dye into the fiber. Hence, the sample dyed at higher temperature and increased time bestowed a better K/S value. The color yield in the case of protein fibers was higher compared to cellulosic (cotton) as the functional groups like thiol and amino in the case of protein fibers are easier to activate compared to the hydroxyl groups of cellulosic fibers. Man-made fibers can also be dyed in the scCO$_2$ with dyed samples having good color depth and fastness properties (Schmidt, Bach, and Schoomeyer, 2003). In a recent study, the polyester fabrics were dyed in scCO$_2$ medium with some commercial photochromic dyes to produce UV sensing fabrics. The photochromic dyes were two types, namely, spirooxazine based and naphthopyran based. In order to evaluate the durability of the photochromic dyes, they were subjected to repeated exposure of UV radiation. The naphthopyran-based dyes showed better durability on repeated exposure compared to spirooxazine based dyes. This can be attributed to the better fatigue resistance of the naphthopyran-based dyes. The color value of the dyed samples with the two different dyes was almost similar; however, inkjet-printed samples with these dyes had greater color value compared to the ones dyed in scCO$_2$ medium due to less penetration of the dye in case of inkjet-printed samples and hence more surface color. The spirooxazine-based dyes showed higher coloration and rapid decoloration compared to the napthopyran dyes. This is because of the inherent features of the dye structures. Spirooxazine-based dyes have an inherent quick switching mechanism and are more temperature sensitive. The napthopyran-based dyes on the other hand have lower fading rates after showing coloration. With respect to the washing fastness of the dyed samples, the naphthopyran-based dyed samples showed better wash fastness compared to the spirooxazine-based dyed samples. This is due to the better stability, improved penetration, and higher affinity of the naphthopyran-based dye compared to the spirooxazine-based dye (Abate et al., 2020).

2.5.2 Plasma-Assisted Dyeing

With the aim of sustainable and eco-friendly dyeing technologies, plasma-assisted dyeing is gaining attention. Plasma is referred to as the fourth state of matter. The active ionic species of plasma have the capacity to break chemical bonds that result in surface modification of textiles. Hence, plasma is a dry technology that is utilized in dyeing as well as finishing treatments. Based on the gas used and the process conditions, plasma can be used to get different effects like cleaning, etching, and other surface functionalization properties. Plasma can be of two types according to the working pressure of the system: hot plasma and cold plasma. Owing to the degradation of textiles at elevated temperatures, cold plasma is used for the plasma treatment of textiles (Haji and Naebe, 2020).

Wool shows a hydrophobic character due to the lipid content in its outermost epicuticle layer. This provides a hindrance to the penetration of water and chemicals during the dyeing process. Hence, surface modification is done by plasma treatment to increase the wettability and absorbency of the fiber. Dielectric barrier discharge plasma was used for this purpose. The reactive dye exhaustion on the fiber increased after the plasma treatment due to the introduction of nanoscale surface roughness that provided a pathway for the diffusion of the dye into the fiber interior. The fixation percentage of the dye also improved due to the inclusion of functional groups on the fiber that facilitated dye attachment. However, the color uniformity of the plasma-treated wool dyed samples was lower due to the rapid adsorption of the dyes at the surface. The wash and light fastness of the samples were better than its rubbing fastness properties. The wet rubbing fastness of the plasma-treated samples were inferior due to the lubricating action of water that broke the hydrogen bonds between the dye and fiber causing staining (Navik, Shafi, and Alam, 2018). Silk, a continuous protein fiber has also been treated in oxygen and argon plasma and the effects on lac dyeing thereafter has been studied. The treatment of plasma brings about structural changes in silk. It changes the random coil conformation in silk to amide-I β-sheet structure due to the oxidation and deoxidation reactions during treatment. The treatment also introduces surface roughness in silk, which is more in oxygen than argon plasma treatment due to the enhanced chemical etching characteristic of oxygen. However, too long plasma treatment reduces the lac dye uptake by silk fibers. The dye uptake increases with the increase in dyeing temperature until the dyeing equilibrium is reached, after which it decreases. This proves the exothermic nature of dyeing (Boonla and Saikrasun, 2013).

Cellulosic fibers like cotton have been treated with plasma and dyed with natural dyes that added functional properties to the dyed fabrics. Cotton was dyed with 17 different types of plant extracts which were checked for their antimicrobial properties as well (Chen, Fan, and Wang, 2007). Out of the 17 dyes applied, 10 showed antimicrobial properties, while others did not. However, after oxygen plasma pretreatment followed by dyeing, the 13 different types of dyed cotton samples showed antimicrobial properties. This was due to the fact that oxygen plasma treatment introduced hydrophilic functional groups on the cotton substrate which could interact with the antimicrobial component of the dye structure and hence showed the expected functionality. However, the washing fastness of the dyed samples was not improved even after oxygen plasma pretreatment.

Polyester is the most popular fiber in the category of synthetic fibers. Due to its hydrophobic nature, it becomes difficult to dye polyester with natural dyes. It has also been treated with plasma through dielectric barrier discharge in air at atmospheric pressure. This treatment has introduced changes in the polyester surface that helped to reduce its hydrophobicity and hence increased its dyeability with natural dyes. The introduction of polar groups on the polyester surface due to plasma treatment increased its wettability confirmed through the reduction in contact angle. The presence of polar groups and the amorphization on plasma treatment increased the natural dye uptake of polyester (Dave et al., 2012). Nylon after polyester is the most widely used synthetic fiber. Knitted fabric made of polyamide and elastane was plasma pretreated and its dyeability was studied. The treatment introduced physical changes in the fabric and chemical modifications in the fiber structures. The plasma treatment introduces groups like C-H, N-H, and N-O which helped to increase the dye uptake by the fabric and also facilitated better fixation of the dye (Gasi et al., 2020).

2.5.3 Air Dyeing

Water is used extensively in textile wet processing. To make wet processing sustainable, waterless technologies are being discovered. One such technology is air dyeing. It colors textiles with negligible or no use of water. The air flow dyeing technology involves atomization of the dyeing liquor followed by mixing with high-pressure air flow. The fabric passes through a mixture of air streams that is emitted from the blower. The movement of fabric takes place with the help of air or a mixture of air and steam. This pressurized air moves the fabric forward. The blower jets the dye-air mixture onto the fabric. Unlike jet dyeing machines where the jet ejects a mixture of dye and water mixture and exerts pressure on the fabric, the pressure on fabric in case of air dyeing of the air dyeing is negligible. This can hence be used as a technique to dye sensitive textiles (Vignesh Dhanabalan, Sukanya, and Lokesh, 2015).

The fabric absorbs the dye from the moisture saturated air flow, and hence, it is lighter than its weight in conventional dyeing. The light weight of the fabric helps to achieve high processing speed, without causing any damage to the substrate. The strain experienced by the fabric during processing is minimal and hence elastane fibers can be processed through this technology. Polyamide dyeing through air dye technology is also used for end-use applications like automotive, swimwear, etc. in other words, delicate fabrics can be dyed effectively with minimal use of water and energy. Moreover, the high processing speed of this technology helps to increase production with short-run cycles.

The high processing speed and minimal liquor ratio make this process sustainable. The reduction in chemicals also supports in overall cost reduction of the process. However, issues may crop up when dye solubility is a concern. Sometimes, the fabric tends to pack in on the lower end of the machine owing to the reduced liquor ratio. This may lead to the development of permanent wrinkles on the fabric. Hence the synthetic fibers should be heat set before they are dyed using this technology. The air dye technology opens up multiple designs of dyeing fabric. When two sides of the fabric are differently colored, it is called dye-to-dye contrast. When one side of the fabric is printed while the other face is solid colored, it is called dye-to-print. When both sides of the fabric are of the same color, it is called dye squared. When both faces of the fabric are printed differently, it is called print-to-print.

The air dye technology claims to reduce the use of resources like water and energy and hence is a sustainable approach. It is a cleaner way of producing dyed fabrics with rich shades. There are no limitations relating to the washing of the air dyed textiles. This technology is, therefore, being increasingly accepted as a sustainable alternative to the conventional dyeing of textiles.

2.5.4 Ultrasound-Assisted Dyeing

Ultrasound-assisted dyeing was first discovered in the mid-1900s. Ultrasound has a frequency range that is inaudible to humans. Ultrasound provides some distinguished advantages in dyeing compared to conventional dyeing techniques. It helps to accelerate the process of dyeing and gives good results in mild processing conditions. The effect of ultrasound on the solubility of dyes and dye uptake by textiles has been studied by various research groups. The primary reason for the improvement in the dyeing process is due to the cavitation phenomenon and its consequent effects (Vajnhandl, Majcen, and Marechal, 2005).

Nylon-6 fabrics were dyed with reactive dyes with the help of ultrasound. The use of ultrasound increased the color depth of the samples. The reasons may be summed to three phenomena of dispersion, degassing, and diffusion. Dispersion resulted in the breaking up of the larger micelles into smaller molecules to give a uniform dye dispersion. Degassing removed the air entrapped in the fiber interstices, thus increasing dye-fiber interaction. The accelerated rate of dye uptake was due to the better diffusion and fixation of the dye. Moreover, the agglomeration tendency of the dye was also reduced at higher concentrations due to the use of ultrasound. The dyeing at different pH using ultrasound gave uniform color depth results of the dyed samples. However, fixation of the dye was much better in the case of neutral pH. The use of ultrasound resulted in improvement in washing fastness of the dyed samples due to the better penetration and fixation of the dye. However, the rubbing fastness of the samples was not improved (Kamel et al., 2003). To compensate for the high temperature required for polyester dyeing, ultrasound-assisted dyeing method was implemented. The K/S values of the samples dyed through the ultrasound-assisted dyeing technique showed an improvement in the increase of K/S value with the increase in dyeing temperature compared to that of the conventionally dyed samples. This was due to the increased rate of dye diffusion and dye-fiber interaction as a result of the cavitation phenomenon. Increasing the time of ultrasound treatment reduced the tendency of the dyes to agglomerate and hence improved the color yield. The better dye penetration due to the ultrasound treatment also helped to obtain better fastness properties of the ultrasound-assisted dyed polyester samples (Youssef and Haggag, 2017).

Not only synthetic fiber dyeing but also dyeing of natural protein fiber like wool through the ultrasound-assisted technique was investigated for its dyeing and fastness properties. It was found that acidic pH rendered better dyeing results compared to dyeing at higher pH values. This was due to the changes in the dye-fiber interactions at low and high pH values. Salt addition in wool dyeing functioned as a retarding agent and provided better results in ultrasound dyeing compared to conventional dyeing techniques. The effect of dyeing temperature and dyeing time was similar as observed in the case of synthetic fiber dyeing through ultrasound. As the power of ultrasound was increased, a phenomenon like dispersion, degassing, and diffusion was improved that resulted in better K/S values of the dyed samples. The fastness properties of the ultrasound-assisted dyed samples were comparable to the of conventional dyeing method (Kamel et al., 2005). Silk filament yarn was dyed with cationic dyes in the presence of ultrasound. The presence of ultrasound increased the dye uptake by silk yarn due to the cavitation phenomenon. Hence, better results were obtained at lower dyeing temperatures. A temperature of 50°C was found to show maximum dye uptake using the ultrasound dyeing technique. The effect of ultrasound was more pronounced with a dye having higher molecular weight than the low molecular weight dyes. Ultrasound helped to break down the high molecular weight dyes into smaller fragments at low temperatures. This eased the dye absorption and diffusion phenomenon (Mathur, 1995).

2.5.5 Use of Natural Mordants in Dyeing

Mordants are an important element of textile dyeing. Natural dyes generally do not possess an affinity toward the textile substrate. Mordants act as a bridge between the fiber and the dyes. They are substantive to dye as well as fiber. Hence, dyes with little or no affinity toward textile can easily be fixed on the substrate with the help of mordants. Depending on the time of mordant application in the sequence of dyeing, the dyeing process can be classified as premordanting, meta/simultaneous mordanting, and postmordanting. The treatment of textile with mordant prior to dyeing is called premordanting, while the treatment of textile with mordant after dyeing is called post mordanting. Simultaneous mordanting involves the dissolution of mordant as well as a dye in the dyebath during dyeing operation. In the view of sustainability in textile processing techniques, biomaterials are being increasingly researched for mordanting properties (Yusuf et al., 2017). Some of the naturally occurring compounds that are used as mordants are tannins and tannic acid. Myrobolans and sumach are natural substances that come under this category. The fruit of *Terminalia chebula* is called myrabalan. The tannin in myrobalan is ellagitannic acid which is contained in the peel of the fruit. It is used in dyeing cotton and black dyeing of silk. Sumach contains gallotannic acid which is not used in dyeing light shades due to the red coloring matter present in it (Mohammad and Mohammad, 2017).

Dyeing of wool has been carried out using root extract from Rubia cordifolia, and the use of Acacia catechu as a biomordant has been assessed. The mordanting was carried out by pre-, meta-, and postmordanting treatments. The FTIR analysis shows the presence of the anthraquinone group in the naturally extracted dye. The mordanting with the biomordant did not lead to changes in the characteristics of wool fiber. The lightness value of the mordant dyed samples was less compared to the one dyed without the use of mordant. The premordanted samples had darker shades followed by post- and meta-mordanted samples. The interaction between the dye and the mordant in the dye bath in case of meta-mordanting led to decrease in darkness of shade. The mordant acts as a bridge between the dye and the substrate and hence helps to obtain better color fastness properties of the dyed samples (Yusuf et al., 2016). In another study, the effect of valex, pomegranate rind, rosemary, and thuja (*Thuja orientalis*) as mordants in the dyeing of wool dyeing with an outer green shell of the almond extract was investigated. Mordanting was done by three methods of pre-, meta- and postmethods. The use of rosemary as mordant led to changes in the color gamut of the dyed sample due to the inherent coloring nature of rosemary. Different mordants showed different effects with different mordants used. These natural mordants showed comparable and in some cases better performance than metallic mordants. The color fastness properties of the biomordanted dyed samples were based on the fixing mechanism of the different plant parts. Thuja leaves however did not have any effect on the fastness properties (Leyla and Özdo, 2014). Silk was dyed using lac dye in the presence of mordants like tamarind. Other chemical mordants were also used in dyeing and their performance were compared. The post mordanting method yielded better results compared to the pre- and meta- mordanting methods due to better adsorption of dye. The type of mordants used influenced the color of the dyed sample. The method of mordanting had an insignificant effect on the breaking strength of the dyed silk samples (Kongkachuichay et al., 2002). In a study, pigments were extracted from different parts of the plant using eco-friendly solvents, ethanol and water. To compare the effect of different mordants, mordanting was carried out using biomordant (tannin) extracted from *Emblica officinalis* G and a metallic mordant copper sulfate. Cotton and silk fabrics were dyed using the natural dyes and the mordants. All the natural extracts and the mordant showed high phenol content which proved them as effective textile colorants. The improved color depth of the samples was observed using natural mordant due to the complex formed between the hydroxyl group of phenol and the dye molecules. Premordanting enhanced the color improvement of the naturally dyed samples. Due to the presence of more functional groups in silk as compared to cotton, the shades were darker in silk than in cotton. The treatment also helped to provide additional functionality of UV protection. It was more in the case of cotton than in silk samples (Chao et al., 2017).

2.5.6 Nanotechnology in Dyeing

Nanotechnology has made a cutting-edge entry in the field of textile wet processing. Fibers having varied dimensions can be dyed using natural dyes. In one such study, natural dye extracted from roselle calyx was used to color nanofibers during bubble electrospinning. Polyvinyl alcohol (PVA) nanofibers were dyed using this naturally extracted dye. The dye could effectively penetrate and color the nanofibers (Ning et al., 2016).

Isotactic polypropylene (PP) is also a fiber that is difficult to dye because of its highly regular arrangement of monomer groups resulting in high crystallinity. Hence, nanotechnology has been utilized to dye the otherwise difficult to dye isotactic PP fiber. The nanocomposites of PP are dyeable as it is believed that the dye sites are formed at the places occupied by the nanoparticles (Fan and Mani, 2007). Nanoclays are generally used as nanoparticles. They are sometimes surface-modified to increase the dye affinity of the composite.

The effect of nanoclay addition to a blend of polypropylene (PP) and polytrimethylene terephthalate (PTT) was studied in terms of its structural properties and dyeability with acid and disperse dyes. The fastness properties of the dyed samples were also investigated. On the other hand, the nanoclay acts as a nucleating effect and hence helps to improve the overall crystallinity of the fibers. The disperse dyeability of the fibers is improved as observed from the enhanced K/S values. This is mainly due to the nanoclays acting as sorption sites because of its high surface area leading to van der Waals' and ionic interactions with the dyes. The presence of quaternary ammonium salts in the nanoclays serves as points of interaction with the acid dyes due to which the acid dyeability of the fibers was improved (Teli and Desai, 2015).

2.6 CONCLUSION

This chapter gave a relatively detailed account of the huge volume of fresh water consumed by the textile industry and the generation of wastewater which directly affects the environment. Various sustainable approaches, such as dyeing of textile materials with natural dyes, radiation tools to extract natural dyes and assist in dyeing, biomordants to fix natural dyes, nanofibrillated cellulose dyeing techniques, etc. can be adopted to minimize the pollution load.

Many researchers have attempted to use electrochemical techniques for bleaching, dyeing, and effluent treatment. Furthermore, the sustainable approaches to the coloration of cellulose, protein, and manmade fibers were elaborately described. New techniques such as plasma treatment, supercritical carbon dioxide dyeing technique, ultrasound, and microwave-assisted methods are reported. Moreover, air dyeing techniques and dyeing using nanotechnology – principles, mechanisms, merits, and limitations – were emphasized.

REFERENCES

Abate, M. T., Seipel, S., Yu, J., Viková, M., Vik, M., Ferri, A., Guan, J., Chen, G., and Nierstrasz, V. (2020). "Supercritical CO_2 dyeing of polyester fabric with photochromic dyes to fabricate uv sensing smart textiles." *Dyes and Pigments* 183, 108671. https://doi.org/10.1016/j.dyepig.2020.108671.

Abou Elmaaty, T., and Abd El-Aziz, E. (2018). "Supercritical carbon dioxide as a green media in textile dyeing: a review." *Textile Research Journal* 88(10): 1184–1212. https://doi.org/10.1177/0040517517697639.

Adeel, S., Bhatti, I. A., Kausar, A., and Osman, E. (2012). "Influence of UV radiation on the extraction and dyeing ofcotton fabric with Curcuma longa L." *Indian Journal of Fiber and Textile Research* 37: 87–90.

Adeel, S., Khan, S. G., Shahid, S., Saeed, M., Kiran, S., Zuber, M., Suleman, M., and Akhtar, N. (2018). "Sustainable dyeing of microwave treated polyester fabric using disperse yellow 211 dye." *Journal of the Mexican Chemical Society* 62(1): 1–9. https://doi.org/10.29356/jmcs.v62i1.580.

Adeel, S., Muneer, M., Ayub, S., Saeed, M., Zuber, M., Iqbal, M., Haq, E., and Kamran, M. (2017a) "Fabrication of UV assisted improvement in dyeing behavior of polyester fabric using disperse orange 25." *Oxid Communication* 40(2): 925–935.

Adeel, S., Rafi, S., Salman, M., Rehman, F. U., and Abrar, S. (2017b). Potential resurgence of natural dyes in applied fields. In Shahid-ul-Islam (Ed.), *Plant-Based Natural Products*, . Wiley Scrivener Publishing LLC, Hoboken, NJ, USA. 1–26. ISBN 978-1-119-42383-6.

Adeel, S., Rehman, Fazal-ur-, Hameed, A., Habib, N., Kiran, S., Zia, K. M., and Zuber, M. (2020). "Sustainable extraction and dyeing of microwave-treated silk fabric using arjun bark colorant." *Journal of Natural Fibers* 17(5): 745–758. doi:10.1080/15440478.2018.1534182

Ahmed, N., Nassar, S., Kantouch, F., and El-Shishtawy, R. M. (2020). "A novel green continuous dyeing of polyester fabric with excellent color data." *Egyptian Journal of Chemistry* 63(1): 1–14. https://doi.org/10.21608/ejchem.2020.22055.2318.

Allegre, C., Moulin, P., Maisseu, M., and Charbit, F., (2006). "Treatment and reuse of reactive dyeing effluents. " *Journal of Membrane Science* 269: 15e34.

Amano Y., and Tanaka, Y. Y. (1990) "Treating agent for bleachprocessing," Japanese Patent: Application number "JP1988000226387".

Amin, M. N., and Blackburn, R. S. (2015). "Sustainable chemistry method to improve the wash-off process of reactive dyes on cotton. " *ACS Sustainable Chemistry & Engineering* 3 (6): 1039e1046.

Baaka, N., Haddar, W., Ben Ticha, M., Amorim, M. T. P., and M'Henni, M. F. (2017). "Sustainability issues ofultrasonic wool dyeing with grape pomace colourant. " *Natural Product Research* 31(14): 1655–1662. doi:10.1080/14786419.2017.1285303.

Banat, I. M., Nigam, P., Singh, D., and Marchant, R. (1996). "Microbial decolorization of textile-dyecontaining effluents: a review. " *Bioresource Technology* 58: 217–227.

Bechtold, T., Maier, P., and Schrott, W. (2005). "Bleaching of indigodyeddenim fabric by electrochemical formation of hypohalogenitesin situ," *Coloration Technology*, 121 (2): 64–68.

Bechtold, T., Turcanu, A., Campese, R., Maier, P., and Schrott, W. (2006). "On-site formation of hypochlorite for indigooxidation—scale-up and full scale operation of an electrolyser for denim bleach processes." *Journal of AppliedElectrochemistry* 36(3): 287–293.

Beisson, F., Li-Beisson, Y., and Pollard, M. (2012). "Solving the puzzles ofcutin and suberin polymer biosynthesis. " *Current Opinion In Plant Biology* 15: 329–337.

Berglund, L. A., and Peijs, T. (2010). "Cellulose biocomposites—from bulk moldings to nanostructured systems." *MRS Bulletin* 35: 201–207.

Boonla, K., and Saikrasun, S. (2013). "Influence of Silk surface modification via plasma treatments on adsorption kinetics of lac dyeing on silk." *Textile Research Journal* 83(3): 288–297. https://doi.org/10.1177/0040517512458344.

Bozic, M., and Kokol, V. (2008). "Ecological alternatives to thereduction and oxidation processes in dyeing with vat andsulphur dyes," *Dyes and Pigments* 76(2): 299–309.

Carneiro, P. A., Umbuzeiro, G. A., Oliveira, D. P., and Zanoni, M. V. B. (2010). "Assessment of water contamination caused by amutagenic textile effluent/dyehouse effluent bearing disperse dyes." *Journal of Hazardous Materials* 174(1–3): 694–699.

Chao, Yu-chan, Ho, Tsung-han, Cheng, Zhi-jiao, Kao, Li-heng, and Tsai, Ping-szu. (2017). "A study on combining natural dyes and environmentally-friendly mordant to improve color strength and ultraviolet protection of textiles" *Fibers and Polymers* 18(8): 1523–1530. https://doi.org/10.1007/s12221-017-6964-7.

Chavan, R. B. (1976). "Solvent dyeing of cotton with a reactive dye." *Journal of the Society of Dyers and Colourists* 92: 59.

Chen, C., Fan, L., and Wang, H. (2007). "Antibacterial evaluation of cotton fabric pretreated by microwave plasma and dyed with taiwan folkloric medicinal plants." *Journal of Fiber Science and Technology* 63(11): 252–255. https://doi.org/10.2115/fiber.63.252.

Chen, L., Wang, B., Ruan, X., Chen, J., and Yang, Y. (2015). "Hydrolysis-free and fully recyclable reactive dyeing of cotton in green, non-nucleophilic solvents for a sustainable textile industry." *Journal of Cleaner Production*, 107, 550–556.

Chen, R. X., Li, Y., and He, J. H. (2014). "Mini-review on bubbfil spinning process for massproduction of nanofibers." *Materia* 19(4): 325–343.

Chen, R., Wan, Y., Si, N., He, J.H., Ko, F. and Wang, S.Q.. (2015). "Bubble rupture in bubble electrospinning." *Thermal Science* 19(4): 1141–1149.

Chhipa, M. K., Srivastav, S., and Mehta, N. (2017). "Study of dyeing of cotton fabric using peanut pod natural dyes using Al_2SO_4 $CuSO_4$ and $FeSO_4$ mordanting agent." *International Journal of Environmental and Agriculture Research*, 3: 36–44.

Chong, C. L., and Chu, P. M. (1998). "Bleaching cotton based onelectrolytic production of hydrogen peroxide." *AmericanDyestuff Reporter* 87(4): 13–19.

Chung, K. T., and Cerniglia, C. E. (1992). "Mutagenicity of azo dyes: structure-activity relationships." *Mutation Research/Reviews in Genetic Toxicology* 277: 201–220.

Cid, M. F., Gerstner, K. N., van Spronsen, J., van der Kraan, M., Veugelers, W., Woerlee, G. F., and Witkamp, G. J. (2007). "Novel process to enhance the dyeability ofcotton in supercritical carbon dioxide." *Textile Research Journal* 77: 38e46.

Daneshvar, N., Khataee, A. R., Amani Ghadim, A. R., and Rasoulifard, M. H. (2007). "Decolorization of C.I. Acid Yellow 23solution by electrocoagulation process: investigation of operational parameters and evaluation of specific electrical energyconsumption (SEEC)." *Journal of Hazardous Materials* 148(3): 566–572.

Dasgupta, J., Sikder, J., Chakraborty, S., Curcio, S., and Drioli, E. (2015). "Remediation of textile effluents by membrane based treatment techniques: a state of the art review. " *Journal of Environmental Management* 147: 55e72.

Datta, S., Uddin, M. A., Afreen, K. S., Akter, S., and Bandyopadhyay, A. (2013). "Assessment of antimicrobial effectivenessof natural dyed fabrics." *Bangladesh Journal of Scientific and Industrial Research* 48: 179–184 doi:10.3329/bjsir.v48i3.17327.

Dave, H., Ledwani, L., Chandwani, N., Kikani, P., Desai, B., Chowdhuri, M. B., and Nema, S. K. (2012). "Use of dielectric barrier discharge in air for surface modification of polyester substrate to confer durable wettability and enhance dye uptake with natural dye eco-alizarin." *Composite Interfaces* 19(3–4): 219–229. https://doi.org/10.1080/1568554 3.2012.702594.

Dhanabalan, V., Sukanya, and Lokesh, K.V. (2015) "Air dyeing technology-a review." October 14, 2015. https://www.textiletoday.com.bd/air-dyeing-technology-a-review/. Accessed February 23, 2021.

Drivas, I., Blackburn, R. S., and Rayner, C. M. (2011). "Natural anthraquinonoid colorants as platform chemicals in the synthesis of sustainable disperse dyes for polyesters." *Dyes and Pigments* 88(1): 7–17. https://doi.org/10.1016/j.dyepig.2010.04.009.

Dufresne, A. (2013). "Nanocellulose: a new ageless bionanomaterial." *Material Today* 16: 220–227.

Elmaaty, T. A., El-Aziz, E. A., Ma, J., El-Taweel, F., and Okubayashi, S. (2015). "Eco-friendly disperse dyeing and functional finishing of nylon 6 using supercritical carbon dioxide." *Fibers* 3(3): 309–322. https://doi.org/10.3390/fib3030309.

Fan, Q., and Mani, G. (2007). "Dyeable polypropylene via nanotechnology." In: *Nanofibers and Nanotechnology in Textiles* (Eds. Brown, P. J. and Stevens, K.), 320–350. Woodhead Publishing Limited, Cambridge, England. https://doi.org/10.1533/9781845693732.3.320.

Farizadeh, K., Montazer, M., Yazdanshenas, M. E., Rashidi, A., and Malek, R. M. A. (2009). "Extraction, identification andsorption studies of dyes from madder on wool." *Journal of Applied Polymer Science* 113(6): 3799–3808. doi:10.1002/app.30051.

Gao, A., Hu, L., Zhang, H., Fu, D., Hou, A., and Xie, K. (2018). "Silicone Nanomicelle Dyeing Using the Nanoemulsion Containing Highly Dispersed Dyes for Polyester Fabrics." *Journal of Cleaner Production* 200: 48–53. https://doi.org/10.1016/j.jclepro.2018.07.312.

Gasi, F., Petraconi, G., Bittencourt, E., Lourenço, S. R., Castro, A. H. R., Miranda, F. De S., Essiptchouk, A. M., et al. (2020). "Plasma treatment of polyamide fabric surface by hybrid corona-dielectric barrier discharge: material characterization and dyeing/washing processes." *Materials Research* 23(1): 1–9. https://doi.org/10.1590/1980-5373-MR-2019-0255.

Ghoranneviss, M., Shahidi, S., Anvari, A., Motaghi, Z., Wiener, J., et al. (2011). "Influence of plasma sputtering treatment on natural dyeing andantibacterial activity of wool fabrics." *Progress in Organic Coatings* 70(4): 388–393.

Ghosh Dastidar, T., and Netravali, A. N. (2012). "Green crosslinking ofnative starches with malonic acid and their properties." *Carbohydrate Polymers* 90, 1620–1628.

Guaratini, C. C. I., and Zanoni, M. V. B. (2000) "Textile dyes," *Quimica Nova* 23(1): 71–78.

Haji, A., and Naebe, M. (2020). "Cleaner dyeing of textiles using plasma treatment and natural dyes: a review." *Journal of Cleaner Production* 265: 121866. https://doi.org/10.1016/j.jclepro.2020.121866.

Haji, A., Shoushtari, A. M., and Mirafshar, M. (2014). "Natural dyeing and antibacterial activity of atmospheric-plasmatreatednylon 6 fabric." *Coloration Technology* 130: 37–42 doi:10.1111/cote.2013.130.issue–1.

Jalali, S., Rezaei, R., Afjeh, M. G., and Eslahi, N. (2019). "Effect of vanilla as a natural alternative to traditional carriers in polyester dyeing with disperse dyes." *Fibers and Polymers* 20(1): 86–92. https://doi.org/10.1007/s12221-019-8482-2.

Jun, J. H., Sawada, K., and Ueda, M. (2004). "Application of perfluoropolyether reverse micelles in supercritical CO2 to dyeing process." *Dyes and Pigments* 61: 17e22.

Kamel, M. M., El-shishtawy, R. M., Hanna, H. L., and Ahmed, N. S. E. (2003). "Ultrasonic-assisted dyeing : I. nylon dyeability with reactive dyes." *Polymers International* 380(May 2002): 373–380. https://doi.org/10.1002/pi.1162.

Kamel, M. M., El-shishtawy, R. M., Yussef, B. M., and Mashaly, H. (2005). "Ultrasonic assisted dyeing III. Dyeing of wool with lac as a natural dye." *Dyes and Pigments* 65(2): 103–110. https://doi.org/10.1016/j.dyepig.2004.06.003.

Kan, C., Chan, K., Yuen, C. W., and Miao, M. (1998). "The effect of low-temperatureplasma on the chrome dyeing of wool fibre." *Journal of Materials Processing Technology* 82(1): 122–126.

Kang, I.-S., Yang, C. Q., Weishu Wei, W., and Lickfield, G. C. (1998). "Mechanical strength of durable press finished cotton fabrics: part I: effects of acid degradation and crosslinking of cellulose bypolycarboxylic acids." *Textile Research Journal* 68: 865–870.

Khatri, A., Peerzada, M. H., Mohsin, M., and White, M. (2015). "A review on developments in dyeing cotton fabrics with reactive dyes for reducing effluent pollution." *Journal of Cleaner Production* 87: 50e57.

Kilinc, M., Canbolat, S., Merdan, N., Dayioglu, H., and Akin, F. (2015). "Investigation of the color, fastness andantimicrobial properties of wool fabrics dyed with the natural dye extracted from the cone of Chamaecyparislawsoniana." *Procedia--Social and Behavioral Sciences* 195: 2152–2159. doi:10.1016/j.sbspro.2015.06.281.

Kim, Y., McCoy, L. T., Lee, E., Lee, H., Saremi, R., Feit, C., Hardin, I. R., Sharma, S., Mani, S., and Minko, S. (2017). "Environmentally sound textile dyeing technology with nano-fibrillated cellulose." *Green Chemistry* 19: 4031–4035.

Kinnersly-Taylor, J. (2003). *Dyeing and Screen-Printing on Textiles*. A and C Black Publishers, London, UK, 176.

Klemm, D. F., Moritz, K. S., Lindström, T., Ankerfors, M., Gray, D., and Dorris, A. (2011). "Nanocelluloses: a new family of nature-based materials." *Angewandte Chemie International Edition* 50, 5438–5466.

Kongkachuichay, P., Shitangkoon, A., and Chinwongamorn, N. (2002). "Studies on dyeing of silk yarn with lac dye: effects of mordants and dyeing conditions." *ScienceAsia* 28: 161–166.

Kulandainathan, M. A., Muthukumaran, A., Patil, K., and Chavan, R. B. (2007). "Potentiostatic studies on indirect electrochemicalreduction of vat dyes," *Dyes and Pigments* 73(1): 47–54.

Langhals, H. (2004). "Color chemistry. Synthesis, properties and applications of organic dyes and pigments. 3rd revised edition. By Heinrich Zollinger." *Angewandte Chemie International Edition* 43, 5291–5292.

Leyla, Y., and Özdo, E. (2014). "Use of almond shell extracts plus biomordants as effective textile dye." *Journal of Cleaner Production* 70, 61–67.. https://doi.org/10.1016/j.jclepro.2014.01.055.

Liman, Md L. R., Islam, M.T., Hossain, Md M., and Sarker, P. (2020). "Sustainable dyeing mechanism of polyester with natural dye extracted from watermelon and their UV protective characteristics." *Fibers and Polymers* 21(10): 2301–2313. https://doi.org/10.1007/s12221-020-1135-7.

Liyanapathiranage, A., Peña, M. J., Sharma, S., and Minko, S. (2020). "Nanocellulose-based sustainable dyeing of cotton textiles with minimized water pollution." *ACS Omega* 5: 9196–9203.

Long, J. J., Ma, Y. Q., and Zhao, J. P. (2011). "Investigations on the level dyeing of fabrics in supercritical carbon dioxide." *Journal of Supercritical Fluids* 57(1): 80–86. https://doi.org/10.1016/j.supflu.2011.02.007.

Lopes, A., Martins, S., Mora͏̈o, A., Magrinho, M., and Gonc͏̧alves, I. (2004). "Degradation of a textile dye C.I. Direct Red 80by electrochemical process." *Portugaliae Electrochimica Acta* 22: 279-294.

Maamoun, D. (2014). "Utilization of nanodisperse dye in printing polyester/cotton substrate via Burn-out Techniques, *Asian Journal of Textiles* 4(1), 18–29.

Maier, R., Schrott, W., Bechtold, T., and Campese, R. (2004). "Electrochemicalbleaching in the finishing of jeans." *MelliandTextilberichte* 85(11–12): 880–884.

Martinez-Huitle, C. A., and Ferro, S. (2006). "Electrochemical oxidationof organic pollutants for the wastewater treatment:direct and indirect processes." *Chemical Society Reviews* 35(12): 1324–1340.

Minko, S., Sharma, S.,, Hardin, I., Luzinov, I., Wu Daubenmire, S., Zakharchenko, A., Saremi, R., and Kim, Y.S. (2016). "Textile Dyeing Using Nanocellulosic Fibers". *Clemson Patents* 565. https://tigerprints.clemson.edu/clemson_patents/565

Islam, S.., and Mohammad, F. (2017). 5- Ecological Dyeing of Wool with Biomordants. In Muthu, S. S. (Ed.), *Sustainable Fibres and Textiles*. Woodhead Publishing,, England. https://doi.org/10.1016/B978-0-08-102041-8.00005-6.

Moon, R. J., Martini, A., Nairn, J., Simonsen, J., and Youngblood, J. (2011). "Cellulose nanomaterials review: structure, properties and nanocomposites." *Chemical Society Reviews* 40: 3941-3994.

Naim, M. M., and El Abd, Y. M. (2002). "Removal and recovery ofdyestuffs from dyeing wastewaters." *Separation and Purification Methods* 31(1): 171–228.

Navik, R., Shafi, S., Alam, M.M., Farooq, M.A., Lina, L.I.N. and Yingjie, C.A.I. (2018). "Influence of dielectric barrier discharge treatment on mechanical and dyeing properties of wool." *Plasma Science and Technology*, 20(6), 065504.Ning, C.-J., He, C.-H., Xu, L., Liu, F-J. and He, J-H. (2016). "Nano-Dyeing." *Thermal Science* 20(3): 1003–1005.

Pasquet, V., Perwuelz, A., Behary, N., and Isaad, J. (2013). "Vanillin, a potential carrier for low temperature dyeing of polyester fabrics." *Journal of Cleaner Production* 43: 20–26. https://doi.org/10.1016/j.jclepro.2012.12.032.

Pawar, S. S., Maiti, S., Biranje, S., Kulkarni, K., and Ravindra, V. A. (2019). "Heliyon a novel green approach for dyeing polyester using glycerine based eutectic solvent as a dyeing medium." *Heliyon* 5(April): e01606. https://doi.org/10.1016/j.heliyon.2019.e01606.

Raghu, S., and Ahmed Basha, C. (2007). "Chemical or electrochemicaltechniques, followed by ion exchange, for recycle of textiledye wastewater." *Journal of HazardousMaterials* 149(2): 324–330.

Roessler, A., Crettenand, D., Dossenbach, O., Marte, W., and Rys, P. (2002). "Direct electrochemical reduction of indigo." *Electrochimica Acta* 47(12): 1989–1995.

Roessler, A., and Jin, X. (2003). "State of the art technologies and newelectrochemical methods for the reduction of vat dyes." *Dyesand Pigments* 59(3): 223–235.

Rosa, J. M., Fileti, A. M. F., Tambourgi, E. B., Santana, J. C. C. (2015). "Dyeing of cotton with reactive dyestuffs: the continuous reuse of textile wastewater effluent treatedby ultraviolet/hydrogen peroxide homogeneous photocatalysis." *Journal of Cleaner Production* 90: 60e65.

Sala, M., and Gutiérrez-Bouzán, C. (2012). "Electrochemical techniques in textile processes and wastewater treatment." *International Journal of Photoenergy* 2012, 629103. doi:10.1155/2012/629103.

Saleem, M. A., Pei, L., Saleem, M. F., Shahid, S., and Wang, J. (2020). "Sustainable dyeing of nylon with disperse dyes in decamethylcyclopentasiloxane waterless dyeing system." *Journal of Cleaner Production* 276: 123258. https://doi.org/10.1016/j.jclepro.2020.123258.

Saleem, M.A., Pei, L., Saleem, M.F., Shahid, S., and Wang, J. (2021). "Sustainable dyeing of nylon fabric with acid dyes in decamethylcyclopentasiloxane (D5) solvent for improving dye uptake and reducing raw material consumption." *Journal of Cleaner Production* 279: 123480. https://doi.org/10.1016/j.jclepro.2020.123480.

Saus, W., Knittel, D., and Schollmeyer, E. (1993). "Dyeing of textiles in supercritical carbon dioxide." *Textile Research Journal* 63 (3): 135–142. https://doi.org/10.1177/004051759306300302.

Sawada, K., and Ueda, M. (2007). "Optimization of dyeing poly(lacticacid) fibers with vat dyes." *Dyes and Pigments* 74(1): 81–84.

Schmidt, A., Bach, E., and Schoomeyer, E.(2003). "The dyeing of natural fibres with reactive disperse dyes in supercritical carbon dioxide." *Dyes and Pigments* 56(1): 27–35. https://doi.org/10.1016/S0143-7208(02)00108-0.

Shipman, A. J. (1971). "The use of non-aqueous solvents in textile processing." *Review of Progress in Coloration and Related Topics* 2: 42e50.

Shukla, S. R., and Mathur, M. R. (1995). "Low-temperature ultrasonic dyeing of silk." *Journal of the Society of Dyers and Colourists* 111(11): 342–345.

Sun, D., Stylios, G. K. (2006). "Fabric surface properties affected by lowtemperature plasma treatment." *Journal of Materials ProcessingTechnology* 173(2): 172–177.

Tadesse, M., Ferri, A., Guan, J., Chen, G., and Nierstrasz, V. (2019). "The journal of supercritical fluids colouration and bio-activation of polyester fabric with curcumin in super-critical CO_2 : part I - investigating colouration properties." *The Journal of Supercritical Fluids* 152: 104548. https://doi.org/10.1016/j.supflu.2019.104548.

Tang, S. Y., Bourne, R. A., Smith, R. L., and Poliakoff, M. (2008). "The 24 principles of green engineering and green chemistry: "IMPROVEMENTS PRODUCTIVELY"." *Green Chemistry* 10: 268e269.

Teli, M. D., and Desai, P. V. (2015). "Disperse and acid dyeable polypropylene polyblend fibers." *International Journal of Current Engineering and Technology* 5(4): 2567–2571.

Teli, M. D., Rohera, P., Sheikh, J., and Singhal, R. (2009). "Useof Amaranthus (Rajgeera) starch vis-`a-vis wheat starch inprinting of vat dyes." *Carbohydrate Polymers* 76(3): 460–463.

Thetford, D., and Chorlton, A. P. (2004). "Investigation of vat dyes aspotential high performance pigments." *Dyes and Pigments* 61(1): 49–62.

Thomas, B., and Aurora, T. (2006). "Iron-complexes of bis(2-hydroxyethyl)-amino-compounds as mediators for the indirectreduction of dispersed vat dyes—cyclic voltammetry and spectro-electrochemical experiments." *Journal of ElectroanalyticalChemistry* 591(1): 118–126.

Tokino, S., Wakida, T., Uchiyama, H., and Lee, L. (1993). "Laundering shrinkage of wool fabric treated with low-temperature plasmas under atmospheric pressure." *Journal of the Society of Dyers and Colourists* 109(10): 334–335.

Vajnhandl, S., Majcen, A., and Marechal, L. (2005). "Ultrasound in textile dyeing and the decolouration / mineralization of textile dyes." *Dyes and Pigments* 65: 89–101. https://doi.org/10.1016/j.dyepig.2004.06.012.

Van der Kraan, M., Cid, F., Woerlee, G. F., Veugelers, W., and Witkamp, G. J. (2007). "Dyeing of natural and synthetic textiles in supercritical carbon dioxide with dispersereactive dyes." *The Journal of Supercritical Fluids* 40: 470e476.

Wakida, T., Lee, M., Sato, Y., Ogasawara, S., Ge, Y., et al. (1996). "Dyeing propertiesof oxygen low-temperature plasma-treated wool and nylon 6 fibres withacid and basic dyes." *Coloration Technology* 112(9): 233–236.

Wakida, T., Tokino, S., Niu, S., Lee, M., Uchiyama, H., et al. (1993). "Dyeingproperties of wool treated with low-temperature plasma underatmospheric pressure." *Textile Research Journal* 63(8): 438–442.

Xu, S., Chen, J., Wang, B., and Yang, Y. (2016). "An environmentally responsible polyester dyeing technology using liquid paraffin." *Journal of Cleaner Production* 112: 987–994. https://doi.org/10.1016/j.jclepro.2015.08.114.

Yang, C. Q., Chen, D., Guan, J., and He, Q. (2010). "Cross-linking cottoncellulose by the combination of maleic acid and sodiumhypophosphite. 1. Fabric wrinkle resistance." *Industrial & Engineering Chemistry Research* 49: 8325–8332.

Yang, C. Q., Hu, C., and Lickfield, G. C. (2003). "Crosslinking cotton withpoly(itaconic acid) and in situ polymerization of itaconic acid: fabricmechanical strength retention." *Journal of Applied Polymer Science* 87: 2023–2030.

Yang, C. Q., Wang, X., and Kang, I.-S. (1997). "Ester crosslinking of cottonfabric by polymeric carboxylic acids and citric acid." *Textile Research Journal* 67: 334–342.

Yang, C. Q., Zhiping, M., and Lickfield, G. C. (2000). "Ester Crosslinking ofCotton Cellulose by Polycarboxylic Acids: pH-Dependency." *Textile Chemist & Colorist* 32, 43–46.

Yoon, N. S., Jin, L. Y., Tahara, M., and Takagishi, T. (1996). "Mechanical and dyeingproperties of wool and cotton fabrics treated with low temperatureplasma and enzymes." *Textile Research Journal* 66(5): 329–336.

Youssef, B. M., and Haggag, K. M. (2017). "Microwave, ultrasound assisted dyeing- part I: dyeing characteristics of C.I. disperse red 60 on polyester fabric." *Egyptian Journal of Chemistry* 151: 143–151. https://doi.org/10.21608/EJCHEM.2017.1478.1102.

Yusuf, M., Mohammad, F., Shabbir, M., and Khan, M. A. (2016). "Eco-dyeing of wool with rubia cordifolia root extract : assessment of the effect of acacia catechu as biomordant on color and fastness properties." *Textiles and Clothing Sustainability* 2(1): 1–9. https://doi.org/10.1186/s40689-016-0021-6.

Yusuf, M., Shabbir, M., and Mohammad, F. (2017). "Natural colorants : historical, processing and sustainable prospects." *Natural Products and Bioprospecting* 7(1): 123–145. https://doi.org/10.1007/s13659-017-0119-9.

Zhang, B., Wang, L., Luo, L., and King, M. W. (2014). "Natural dye extracted from Chinese gall - the application of colorand antibacterial activity to wool fabric." *Journal of Cleaner Production* 80: 204–210. doi:10.1016/j.jclepro.2014.05.100.

Zheng, H., Xu, Y., Zhang, J., Xiong, X., Yan, J., and Zheng, L. (2017). "An ecofriendly dyeing of wool with supercritical carbon dioxide fluid." *Journal of Cleaner Production* 143: 269–277. https://doi.org/10.1016/j.jclepro.2016.12.115.

3 Aspects of Mordant Dyeing of Textile Fibers

Supriyo Chakraborty
Uttar Pradesh Textile Technology Institute

Lipika Chakraborty
Government Polytechnic

CONTENTS

3.1 Introduction 38
3.2 Traditional Mordant Dyeing Practices..38
   3.2.1 Chrome Mordant Dyes 38
      3.2.1.1 Mechanism of Chrome Mordant Process 39
   3.2.2 Mordant Wool Dyeing with Pre-Metallised Acid Dyes (i.e., Metal Complex Dyes) 39
      3.2.2.1 Chemistry of Metal – Dye Complex Formation 39
      3.2.2.2 Dyeing using Metal Complex Dyes 41
3.3 Coloration of Textiles with Natural Dye..41
   3.3.1 Resources of Natural Dye: Plants, Animal, and Minerals 42
   3.3.2 Categorizing Natural Mordant Dyes as per Chemical Structure 43
      3.3.2.1 Anthraquinone Dyes 43
      3.3.2.2 Indigoid Dyes...44
      3.3.2.3 α-Naphthoquinone Dyes 44
      3.3.2.4 Carotenoid Dyes..44
      3.3.2.5 Flavonoid Dyes 44
      3.3.2.6 Dihydropyran Dyes..45
   3.3.3 Extraction of Natural Dyes 45
      3.3.3.1 Aqueous Extraction..45
      3.3.3.2 Organic Media/Mixed Aqueous and Organic Solvents........45
      3.3.3.3 Acid/Alkali Extraction..45
   3.3.4 Mordants 45
   3.3.5 Application of Mordants on the Textile Material 47
   3.3.6 Types of Mordant..47
      3.3.6.1 Metal-Based Mordants...47
      3.3.6.2 Tannin 48
      3.3.6.3 Oil-Mordants..49
   3.3.7 Recent Trend of Bio-Mordants 49
   3.3.8 Enzymes Mordant 51

DOI: 10.1201/9781003140467-3

3.3.9 General Method of Fabric Mordanting and Natural Dyeing..............52
 3.3.10 Ecological Aspects: Natural Dyeing with Mordants........................52
3.4 Conclusion 53
References..53

3.1 INTRODUCTION

A range of dyes have limited or no affinity for the fibre substrate but when they form complexes with some metal oxides or organic/natural mordant, they can be applied to the textile fibers such as wool, silk, and cellulosic materials. Different mordants, while forming complexes with dyes, can produce different hues; they improve the fastness of the dyes by forming insoluble complexes. Various metal salts used for mordanting are not environmentally friendly; some restrictions are imposed on such chemicals. Chromium is a red-listed chemical under eco-regulations, and copper also falls in the restricted category. The presence of tin is also not a desirable chemical to be present in the effluents. Some biomordants such as tannin and oil mordants used in place of metal mordants are a better options.

3.2 TRADITIONAL MORDANT DYEING PRACTICES

Mordant dyes are capable of combining with metallic oxides to form insoluble colored lakes. Many of the old natural dyes such as alizarin, logwood, and weld were dyed on textiles that had previously been impregnated or mordanted, with hydroxides of chromium, tin, or aluminum. Some of these natural products were replaced with synthetic equivalents. After the structure of alizarin was determined, the range was increased by adding many analogous substances derived from anthraquinone.

This class of dyes can be found in the Color Index, which includes acid dyes only, where the fastness properties of the dyes are noticeably better when they form complexes with some inorganic salts (such as chromium, tin, copper, aluminum, etc.) as a result of insolubilization. These dyes were employed for the dyeing of wool before the premetallized acid dyes were found to serve the purpose. They are affixed with the application of metal salts before (premordanting method), or after (postmordanting method) or applied simultaneously with (co-mordanting method) in the bath, with the acid dye to color the fiber. The mordant dyes are applied mainly to the wool or polyamides, as in the case of the acid dyes.

From the classification of dyes as per the methods of application, mordant dyes may contain characteristic structural unit or functional groups such as nitroso (quinoneoximes) (O=C-C=NOH <----> -C=C-N=O), azo [(trisazo/(R-N=N--)$_3$-R′) and (polyazo/(R-N=N-)$_{>4}$-R′)], triarylmethane, xanthene, oxazine, thiazine, lactone, aminoketone and hydroxyketone [R may not be same as R′] (Vigo, 2013).

3.2.1 CHROME MORDANT DYES

Most of the mordant dyes are similar to non-metalized acid dyes because they are anionic in nature, have sulphonate ($-SO_3H$) and or carboxylic acid ($-COOH$) groups. The mordant dyes possess some added chemical groups (e.g., $-OH$, $-NH_2$, etc.) that

Mordant Dyeing of Textile Fibers

allow them to form more stable, coordination compounds with metal ions (such as chromium ions) inside a fibre. These complexes so formed are often accompanied by an intense change in the shade of the dyeing; at the same time, there is a noticeable increase in the wet fastness and light fastness (Rattee, 1953).

Chrome dyeing methods were developed for the dyeing of wool fiber textile materials. As already mentioned, there are three methods that can be used for their application of these dyes on the wool (specifically pre-, meta-, afterchrome process) depending on at what time the mordant (commonly $Na_2Cr_2O_7$, $K_2Cr_2O_7$ or Na_2CrO_4, K_2CrO_4, etc.) is applied in the dyeing process. After chrome method of application is usual for chrome dyeing of nylon fiber. In general, very poor fastness in the case of un-mordanted dyeing disprove the practice of applying the mordant dyes as an acid dye.

3.2.1.1 Mechanism of Chrome Mordant Process

As wool fibre is processed with a hot solution of dichromate, as in mordanting or after chroming, firstly, there occurs a reaction of dichromate with the basic groups on side chains available on the wool, as by some inorganic acids. Absorption of dichromate H^+ ions takes place and at the same time to adjust the neutrality liberation of some alkali and a small increase in pH occurs. The subsequent process is associated with the reduction of Cr by reaction with some groups of wool at 100°C. The trivalent chromium then forms complex with –COOH groups of wool; then it can form complex with Dye molecule in 1:2 ratio.

3.2.2 Mordant Wool Dyeing with Pre-Metallised Acid Dyes (i.e., Metal Complex Dyes)

In the chrome mordant process, there are some problems associated with fiber damage due to a long time for dyeing; also, matching shades becomes difficult. The crucial factors of the dye chemistry and properties of these metal complex dyes were cultured and established to get rid of this crucial problems of using mordant dyes on wool textile materials. In Color Index, the metal complex dyes are included as the Acid Dye; these dyes are also known as pre-metalized acid dyes. There are two types of these dyes that were introduced,

1. Acid dyeing pre-metalized dyes (1919)
2. Neutral dyeing pre-metalized dyes, (1951) (Bird, 1972).

3.2.2.1 Chemistry of Metal – Dye Complex Formation

Metal complexes formed are coordination compounds where co-ordinate links develop between the dye and metal. Co-ordinate bonds occur when lone pairs of electrons of, say, oxygen or nitrogen atom are shared with another atom. A well-identified case of coordination valencies is the formation of an ammonium ion from ammonia and a proton—resulting $[NH_4]^+$. As shown (Scheme 3.1)

Thus an oxygen atom from the carbonyl group and nitrogen from the azo group will co-ordinate with chromium or the metal. The chromium atom, as it loses its three valency electrons, results in the chromic ion Cr(III), which can readily take up

$$H^+ + :NH_3 \longrightarrow NH_4^+$$

SCHEME 3.1 Simple example of co-ordinate covalent bond in case of formation of Ammonium Ion (Bird, 1972).

six pairs of electrons in its outer shells by the formation of co-ordinate bonds, and acquires a stable structure. The donor molecule, containing more than or one free electron pair, which is known as a ligand. In a $[Cr-(H_2O)_6]^{3+}$ complex, 12 electrons are provided by the oxygen atoms in six water molecules, while the loss of the three valence electrons results in the complex a trivalent ion.

For the dye to be firmly attached to the metal ion, at least two bonds are necessary, so in the case of chromium, a maximum of three dye molecules can co-ordinate with the metallic atom. In the case of alizarin, as an example, the complex formation with a chromic salt may be as follows by both ionic and co-ordinate bonds (Scheme 3.2).

One covalent bond is formed at the first stage, the compound does not ionize. As chromium is trivalent and can replace three hydrogen atoms, and one dye molecule has been represented to form a bond with Cr/3 and the other bond being the co-ordinate bond (Trotman, 1975). In the case of forming co-ordinate bonding with the azo group following scheme can be shown (Scheme 3.3).

For the chelation to occur, essentially, there should be a salt-forming group, which is usually hydroxyl (but maybe carboxyl) and unsaturated atoms such as oxygen or nitrogen, present in such groups as keto and azo (—N=N—) that is acting as an electron donor to the central metal atom forming co-ordinate bonds.

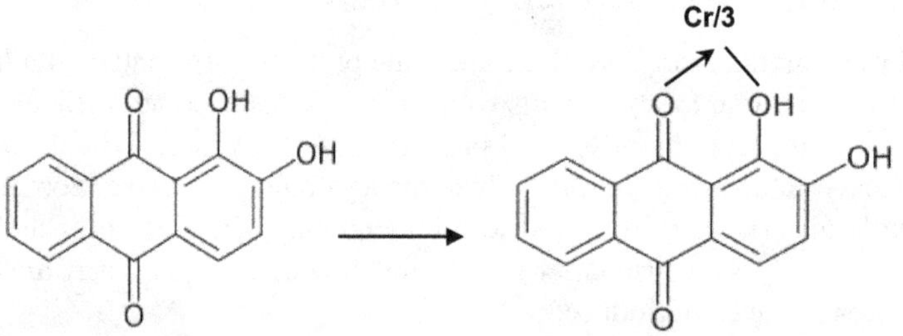

SCHEME 3.2 Mechanism of formation of Chromium complex with Alizarin (Bird, 1972).

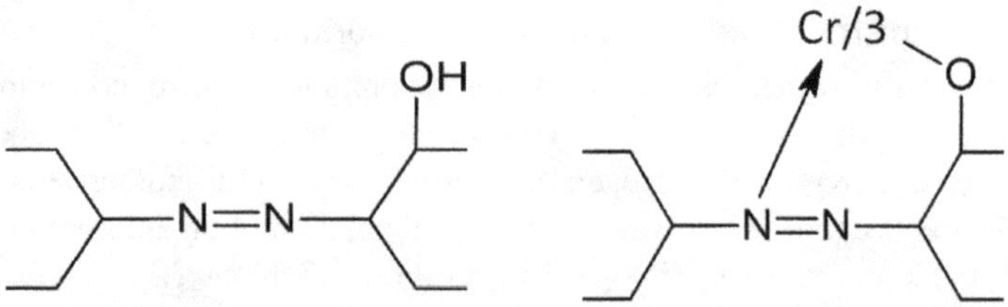

SCHEME 3.3 A figure showing the mechanism of co-ordinate bond formation between Azo group of the Dye molecule and Chromium ion (Bird, 1972).

3.2.2.2 Dyeing using Metal Complex Dyes

Some basic differences are there between the way of application of acid dyes and the application of acid dyeing pre-metalized dyes. Wools are dyed with 1:1 metal complex dyes from a strongly acidic dye bath where dichromate is not added, but with a few exceptions, these chromium-containing dyes require an addition to the dye bath of at least 8% sulphuric acid on the weight of material; with smaller amounts, they give very unsteady dyeing of poor wet-fastness. An extensive boil is indispensable to develop a complete shade. As compared to the acid dyeing process, metal complex dyeing is done without sodium sulfate added to the dye bath because it does no leveling action in this case. Thus the nature of the dyeing process is different from that of the application of acid dyes to wool. In a study on exhaustion of pre-metalized dye (2%, 4 Hours at boil) on wool, it noted that at more than pH – 3, dye bath exhaustion decreases (Burkinshaw et al., 1992). During dyeing of wool with metal complex dyes, four types of linkages are possible, namely following different principles [Dye-SO$_3$$^-$$^+H_3$N–< (acid dying), Dye-Cr$^+$$^-$OOC–< (basic dyeing), Dye-Cr ← NH–< (coordination), Dye---Wool (van der Waals forces of attraction)]. This dyeing process is controlled by the acid concentration and pH. These dyes possess good overall fastness properties and are moderately bright. Neutral dyeing pre-metalized dyes dissoluble as 1:2 Cr or Co metal complex of azo-dye are applied for wool dyes at neutral or slightly acidic dye bath. As being fully chelated, 1:2 dye-metal complexes cannot form coordination with suitable groups in the wool molecule. Previously described acid dyeing mechanism and van der Waals forces of attraction are the key mechanism by which the dyes get attached to the fibres. Dyeing can be done even at pH – 7 (Trotman, 1975).

3.3 COLORATION OF TEXTILES WITH NATURAL DYE

From the historic time for the purpose of coloring textiles, people initiated using nature-based colors; those were extracted from nature are recognized as the natural colorants. They usually yield very special, gentle, and soft shade and also can be regarded as eco-friendly and non-toxic. The natural dyes or colorants include a range of dyes and colors acquired straight from nature (animals, vegetable matters) (Samanta and Konar, 2011). Natural coloration and preparation also involve fewer chemicals, fewer reactions, and at the same time, they are obtained from renewable natural resources, and those are ecofriendly or biodegradable in nature (Evans and McCarthy, 1998).

These dyes usually have less substantivity for textile fibres, and they are applied on textile commonly with a metal salt or other mordants that have an affinity for the natural colorant and also the textile. Natural dyes may be classified as other synthetic dyes from different aspects. The natural dyes do not possess an affinity for them and require chemical assistance to chemically attach the dye molecule to the fiber are classified as mordant dyes, and the substantive dyes can chemically react with the fiber and form a strong bond without any chemical assistance (Sequin-Frey, 1981; Vankar, 2016). These metal salts combine with the dyestuff to produce dye aggregates, which cannot be removed from the cloth easily. Natural dyes used for textile coloration are mainly mordant dyes but can fall in vat, solvent, pigment, direct group of dyes

(NIIR, 2005). The natural dyes can be found to be indexed in the range 75,000–76,999 in the Color Index (Chakraborty, 2015). The invention of synthetic dyes, which easily dyed the textile with bright color and excellent fastness properties, made natural dyes a past. In the recent time, environmental consciousness of consumers as well as various industrial standards related to the environment increases impact on several harmful synthetic dyes, and related processes as they are having a serious threat to mankind; due to this, an interest in natural dyes has gained a lot of momentum (Pargai et al., 2020). It appears that coloration through natural dyes and following an eco-friendly or green process approach is the only harmless alternative (Vankar, 2017). It is very true that all the natural dyes are not necessarily environmentally friendly. In many natural dyes, heavy metals may be present or can contain some other material that may possess toxicity. Hence the natural dyes that are obtained from various natural resources are also essential to be carefully confirmed for any toxicity before they are really taken for use (Samanta and Agarwal, 2009). Different extraction processes are usually carried out for each plant, and subsequently, dyeing can be done with different mordants. Important to mention that some of these plants also have high medicinal values and are truly harmless (Samanta, 2020; Worku, 2018). For the successful commercialization of natural dyeing practices, some suitable and standardized dyeing methods are required to be developed without really loosening the required quality of dyeing. Hence to achieve colors with acceptable fastness and reproducibility, apposite scientific methods need to be developed from systematic experimentation on dyeing procedures, develop understating of the dyeing process variable, dyeing kinetics, and compatibility of natural dyes (Samanta and Agarwal, 2009).

As already mentioned that some transition metals such as chromium, cobalt, tin, etc., can form coordination compounds with dye molecules and act as a bridge or can link the molecule with the textile material natural dyes by chemical attraction or interaction. Dye molecules must possess some mordantable groups (i.e., keto or azo groups) that facilitate the formation of the dye/metal/fiber complexes. According to the activity with the mordant and textile fiber, dyes can be classified based on both their chemical structure and the process of application of the dye. Broadly on the general chemistry of dye as the base for classification, textile dyestuffs are divided into 14 groups: acid, direct, azoic, disperse, sulfur, reactive, basic, oxidation, mordant (chrome), developed dyes, vat, pigments, optical/fluorescent brighteners, and solvent dyes, etc. (Kulkarni et al., 1985). Usually, dyes in a soluble form containing active negative and positively charged groups easily get absorbed in the fiber, forming bonds with the ionic and polar sites available on them. Some exhausting agents can be applied to complete the dye uptake process. On the contrary mordanting (metal ions) are required to get the mordant dyes to get attached to the fiber (Vankar, 2017).

3.3.1 Resources of Natural Dye: Plants, Animal, and Minerals

From various plants, minerals, and animals from which natural coloring matters are extracted. They are extracted from roots, woody stems, plant leaves, fruit, etc., that contain coloring substances that can be used for dyeing textile goods. Some plants have more than one color, depending on the part used. The mordants used in dye-bath can alter the shade and hue of the color. Some natural dye contains

Mordant Dyeing of Textile Fibers

natural mordant too. The dyes can be applied on various textile fibers like wool, silk, cotton, nylon, polyester, etc. Natural resources that are used to extract coloring matter may include the following Jack fruits (*Artocarpus heterophyllus Lam*), Turmeric (*Curcuma longa*), Onion (*Allium cepa*), Hina (*Lawsonia inermis L*), Indigo (*Indigofera tinctoria*), Madder or Manjistha or Rubia (*Rubia tinctorum*), Tea leaves (*Camellia sinensis*), Safflower (*Carthamus tinctorius*), Sappan wood (*Caesalpinia sappan*), Logwood (*Haematoxylon compechianum*), Saffron (*Crocus sativus*), Pomegranate rind (*Punica granatum*), Lac (insect) (*Laccifer lacca Kerr*), Cochineal (insect) (*Dactylopius coccus*). A detailed list of important natural dyes obtained from different sources is given in Table 3.1 (Vankar, 2000). Various mordants used for the application of natural dyes include - alum, stannic chloride, ferrous sulfate, chrome mordants, etc. Natural dyes can also be extracted from colored clays such as mineral ores, red clay, etc. Dyes can be obtained from animal sources such as Lac insect and cochineal, produces good crimson red, scarlet red and pink color for cotton, wool and silk fibers, some other shades can be obtained after mordanting. Natural dyes are obtained from the by-product of lac industry that is extracted from the process effluent. In the same way, a brown dye is obtained as *Cassia tora*, which is a by-product of the gum manufacturing industry. In contemporary research work, a dye extracted from leaves of henna was tried on wool yarn to examine dyeing and antimicrobial activity against some common human pathogens viz. Escherichia coli MTCC 443, Staphylococcus aureus MTCC 902 etc (Yusuf et al., 2012). Biocidal activity of henna dyed wool yarns was then compared with commercial antibacterial and antifungal agents. The dyed woolen yarn samples were found considerably active against tested microorganisms; fastness properties were also found to be quite good. In some other research work with natural dyes extracted from powdered henna (*Lawsonia inermis*) leaves and madder (*Rubiacordifolia*) roots on woolen yarn were dyed with stannous chloride mordant (Yusuf et al., 2015). Various characteristics such as dyeability, shades developed, fastness, and antifungal properties of the dyed woolen yarn. A few shades developed showed considerably good results in terms of color strength determined by K/S value and fastness properties.

3.3.2 Categorizing Natural Mordant Dyes as per Chemical Structure

Mordant dyes can be categorized according to the chemical structure, as azo, anthraquinone, oxazine, xanthene, triphenylmethane, nitroso, etc., in which the azo dyes include a whole variety of range and are of principal importance (Shenai, 1977). Natural dyes can be classified according to the hue or as per chemical nature. Natural dyes include a wide variety of chemical categories, namely anthraquinones, naphthoquinones, indigoids, carotenoids, flavones, dihydropyrans, anthocyanidin, and flavonols can be mentioned. Some representative chemical structures of these dyes are shown in Figure 3.1.

3.3.2.1 Anthraquinone Dyes

Some significant red colorants have a chemical structure obtained from both animal and plant resources. They have marked fastness against light and forming complexes of the metal show better wash fastness properties also. Some examples of these dyes

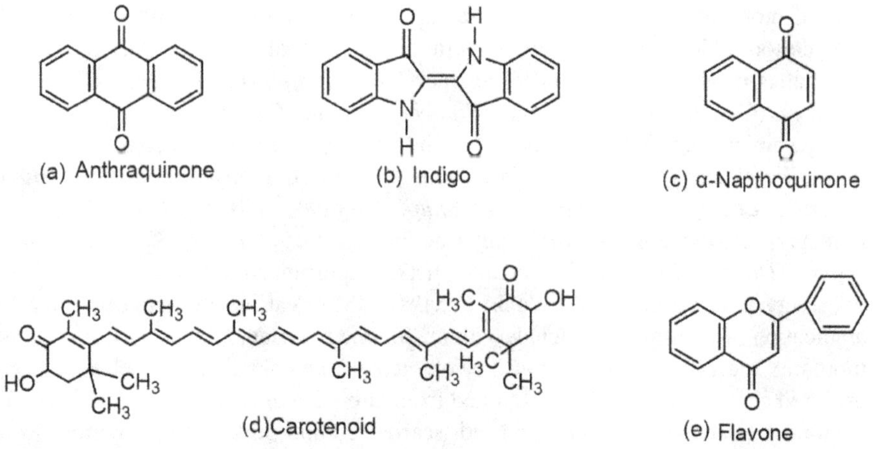

FIGURE 3.1 Chemical structure of some common natural dyes (Chakraborty, 2015).

are alizarins extracted from madder root (*Rubiatinctorum L*), the lac dye and kermesic acid or kermes (obtained from insects), scarlet dyes, cochineal (extracted from crushed bodies of insects, coloring constituent is carmine).

3.3.2.2 Indigoid Dyes

Indigo and purple natural dyes are two significant dyes that have indigoid structures. Indigo dyes are perhaps the oldest natural colorants used by humans. Historically this dye is being used by the Indians, as it was developed first a long 4000 years back. This occurs as *glucoside indican is* obtained from the plant tinctoria.

3.3.2.3 α-Naphthoquinone Dyes

The most noticeable dyestuff in naphthoquinone class is the lawsone or henna. The dye is extracted from *Henna* or *Hina*, which is principally grown in Egypt and India. One more comparable dye molecule is *juglone* (natural brown), extracted from the shell of green walnut.

3.3.2.4 Carotenoid Dyes

"Carotenoid dyes," the class name, results from the orange pigment originating in the carrots. For these chemicals, the color is due to the presence of extended series of the conjugated double bond. Some other common cases of carotenoid structure-based dyes are annatto, saffron, etc., to be mentioned.

3.3.2.5 Flavonoid Dyes

Flavonoids (polyphenolic secondary metabolites), which generate yellow color, are classified as flavones (colorless organic compounds), aurones, chalcones, and anthocyanins. The majority of the natural yellow dyes obtained are by-products of flavones and isoflavones. For instance, dyer's weed or weld that is rich with luteolin pigment results in bright and fast coloration of natural protein fibers.

Mordant Dyeing of Textile Fibers

3.3.2.6 Dihydropyran Dyes

Dihydropyran colors include brazilin that is derived from a Brazilian tree known as Pernambuco wood, also hematoxylin obtained from logwood. Aqueous or alkaline extraction of the dye is in practice can be applied on textile fibers without alums. These dyes can be applied on natural protein fibers, wool, and silk for dark shades.

The dye from Turmeric (a popular natural dye) is derived from the stems of Curcuma longa. The chemical present as the dye is curcumin, which belongs to the diaroylmethane class.

Some other natural dyes include berberine extracted from *Berberis aristata* etc. is isoquinoline alkaloid which has a bright yellow color. Another class of natural dyes is betalains, which are water-soluble tyrosine-derived plant pigments obtained from plant Caryophyllales that comprise yellow betaxanthins and the violet betacyanins [sources are *Opuntia lasiacantha* and *Beta vulgaris* (beetroot)]. Some vegetable products can be found as the source of tannin that can be extracted from, for example, using dry myrobolan (*Terminalia chebula*) fruits that have high tannin content and natural dyes too, can be used for achieving bright yellow shades for textile materials.

3.3.3 EXTRACTION OF NATURAL DYES

The extraction efficiency of coloring ingredients present in natural plant/animal/mineral origins depends on the type of medium (aqueous/organic or acid/alkaline solvents), pH of the medium and extraction conditions such as temperature, time, material-to-liquid ratio, and particle size of the substrate.

3.3.3.1 Aqueous Extraction

The part of plants is dried first and then treated in an aqueous medium under varying conditions to get the desired colorant.

3.3.3.2 Organic Media/Mixed Aqueous and Organic Solvents

Supercritical carbon dioxide fluid can be used for extraction and purifying natural colorants. Organic solvents, ethanol, methanol are common chemicals. In the soxhlet apparatus used for extraction with organic solvents, a mixture of solvent and water was used to extract natural colorants. A reflux condenser was also used in the process.

3.3.3.3 Acid/Alkali Extraction

Hydrogen chloride, sodium carbonate in an aqueous medium can be used to extract the colorant. (Samanta and Agarwal, 2009).

3.3.4 MORDANTS

A mordant is an uncomplicated chemical compound that has an affinity for every fiber further to dye. Mordants act as a bridge that attracts both dye molecules and fibers together by means of forming a complex. These mordants are fundamentally metallic salts and cationic in nature.

In the following table, a detailed list of various dyes is given after(Vankar, 2017), their sources, from where the dyes are extracted, and the mordant used for dyeing textile substrate. There are certain important mordants often found in use, namely aluminum-based mordants, $K_2Cr_2O_7$, $FeSO_4$, $CuSO_4$, $ZnSO_4$, tannin, and tannic acid, etc. (Maulik et al., 2005; Nalankilli, 1997). These metallic mordants are used for developing a wide range of shades as they complexes with the biological coloring ingredients, but almost all of the metals are harmful, and only in bit quantity being present may be safe for the wearer.

TABLE 3.1
Classification of Natural Dyes Based on the Color and Mordants Used

Color	Botanical Name	Parts Used	Mordant
Red Dyes			
Safflower	*Carthamu stinctorious*	Flower	-
Caesalpina	*Caesalpinia sappan*	Wood chips	Alum
Maddar	*Rubia tinctorium*	Wood	Alum
Lac	*Coccus lacca (insect)*	Twigs inhabited Stannic chloride by these insects	Stannic chloride
Yellow Dyes			
Bougainvillea	*Bougainvillea glabra*	Flower	Tin
Golden rod	*Solidago grandis*	Flower	Alum
Teak	*Tectona grandis*	Leaves	Alum
Marigold	*Tagetes species*	Flower	Chrome
Parijata	*Nyetanthes arbortristis*	Flower	Chrome
Blue Dyes			
Indigo	*Indigofera tinctoria*	Leaves	-
Woad	*Isatis tinctoria*	Leaves	-
Suntberry	*Acacia nilotica*	Seed pods	-
Pivet	*Ligustrum vulgare*	Mature berries after frost	Alum and iron
Water lily	*Nymphaea alba*	Rhizomes	Iron and acid
Black Dyes			
Lac	*Coccus lacca (insect)*	Twigs inhabited by these insects	Ferrous sulphate
Alder	*Alnus glutinosa*	Bark	Ferrous sulphate
Rofblamala	*Loranthus pentapetalus*	Leaves	Ferrous sulphate
Custard apple	*Anona reticulata*	Fruit	-
Harda	*Terminalia chebula*	Fruit	Ferrous sulphate
Brown Dyes			
Caesalpina	*Caesalpinia sappan*	Wood chips	Ferrous sulphate
Bougainvillea	*Bougainvillea glabra*	Flower	Ferrous sulphate + acid/neutral Alum

(Continued)

TABLE 3.1 (*Continued*)
Classification of Natural Dyes Based on the Color and Mordants Used

Color	Botanical Name	Parts Used	Mordant
Balsam	*Impatiens balsamina*	Flower	
Marigold	*Tagetes species*	Flower	Chrome
Black berries	*Rubus fructicosus*	Berries	Iron
Green Dyes			
Tulsi	*Ocimum sanctum*	Leaves	Ferrous sulphate
Bougainvillea	*Bougainvillea glabra*	Flower	Alum + Base and Ferrous sulphate + Acid
Canna		Flower	Alum and Ferrous sulphate
Lily	*Convallaria majalis*	Leaves and stalk	Ferrous sulphate
Nettles	*Urtica diocia*	Leaves	Alum
Orange/Peach Dyes			
Bougainvillea	*Bougainvillea glabra*	Flower	Stannous chloride + Acid; Alum + base
Balsam	*Impatiens balsamina*	Flower	Tin
Dahlia	*Dahlia* species	Flower	Alum/chrome
Annatto	*Bixa orellana*	Seeds	-

Source: Vankar, 2000; Yusuf et al., 2018; Sachan and Kapoor, 2007; Saleh et al., 2013; Samanta and Konar, 2011.

3.3.5 Application of Mordants on the Textile Material

In the premordanting process, as a first step, mordant is applied to textile material and then dyeing carried out with dye. In the meta-mordanting process, there is simultaneous application of mordant, and dye is applied in the same dye bath. Whereas in the post-mordanting method, it is done as an after-treatment of the dyed material.

3.3.6 Types of Mordant

Mainly there are three types of mordants that can be found are metal salts, tannic acid, and oil mordants.

3.3.6.1 Metal-Based Mordants

Metal mordants have been mentioned already in this chapter discussion that metal salts of Transition elements can act as mordants. They can be further subdivided as metallic mordants, can be in addition divided into two types, i.e., brightening mordants (viz. alum, $K_2Cr_2O_7$, and tin ($SnCl_2$) fall under the group of brightening mordants) and the dulling mordants such as $CuSO_4$ and $Fe_2(SO_4)_3$ (Prabhu and Bhute, 2012).

3.3.6.2 Tannin

Historically the term 'tannin' was first coined by Seguin in 1796 (Arapitsas, 2012) and used to call the process of converting animal pelts into the leather by using natural plant extracts from different parts of some plants. The name 'tannin' is originated from the French word 'tanin' (tanning substance), and it is a polyphenol biomolecule (Khanbabaee and Van Ree, 2001). Tannins are the most significant chemical component that is used to produce mordanted yellow, black, brown, and grey color. They can modify the affinity of the fibers for different dyes. A post tannin treatment can improve the fastness properties of some dyes, but its application usually makes the dyeing duller. They are chemicals of high molecular weight (500–3000) containing phenolic groups that can form crosslinks with other macromolecules. Tannins have *o-dihydroxy(catechol)* groups by which they may form metal chelates, producing various colors from different metals. Tannins form hydrogen bonds, ionic bonds, and covalent bonds as well with the protein and cellulosic fibers. The stability of these bonds depends upon some factors such as pH, ionic strength, and metallic chelators. The chemical formula of tannin can be $C_{76}H_{52}O_{46}$ with molecular weight ~1701, mp-220°C (Ramakrishnan et al., 2006), and they include a large class of compounds with different chemical structures. Structurally plant based tannins can be divided into two groups, such as hydrolyzable and condensed tannins on the ground of type of phenolic nuclei and nature of complex formation.

3.3.6.2.1 Hydrolysable Tannins

This category of tannins is characterized by having a core of polyhydric alcohol (Arapitsas, 2012). The structure of this tannin contains several molecules of polyphenolic acids, namely gallic and ellagic acids. These tannins are a type of tannin that, on heating with HCl or H_2SO_4, hydrolyses to produce gallic or ellagic acids.

These tannins structurally include numerous polyphenolic acid molecules viz. gallic, hexahydrodiphenic, and ellagic acids (Figure 3.2). Based on the phenolic acids formed as a result of hydrolysis of these tannins, they can be further characterized as gallotannins consisting of gallic acid or ellagitannins that contain hexahydrodiphenic acid. The group of hydrolyzable tannins is water-soluble, produces blue color with ferric chloride ($FeCl_3$) (Mueller-Harvey, 2001). Few important raw material sources for these tannins are myrobolan fruit, oak bark, sumac leaves, gallnuts, pomegranate rinds, etc.

3.3.6.2.2 Condensed Tannins

This class of tannins is not easily hydrolyzable using mineral acids or enzymes, so they are also mentioned as non-hydrolyzable tannins too. They are polyhydroxy phenols soluble in water, alcohol, and acetones, able to coagulate proteins.

The chemical structure of condensed tannin is shown in Figure 3.3 (Pizzi, 2014). The term proanthocyanidins are sometimes alternatively used for these tannins. The compounds containing condensed tannins contain only phenolic nuclei, which are bio-synthetically plant-derived flavonoids. Catechin that is present in tannins is flavan-3-ol, whereas leucoanthocyanidins are flavan-3,4-diol structures. These

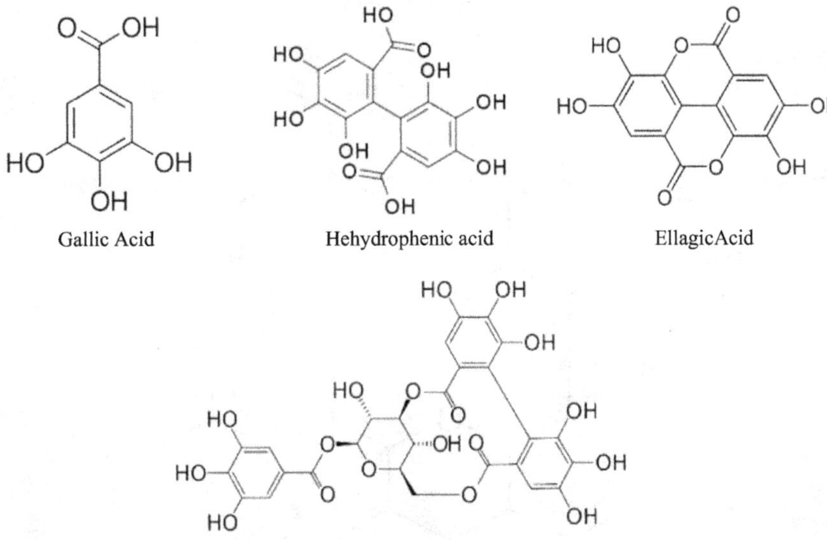

Mordant Dyeing of Textile Fibers

FIGURE 3.2 Chemical structures of Gallic Acid, Hehydrophenic acid, Ellagic Acid, Corilagin(Gallotamine) (Prabhu and Bhute, 2012).

phenolics are often bonded to the carbohydrates or protein molecules to form further complicated tannin compounds. When reacted with acids or enzymes, they have a tendency to polymerize, yielding insoluble red-colored chemicals called phlobaphenes. The phlobaphenes provide typical red colors to many medicines, for example as cinchona and wild cherry bark. Following the dry distillation process, they produce catechol derivatives. Condensed quality tannins are also water-soluble and result in eco-friendly textile coloration using $FeCl_3$.

3.3.6.3 Oil-Mordants

Oil mordants are commonly used to dye using Turkey Red color extracted from madder root, which produces a very fast color. Oil mordant combine with the principal mordant, alum, to form a complex. Due to the presence of sulfonic acid groups ($-SO_3H$), sulfonated oils have a higher metal binding capacity, allowing them to form complexes with metal ions and improving dye fastness.

3.3.7 Recent Trend of Bio-Mordants

Usually, some metallic mordants based on Fe, Al, Cu, Pb, and Sn are applied in dyeing. These metals can be present in minute amounts in plants and act as a nutrient in plant growth. These metal ions can function like a chelating agent that causes dye fixation. Nowadays, it is well known that metallic mordant may cause some serious issues from an ecological point of view (Rahman et al., 2013). In a relatively recent approach, plants with tannins and metal have been utilized as biomordant.

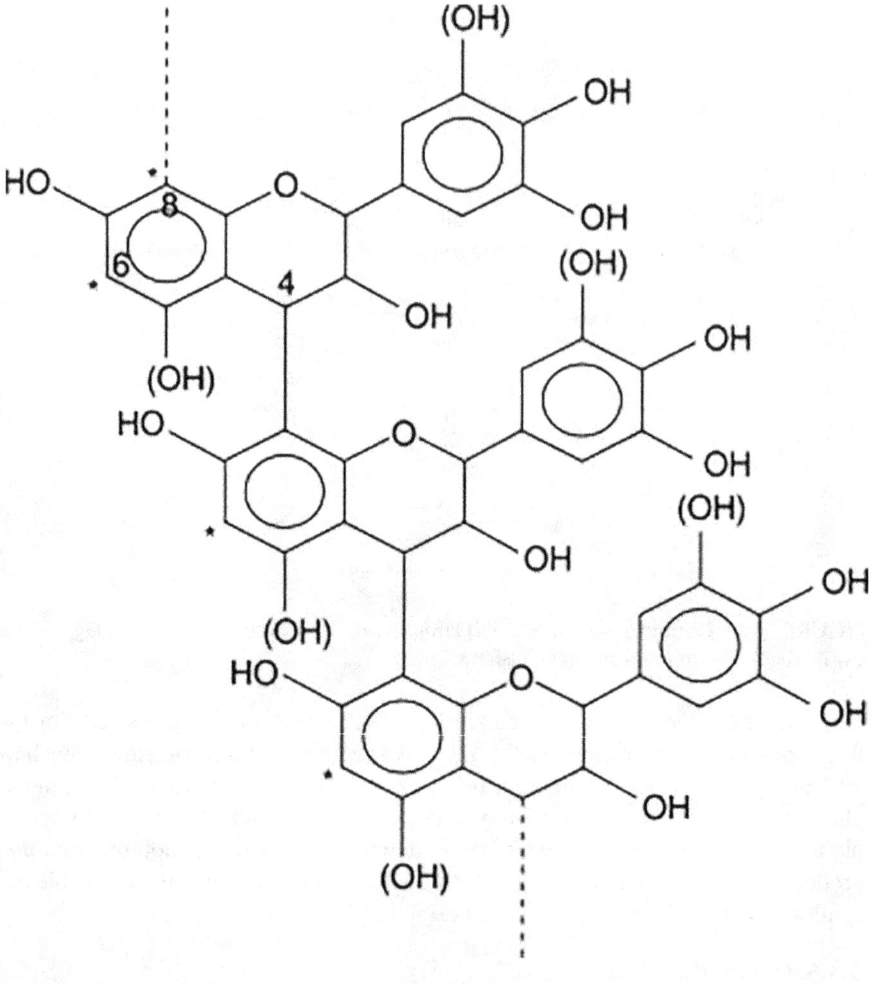

FIGURE 3.3 Schematic representation of Structure of Condensed Tannin (Pizzi, 2014).

Biomordant *Eurya acuminata* (Nausankhee) has been attempted to discard the use of metal mordant, with *Rubia cordifolia,* an anthraquinone dye obtained from its roots, stems, and leaves, for dyeing of cotton fibers (Vankar et al., 2008). It has been shown that *Pyruspashia* can be applied as a source of biomordant. Many *Pyrus* and *Prunus*species contain copper, including *Pyrusdomestica L.* (0.33–34 ppm copper), *Prunusserotina* or black cherry (1.3–378 ppm copper), and *Prunuspersica (L.)* or peach (fruit) (0.3–30 ppm copper); in addition, willow oak(stem), *Liquidambar styraciflua L., Brassica oleracea L. var. capitata L., Corylusavellana L.,* and *Sassafras albidum* [http://www.levity.com/alchemy/metals_i.html]. Bio-mordants have also been applied for many natural dyes in the case of wool dyeing (Jayalakshmi and Amsamani, 2007).

Anthocyanin-metal complexes are a practical option for color stabilization, especially when the metals involved do not cause a risk of polluting the environment.

Mordant Dyeing of Textile Fibers

According to some studies on the stability of plant color, the blue colors have been proposed to be caused by a complex formation of anthocyanin with metals such as Al, Fe, Cu, and Sn (Muhan and Sibiao, 1991).

Though anthocyanins are the natural colorants available in nature with a wide range of colors, orange, pink, red, violet, blue, etc., their use is restricted due to lower extraction percentages and poor stability. Nevertheless, anthocyanins extracted from black carrots found to be more stable over a wide pH range than that is extracted from other fruit, thus causing them ideal for use (Shukla and Vankar, 2013; Castañeda-Ovando et al., 2009).

The higher copper content shows stronger and more suitable chelation with the colorants for better dye association. The existence of the 4-oxo groups in quercetin (plant flavonol) in combination with the −OH group facilitates chelation in flavonoids. Chelation of copper between the 4-oxo group and −OH group in flavonol and flavones was suggested by concerned researchers (Mira et al., 2002; Brown et al., 1998). The frequency of OH groups is very important; the more they are in number, the better will be chelation. Similarly, the copper ion in *Pyrus* facilitates the chelation of Cu (II) with the colorant ingredients such as flavones and flavonols. As per the previous research observations, the probable mode of chelation of copper in *Pyrus* with *quercetin* has been proposed, which is similar to the metal-dye complex formation. In some recent studies, anthraquinone-based colorants extracted from the powder of *Rubia cordifolia* roots have been applied on wool which was pre-mordanted with biomordant *Acacia catechu*, as a replacement of metallic mordants. On dyeing with wool shades obtained, dye uptake, color, and fastness properties were assessed to be very good (Yusuf et al., 2016). In another experimental study, woolen yarn substrate was dyed with natural anthraquinone colors obtained from madder roots using gallnut (*Quercusinfectoria*) extract as a mordant, and a broad range of attractive and colorfast shades (located in the red-yellow quadrant of the CIELab color space) was obtained. Using gallnut extract as the anchoring agent (5%) resulted in increase in the fastness properties, and better results were obtained using it as a pre-treatment (Yusuf et al., 2016).

3.3.8 ENZYMES MORDANT

Apart from the previously discussed mordants such as tannin and oil mordants, various metals viz. iron, aluminum, copper, tin are frequently used in dyeing, but some serious ecological constraints are associated with these applications (Vankar et al., 2007). Different metal complexes result in different colors from the same natural dyes, and there is an inclination to use various metals for producing different shades. However, there are restrictions to the use of different metals from an ecological point of view. Maximum allowable amounts of various metals in the final product are as per the well-known "German ban" as follows: As (1.0 ppm), Pb (1.0 ppm), Cd (2.0 ppm), Cr (2.0 ppm), Co (4.0 ppm), Cu (50 ppm), Ni (4.0 ppm), and Zn (20 ppm) (Gulrajani, 2001). The upper limits of the presence of various metals vary from product to product; for example, for iron, tin, and aluminum, there is no upper limit, and it is fairly high for copper. These metals can be used for complexing and mordanting purposes. It has been found that Natural dye extracted from the *Delonix* flower can be used combined using some enzymes—such as protease, lipase, etc. for dyeing of silk, and the source of biomordant is *Pyruspashia* (Vankar and Shanker, 2009). Wool fibers have been reported to be dyed well using casein enzymes with some natural dyes, namely juglone, lawsone, berberine, and quercetin (Doğru et al., 2006). More recently, natural pigment from *Sappan* has been reported to use for wool fibers dyeing after being treated with enzymes such as *protease* and *transglutaminase* (Zhang and Cai, 2011). In that sense, anionic and cationic sites can similarly be created by various means to facilitate or create affinity for the dyes can also be considered as a mordanting process (Gulrajani et al., 2001).

3.3.9 General Method of Fabric Mordanting and Natural Dyeing

Usually, as a first step fabric sample to be dyed is weighed with accuracy and treated with some suitable metal salts; in general, a premordanting is done. The mordant solution may be prepared with water, typically to make an M:L ratio of 1:50. The soaked sample is put into the solution of the mordant, followed by heating. The temperature of the mordanting bath is to be raised to 60°C for over thirty mins and kept in that stage for extra thirty mins after which treated textile material is squeezed and dried. Immediate dyeing is required for the mordant dyeing of silk and wool, as some mordants can be very sensitive to light (Vankar, 2017).

For dyeing with natural dyes, prewashed and clean cotton fabric is to be pre-treated with Tannic acid. Subsequently, alum, tin, and copper mordants can be prepared, where the cream of tartar can be used suitably as per demand. Chrome mordants are considered best and can be applied with lukewarm water. Photosensitivity of the solution needs to be taken care of, as exposure to light may result in unevenness in dyeing. Some health-related issues are also of serious concern with these mordants. In a similar process, iron-based mordants can also be applied to cotton fabric substrates. Silk and wool fabric can also be mordanted following some similar ways. Metal ions- dye complexes form bonds with the available sites on the fibers.

3.3.10 Ecological Aspects: Natural Dyeing with Mordants

It is mentioned that dyes which are extracted from nature are non-carcinogenic, non-allergic, and easily biodegradable. Concerned Textile dye houses are consistently involved in extraction, application and improving the fastness of the natural dyes (Blackburn, 2004; Saleh et al., 2013). It is well known now that these dyes have numerous disadvantages like low exhaustion, the difficulty of the dyeing process, reproducibility of shades and standard extraction process, etc. (Sachan and Kapoor, 2007). Depending on the concentration level of metals present in finished textiles, metallic salts of aluminum, potassium, tin, iron, or copper that forms a complex with the natural dye molecules, can cause numerous skin and other ailments including dermatitis, irritation, cancer, etc (Leme et al., 2014; Rovira et al., 2015; Rovira and Domingo, 2019; Vankar, 2000; Bechtold et al., 2003; Yusuf et al., 2017). For ecological risk analysis, the presence of metals can be calculated from the remaining mordant bath and the wastewater and levels of metal traces present in the finished fabric (Repon et al., 2017). The textile industry causes environmental pollution due to effluents or wastewater produced from chemical processing plants. A way of reducing pollution is to use natural/biodegradable materials. Natural dyes have a low affinity to the fiber and considering the biocompatibility issues of the dyeing oak may be applied as a source of tannin mordant (Hosseinnezhad et al., 2021). Market and industrially natural dye is promising and is still growing, but its contribution to the textile business is very small. The greater part of it involves coloring food items because the impetus on healthier diets is expanding (Downham and Collins, 2000; Křížová, 2015). Improving the traditional mordant dyeing processes and choosing new mordants in place of traditional heavy metal based process should be an important part of the development of natural textile dyeing. If the biomordants are used for improving the fastness properties of natural dyes, then bulk production can be possible. Standardization of different biomordant for different categories of natural dyes is to be done in the latest years to make our environment and human beings safe from the toxic synthetic dyes and metals salts.

3.4 CONCLUSION

Though the application of mordant dyes with different metal complexes sounds attractive from the dyers and experimenter's point of view, many serious environmental aspects and challenges are associated with it. Metal mordants or complex-forming chemicals are often toxic in use, or they may affect the environment by discharging toxic effluents. As alternatives, some natural mordants such as Tannins and oil mordants can be applied along with some safe metal oxide such as *alum* to dye a textile material. Natural dyes obtained from various natural resources often show some promise as being environmentally friendly. Apart from the limitations of poor fastness of natural mordanted dyeing products, some possibilities in it can be further explored.

REFERENCES

Arapitsas, P. 2012. Hydrolyzable tannin analysis in food. *Food chemistry* 135(3): 1708–1717.

Bechtold, T., Turcanu, A., Ganglberger, E. and Geissler, S. 2003. Natural dyes in modern textile dyehouses—how to combine experiences of two centuries to meet the demands of the future? *Journal of Cleaner Production* 11(5): 499–509.

Bird, C. L. 1972. *The theory and practice of wool dyeing*. Bradford: Society of Dyers and Colorists.

Blackburn, R. S. 2004. Natural polysaccharides and their interactions with dye molecules: applications in effluent treatment. *Environmental Science & Technology* 38(18): 4905–4909.

Brown, E. J., H. Khodr, C. R. Hider, and C. A. Rice-Evans 1998. Structural dependence of flavonoid interactions with Cu^{2+} ions: implications for their antioxidant properties. *Biochemical Journal* 330(3): 1173–1178.

Burkinshaw, S. M. 1992. Dyeing Wool with Metal-Complex Dyes, in *Wool Dyeing*, ed. D. M. Lewis, 229–247. Bradford, England: Society of Dyers and Colorists.

Castañeda-Ovando, A., de Lourdes Pacheco-Hernández, M., Páez-Hernández, M.E., Rodríguez, J.A. and Galán-Vidal, C. A. 2009. Chemical studies of anthocyanins: a review. *Food Chemistry* 113(4): 859–871.

Chakraborty, J. N. 2015. *Fundamentals and practices in coloration of textiles.* New Delhi: CRC Press.

Doğru, M., Baysal, Z. and Aytekin, Ç. 2006. Dyeing of wool fibers with natural dyes: effect of proteolytic enzymes. *Preparative Biochemistry & Biotechnology* 36(3): 215–221.

Downham, A. and Collins, P. 2000. Coloring our foods in the last and next millennium. *International Journal of Food Science & Technology* 35(1): 5–22.

Evans, E. and McCarthy, B. 1998. Biodeterioration of natural fibers. *Journal of the Society of Dyers and Colorists* 114(4): 114–116.

Gulrajani, M. L. 2001. Present status of natural dyes. *Indian Journal of Fiber and Textile Research* 26: 191–201.

Gulrajani, M. L., Srivastava, R. C. and Goel, M. 2001. Color gamut of natural dyes on cotton yarns. *Coloration Technology* 117(4): 225–228.

Hosseinnezhad, M., Gharanjig, K., Jafari, R., Imani, H. and Razani, N. 2021. Cleaner colorant extraction and environmentally wool dyeing using oak as eco-friendly mordant. *Environmental Science and Pollution Research* 28(6): 7249–7260.

http://www.levity.com/alchemy/metals_i.html

Jayalakshmi, I. and Amsamani, S. 2007. Bio-mordant for wool. *Man-Made Textiles in India* 50(7): 267–270

Khanbabaee, K. and T. Van Ree. 2001. Tannins: classification and definition. *Natural Product Reports* 18(6): 641–49.

Kulkarni, S. V., Blackwell, C. D. Blackard, A. L. Stackhouse, C. W. and Alexander, M. W. 1985. *Textile dyes and dyeing equipment: classification, properties and environmental aspects.* Research Triangle Park, NC: US Environmental Protection Agency,

Leme, D. M., de Oliveira, G. A. R., Meireles, G., dos Santos, T.C., Zanoni, M. V. B. and de Oliveira, D. P. 2014. Genotoxicological assessment of two reactivedyes extracted from cotton fibers using artificial sweat. *Toxicology in Vitro* 28(1): 31–38.

Maulik, S. R. and S. C. Pradhan. 2005. Dyeing of wool and silk with Hinjal bark, Jujube bark and Himalayan rhubarb. *Colorage* 52(9): 67–71.

Mira, L., Tereza Fernandez, M., Santos, M., Rocha, R., Helena Florêncio, M. and Jennings, K.R. 2002. Interactions of flavonoids with iron and copper ions: a mechanism for their antioxidant activity. *Free Radical Research* 36(11): 1199–1208.

Mueller-Harvey, I. 2001. Analysis of hydrolysable tannins. *Animal Feed Science and Technology* 91(1–2): 3–20.

Muhan, Z. and Sibiao, L. 1991. The effect of metal ions on mulberry and Rhododendron pigments. Shipin Yu Fajiao Gongye 6: 82.

Nalankilli, G. 1997. Application of tannin in the coloration of textiles. *Textile Dyer & Printer* 30(27): 13–15.

NIIR Board of Consultants and Engineers. 2005. *The Complete Book on Natural Dyes and Pigments.* New Delhi: Asia Pacific Business Press.

Pargai, D. S. Jahan, and Gahlot, M. 2020. Functional properties of natural dyed textiles, in *Chemistry and technology of natural and synthetic dyes and pigments*, ed. A. K. Samanta, 1–19. London: Intech.

Pizzi, A. 2014. *Types, processing and properties of bioadhesives for wood and fibers. in Advances in Biorefineries*, ed. K. Waldron, Amsterdam: Woodhead Publishing, (https://doi.org/10.1533/9780857097385.2.736).

Prabhu, K. H. and Bhute, A. S. 2012. Plant based natural dyes and mordants: a review. *Journal of Natural Product and Plant Resources* 2(6): 649–664.

Rahman, A., Urabe, T. and Kishimoto, N. 2013. Color removal of reactive procion dyes by clay adsorbents. Procedia Environmental Sciences 17: 270–278.

Ramakrishnan, K. S. R. Selvi, and Shubha, R. 2006. Tannin and its analytical techniques. *Indian Chemical Engineer* 48(2): 88–89.

Rattee, I. D. 1953. The use of non ionic level dyeing assistants with chrome complex dyes. *Journal of the Society of Dyers and Colorists* 69: 288–295.

Repon, M. R., Islam, M. T. and Al Mamun, M. A. 2017. Ecological risk assessment and health safety speculation during color fastness properties enhancement of natural dyed cotton through metallic mordants. *Fashion and Textiles* 4(1): 1–17.

Rovira, J. and Domingo, J. L. 2019. Human health risks due to exposure to inorganic and organic chemicals from textiles: a review. *Environmental Research* 168: 62–69.

Rovira, J., Nadal, M., Schuhmacher, M. and Domingo, J. L. 2015. Human exposure to trace elements through the skin by direct contact with clothing: Risk assessment. *Environmental Research* 140: 308–316.

Sachan, K. and Kapoor, V. P. 2007. Optimization of extraction and dyeing conditions for traditional turmeric dye. *Indian Journal of Traditional Knowledge* 6(2): 270–278.

Saleh, S. M., Abd-El-Hady, Y. A. and El-Badry, K. 2013. Eco-friendly dyeing of cotton fabric with natural colorants extracted from banana leaves. *International Journal of Textile Science* 2(2): 21–25.

Samanta, A. K. and Agarwal, P. 2009. Application of natural dyes on textiles. *Indian Journal of Fiber and Textile Research* 34: 384–399.

Samanta, A. K. and Konar, A. 2011. Dyeing of textiles with natural dyes, in *Natural dyes*, ed. E. P. A. Kumbasar, 30–56. Rijeka, Croatia: InTech,.

Samanta, P. 2020. A review on application of natural dyes on textile fabrics and its revival strategy, in *Chemistry and technology of natural and synthetic dyes and pigments*, ed. A. K. Samanta and N. Awwad, 1–25, London: InTech.

Sequin-Frey, M. 1981. The chemistry of plant and animal dyes. *Journal of Chemical Education* 58(4): 301–305.

Shenai, V. A. 1977. *Technology of textile processing.* Vol. 6. Ahmedabad: Sevak Publisher.

Shukla, D. and Vankar, P. S. 2013. Natural dyeing with black carrot: new source for newer shades on silk. *Journal of Natural Fibers* 10(3): 207–218.

Trotman, E. R. 1975. *Dyeing and chemical technology of textile fibers.* London: Charles Griffin.

Vankar, P. S. 2000. Chemistry of natural dyes. *Resonance* 5(10): 73–80.

Vankar, P. S. 2016. *Handbook of Natural Dyes for Industrial Applications.* New Delhi: NIIR Project Consultancy Services.

Vankar, P. S. 2017. *Natural dyes for textiles: Sources, chemistry and applications.* Cambridge: Woodhead Publishing.

Vankar, P. S., Shanker, R. and Verma, A. 2007. Enzymatic natural dyeing of cotton and silk fabrics without metal mordants. *Journal of Cleaner Production* 15(15): 1441–1450.

Vankar, P.S. and Shanker, R. 2009. Potential of Delonix regia as new crop for natural dyes for silk dyeing. *Coloration Technology* 125: 155–160.

Vankar, P.S., Shanker, R., Mahanta, D. and Tiwari, S.C. 2008. Ecofriendly sonicator dyeing of cotton with Rubia cordifolia Linn. using biomordant. *Dyes and Pigments* 76(1): 207–212.

Vigo, T. L. 2013. *Textile processing and properties: Preparation, dyeing, finishing and performance.* Amsterdam: Elsevier.

Worku, A. 2018. Extraction and application of natural dye on tanned leather an eco-friendly approach. *International Research Journal of Engineering and Technology* 5: 431–434.

Yusuf, M., Ahmad, A., Shahid, M., Khan, M. I., Khan, S. A., Manzoor, N. and Mohammad, F. 2012. Assessment of colorimetric, antibacterial and antifungal properties of woollen yarn dyed with the extract of the leaves of henna (Lawsonia inermis). *Journal of Cleaner Production* 27: 42–50.

Yusuf, M., Mohamad, F., Shabbir, M. and Khan, M. A. 2017. Eco-dyeing of wool with Rubia cordifolia root extract: Assessment of the effect of Acacia catechu as biomordant on color and fastness properties. *Textiles and Clothing Sustainability* 2(1): 1–9.

Yusuf, M., Shahid, M., Khan, M. I., Khan, S. A., Khan, M. A. and Mohammad, F. 2015. Dyeing studies with henna and madder: A research on effect of tin (II) chloride mordant. *Journal of Saudi Chemical Society* 19(1): 64–72.

Zhang, R. P. and Cai, Z. S. 2011. Study on the natural dyeing of wool modified with enzyme. *Fibers and Polymers* 12(4): 478–483.

4 Insights into Electrochemical Technology for Dyeing of Textiles with Future Prospects

Amit Madhu
The Technological Institute of Textile and Sciences

CONTENTS

4.1 Introduction 58
4.2 Electrochemical Dyeing: State of the Art... 60
4.3 Types of Electrochemical Dyeing Processes 61
    4.3.1 Direct Electrochemical Reduction... 61
        4.3.1.1 Direct Electrochemical Reduction in Sulfur Dyeing........... 62
        4.3.1.2 Direct Electrochemical Reduction in Vat Dyeing 64
        4.3.1.3 Direct Electrochemical Reduction of Dye Pigment
                via the Radical Process 64
        4.3.1.4 Direct Electrochemical Reduction on Graphite Electrode.... 65
    4.3.2 Indirect Method or Mediator-Enhanced Electrochemical Reduction 69
        4.3.2.1 Mediator Systems... 70
        4.3.2.2 Organic Mediator Systems.. 70
        4.3.2.3 Inorganic Mediator Systems 72
        4.3.2.4 Indirect Electrochemical Reduction of Dyes...................... 74
        4.3.2.5 Requirements for a Mediator System................................. 76
        4.3.2.6 Reducing Power of Mediator Systems 77
        4.3.2.7 Indirect Electrochemical Dyeing Process........................... 77
    4.3.3 Electrocatalytic Hydrogenation 79
4.4 Advantages of Electrochemical Dyeing Process 83
    4.4.1 Environment 83
    4.4.2 Economy 83
    4.4.3 Health.. 83
    4.4.4 Dyeing Quality 83
4.5 Future Prospects 84
4.6 Conclusion 84
References... 85

DOI: 10.1201/9781003140467-4

4.1 INTRODUCTION

In the coloration of cellulosic textiles, the vat and sulfur dyes are the most commercial dyes after the reactive ones. Vat dyes produce colored textiles that are extremely resistant to washing, light, rubbing, and chlorine bleaching. Sulfur dyes are very significant for the manufacturing of low-cost items with moderate fastness needs. The dyes are light and washing resistant but not chlorine resistant. Therefore, these dyes consume a relatively large segment (35%) of the dyestuff market, and in the near future, their contribution even can be increased. Both vat and sulfur dyes are water-insoluble, and for the application, these are converted to their leuco form on reduction followed by solubilization. In their reduced or leuco form, they become water-soluble and exhibit substantivity for cellulosic textiles. After dyeing, when the dye gets adsorbed into the fiber, these are reconverted to their original pigment form by the process of oxidation.

Chemically, vat dyes are either indigoid or anthraquinone-based structures having carbonyl groups (–C=O) as the chromophores in them, and during reduction, these carbonyl groups are changed to hydroxyl (–C–OH) groups and then to soluble sodium salt derivatives (–C–ONa) in the presence of alkali. In chemical structures of sulfur dyes, the sulfur linkages are an essential component of the chromophore, and a significant amount of sulfur exists in the form of sulfide (–S–), disulfide (–S–S–), and polysulfide (–Sn–) links in heterocyclic rings. Reduction of sulfur dyes involves the conversion of these sulfide linkages into leuco thiols followed by the formation of sodium salt of thiol or thiolates, which are soluble in water and substantive toward cellulosic materials (Chakraborty 2014). The application mechanisms (reduction and oxidation) of vat and sulfur dyes are shown in Figure 4.1.

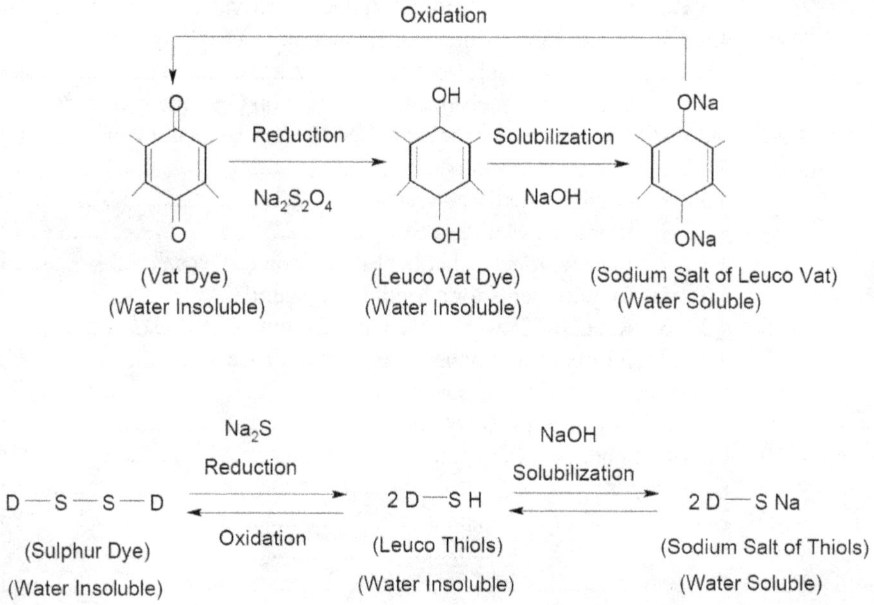

FIGURE 4.1 Application mechanisms (reduction/oxidation) of vat and sulfur dyes.

Sodium hydrosulfite or sodium dithionite ($Na_2S_2O_4$) is the universally accepted reducing agent for the reduction of vat dyes along with sodium hydroxide (NaOH) alkali. When sodium dithionite is dissolved in water in the presence of sodium hydroxide (NaOH), it releases nascent hydrogen, giving it a reducing characteristic. The amount of $Na_2S_2O_4$ and NaOH needed to reduce a particular vat dye is proportional to the number of reducible (−C=O) groups, but we always add an excess of these to ensure a reduced state till the end of dyeing.

Generally, indigoid vat dyes require −700 to −750 mV reduction potential for their complete reduction, while anthraquinoid vat dyes require reduction potential of −800 to −1000 mV for complete reduction depending on their class, i.e., IK, IN, IW, and IN special; IK dyes require the lowest, while IN special dyes require the highest range of reduction potential. A very attractive feature of $Na_2S_2O_4$ as a reducing agent is that it can generate −700 to −1000 mV of reduction potential, and required reduction potential can be varied by changing its concentration, alkali, or temperature. All vat dyes are reduced at 30°C–60°C and above using $Na_2S_2O_4$. Sulfurized vat dyes contain both the sulfur linkage and the carbonyl (C=O) group as a chromophore and can be reduced with $Na_2S_2O_4$ and NaOH beyond a certain temperature but not with Na_2S (Ahmed and El-Shishtawy 2010; Chakraborty 2014).

However, $Na_2S_2O_4$ is highly unstable, and it decomposes during reduction into a variety of hazardous sulfur compounds by hydrolysis, thermal breakdown, oxidative decomposition, and other methods. A few byproducts (Na_2S, NaHS, etc.) pollute the air by generating H_2S and sulfur salts (sulfates and sulfites; Na_2SO_3, $NaHSO_4$, Na_2SO_4, and $Na_2S_2O_3$), contaminate sewage, lower its pH, and corrode concrete pipelines. Other issues with using $Na_2S_2O_4$ are its expense and storage stability. Similarly, sodium sulfide (Na_2S) is the worldwide used, reducing agent for sulfur dyes, and almost 90% of commercial sulfur dyes are reduced with sodium sulfide in alkaline solution. As sodium dithionite, sodium sulfide as a reducing agent also contaminates effluent with sulfur, liberate H_2S gas, and adds toxicity in nature.

Other sulfur-based compounds such as hydroxyalkyl sulfinate, thiourea, and others also have been suggested for the reduction of dyes. These compounds have low sulfur content as well as a lower equivalent mass, resulting in a lower sulfur-based salt in wastewater. However, sulfur-related issues cannot be completely avoided in these situations as well.

A variety of alternative approaches like the fermentation method, zinc-lime, bisulfite zinc–lime, thiourea–dioxide, sodium borohydride, $FeSO_4$-lime, and other possibilities have been investigated by many researchers. However, due to a variety of difficulties, these alternative reducing agents were not proven to be commercially viable. As a result, sodium dithionite is currently the sole widely used reducing agent for the reduction of vat dyes to convert their pigment form to water-soluble leuco form.

Several studies have been done for the substitution of sodium dithionite by an organic reducing agent (hydroxyketones, glucose-NaOH, Fe(II)-ligand complex, electrochemical reduction), which meets the efficient reduction and biodegradability simultaneously. Some chemicals, however, are costly, and the use of some, like hydroxyketones, is limited to closed systems due to the formation of obnoxious smelling condensation byproducts (Božič and Kokol 2008).

Fe(OH)$_2$ is a well-known powerful reducing agent in an alkaline medium. When iron (II) salts react with NaOH, they form Fe(OH)$_2$, which is insoluble in alkaline environments and precipitates. Therefore, to keep Fe(OH)$_2$ in solution and to employ it as a powerful reducing agent at room temperature, it needs to be complexed with weaker ligands such as citric acid, tartaric acid triethanolamine, etc. These single-ligand systems work well for indigo dyeing but not for coloring cotton with other vat dyes. Hence, Fe(II) salt complexed with citric acid and triethanolamine or Fe(II) salt complexed with tartaric acid and triethanolamine has been proposed as a two-ligand system. The two-ligand systems also proved the suitability of successful dyeing with a wide range of vat dyes due to an increase in Fe(OH)$_2$ solubility when compared to the single-ligand system (Saïd et al. 2008).

For sulfur dyes, glucose is a well-known eco-friendly reducing agent. The reduction potential (−600 to −750 mV) of glucose in conjunction with NaOH is usually sufficient for the reduction and solubilization of sulfur as well as indigo colors at the boil. Molasses, a byproduct of sugar production, can be utilized as a source of reducing sugars because it contains approximately 60% of total sugar (glucose, fructose, and sucrose). The combination of molasses and NaOH provides a dye bath that is very stable and has a high dye absorption. These can be a less expensive option to sodium sulfide and do not require any capital investments. β-mercaptoethanol can also be used as a replacement for sodium sulfide, where sulfur dyes can be reduced using mercaptoethanol and NaOH. Its limited application, excluding ready-to-use liquid sulfur dyes, and high overall dyeing costs are disadvantages. Furthermore, many of the water-insoluble sulfur dyes do not completely dissolve and leave some residue (Božič and Kokol 2008; Chakraborty 2014).

Thus, the use of traditional reducing agents in the coloration of vat and sulfur dyes results in non regenerable oxidized byproducts that remain in the dye bath. Also, the non-eco-friendly nature of the decomposition products poses a range of challenges when it comes to disposing of dye baths and wastewater. Additionally, these reducing agents cannot be recycled since they will eventually be oxidized into species that are difficult to regenerate, and the reducing power of these molecules will be lost. Therefore, many attempts have been made to replace the environmentally unfavorable sulfur-based reducing agents with ecologically more attractive alternatives.

This chapter gives a detail of the electrochemical alternatives in which the electric current is used as a reducing agent for the reduction of vat and sulfur dyes. The electrochemical reduction can be direct or indirect, and various electrochemical reducing methods, such as direct electrochemical reduction of dye via radical ion, reduction of indigo itself on graphite electrodes, indirect electrochemical reduction using a redox mediator, and electrocatalytic hydrogenation, have been described in this technique. These approaches have significant environmental benefits since they reduce chemical use and effluent discharge.

4.2 ELECTROCHEMICAL DYEING: STATE OF THE ART

Electrochemical dyeing is a cost-effective and ecological alternative to traditional chemical reduction processes. Daruwalla was the first to use electrochemistry in vatting (vat dye reduction), attempting to reduce the amount of sodium dithionite

$$S_2O_4^{2-} \rightarrow 2SO_2^-$$

SCHEME 4.1 Conversion of sodium dithionate to active species.

required for the reduction of vat dyes by using a direct voltage. Sodium dithionate at the cathode is transformed into a strong reducing species; hence electrochemical reduction was not directly involved in this. A highly active species (SO_2^-) with a redox potential higher than sodium dithionate can be generated under the right cathodic reduction circumstances (potential, sodium dithionate concentration, temperature, and pH) (Scheme 4.1). However, after dyeing, these compounds cannot be regenerated from the cathode by applying voltage, making bath liquor recycling impossible (Božič and Kokol 2008).

$$S_2O_4^{2-} \rightarrow 2SO_2^-$$

The same concept is used in the electrochemical dyeing procedure, which takes it a step further by allowing liquid recycling. Grotthuss proposed the first discovery that the electric current itself can reduce the vat dyes, in which electrons from the electric current were employed to convert dispersed vat dye pigments to their leuco form. It appears to be a more environmentally friendly approach to reducing chemical consumption (Roessler and Jin 2003).

4.3 TYPES OF ELECTROCHEMICAL DYEING PROCESSES

In electrochemical dyeing, the conventional sulfur-based reducing agents are replaced by electrons from the electric current via special cathodes. Electrochemical dyeing can be carried out by two methods depending on the reduction process; either direct or indirect approaches are used. In a direct electrochemical process, an electron is transmitted directly from the cathode surface to the dispersed dye pigment, whereas in an indirect electrochemical process, an electron is carried from the cathode to the insoluble dye through a redox mediator system. This reversible redox mediator-system is constantly regenerated at the cathode, allowing the reducing agent to be renewed. This method allows for dye bath recycling in its entirety, including the reuse of reducing agents. Another benefit of indirect dye reduction is that the dye-reducing bath's parameters may be regulated using electrochemical methods.

4.3.1 Direct Electrochemical Reduction

In the direct electrochemical dyeing technique, the dyestuff is directly reduced by the contact between the electrode surface and the dye molecules. In the mechanism of direct electrochemical reduction, the dye in their pigment form comes in contact with the cathode surface and negative potential at the electrode surface, transfer the electron, and dye gets reduced. The direct reduction process also needs some reducing agent to start the process, and the pH of the dye solution is kept alkaline to keep the dye in its solubilized form. The mechanism of electron transfer from the cathode for the reduction of dyes is shown in Figure 4.2.

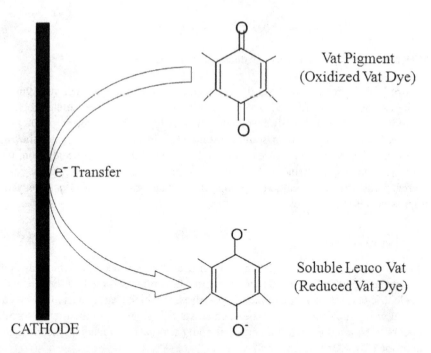

FIGURE 4.2 Mechanism of direct electrochemical reduction.

In direct electrochemical reduction, the electrons have to be transferred directly between cathode surfaces to the surface of the dispersed vat dye pigment. Vat, indigo, and sulfur dyes are water-insoluble and present as dispersed pigments, which results in a very low probability of contact between dye particles and the surface of the cathode. Thus, the rate-limiting step of direct electrochemical reduction is the electron transfer from the cathode surface, and a direct cathodic reduction is very difficult to achieve. Hence, an alternate technique for electron transfer for reduction is necessary to improve the rate of electron transfer (Ahmed and El-Shishtawy 2010; Holme 2002).

4.3.1.1 Direct Electrochemical Reduction in Sulfur Dyeing

The application of sulfur dyes is similar to that of dyeing with vat and indigo dyes. Due to the lower reduction potential, the reduction of sulfur dyes can be achieved easily in comparison to vat and indigo dyes. According to the electrochemical properties of the sulfur dyes, the prereduced sulfur dye can easily act as self-medicating for oxidized sulfur dye. Therefore, the prereduced leuco sulfur dye gets oxidized by oxygen present in the dye bath during dyeing. The oxygen is further recycled by direct self-mediated cathodic reduction. Thus the leuco form of sulfur dye produced by cathodic reduction act as in-situ reducing agents and as dyestuff also (Figure 4.3).

According to the investigation, the redox potential plays an important role in the conventional chemical reduction process, and the color depth increases with higher reduction potentials as a high degree of reduction is achieved. In the same way, the color values and shade of dyed material can be influenced and controlled by varying current densities applied during electrochemical dyeing. As in electrochemical

Electrochemical Technology for Dyeing

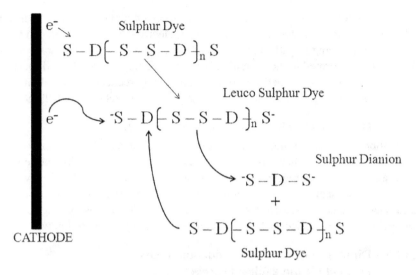

FIGURE 4.3 Direct electrochemical reduction of sulfur.

dyeing, the composition of leuco sulfur dyes is influenced by the current density and as it serves as reducing agent itself and thus results in maximum color yield. Thus, considerable ecological and economic advantages are expected from this sulfide-free cathodic reduction (Bechtold and Brunner 2005; Miled et al. 2014).

In 2008 Bechtold et al. substitute reducing agents with direct cathodic reduction of CI Sulphur Black 1 dye. The direct combination of a multicathode electrochemical cell to a dyeing unit allows for the instant generation of a reduced dye bath and easy control of the dyeing conditions via cell current and dye bath redox potential measurements. Charge flow applied to reduce a dye bath could be used as a measure to predict color depth, and analysis of the color depth by measurement of K/S and CIE Lab-coordinates indicates that the color depth of the dyeing depends on the redox potential in the dye bath. However, cell efficiency and dyestuff re-oxidation due to air-oxidation have to be considered and kept constant. Therefore, both redox potential measurement and applied charge flow are useful factors for achieving repeatable dyeing conditions in a practical application. While redox potential is essential in describing dyestuff reduction and charge flow measurement with attained redox potential is useful in describing the electrochemical equipment efficiency.(Bechtold, Turcanu and Schrott 2008)

In continuation to the above work, Bechtold and his team also established a mathematical model for optimization of a cell with multiple porous electrodes concerning dimensions, cell costs, and energy consumption. Multicathode electrochemical cells are a promising concept for electrochemical reduction at low current densities. However, in the multicathode system, the first electrodes are a prominent risk for bipolar reaction at high cell current. In bipolar behavior, the rear face of a cathode exhibits anodic dyestuff oxidation, while cathodic water reduction is observed at the front side. Thus it limits an increase in cell current of a multicathode electrolyzer by increasing the number of cathodes used. Therefore, in a multicathode stack, the cathode thickness has to be reduced to avoid bipolar behavior to increase cell current. Optimization of electrode number, electrode thickness, and cell voltage was performed using a mathematical model. The cathodic reduction of CI Sulphur Black-1 was performed in these multicathode cells equipped with three-dimensional electrodes, and the dyeing behavior of the cathodically reduced dye was found to be comparable to dyeing results with commercial samples (Bechtold et al. 2009).

4.3.1.2 Direct Electrochemical Reduction in Vat Dyeing

As mentioned earlier that the vat dyes require complicated and sophisticated dyeing procedures because of their insolubility in aqueous media; normally, the reduction of vat and indigo dyes requires a higher reduction potential than sulfur dyes. The first electrochemical reduction of indigo was studied in 1807, and after that, many researchers extensively studied the cathodic reduction of vat dyes. A series of papers were published on the electrochemical reduction of indigo dispersed in water.

4.3.1.3 Direct Electrochemical Reduction of Dye Pigment via the Radical Process

It is a unique direct electrochemical reduction of vat and sulfur dyes that do not require the presence of a soluble reducing agent or a redox-mediator system. In the reaction mechanism of this process, a radical anion is generated as a result of a comproportionation reaction between the dye and the leuco dye, and this radical is then electrochemically reduced to act as an intermediate species. In order to start the process, an initial amount of the leuco dye has to be generated by a conventional reaction, e.g., by adding a small amount of a soluble reducing agent. Further electrochemical reduction is self-sustaining after the process has started, as shown in Figure 4.4 for indigo (Roessler et al. 2001).

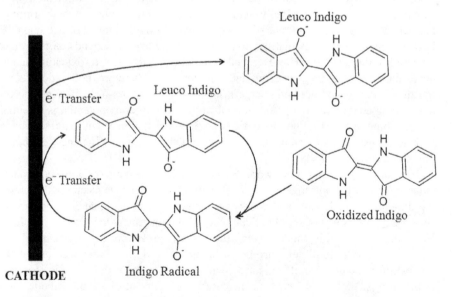

FIGURE 4.4 Direct electrochemical reduction of indigo via radical process.

Electrochemical Technology for Dyeing

This technology was patented by Morte in 2000, and further, Rosseller and his coworkers studied more insight into the mechanism and established the process for laboratory scale. Electrochemical investigations concluded that the electrochemical reaction rate is limited by the diffusion transport of the intermediate radical anion. In fact, as the current density was increased to improve the reaction rate, the experimental current efficiency rapidly declined. Current efficiency is a very important parameter in electrochemical reduction and represents the fraction of electric current used for the desired reaction. With the increase in current density, the electrode potential becomes more negative and results in the liberation of hydrogen as a side reaction (at potentials more negative than −1100 mV, the cathodic liberation of hydrogen as side reaction starts). Thus the current efficiency was limited to a maximum of 80% because of side reactions in these experiments (Roessler et al. 2002, 2003; Roessler and Jin 2003).

Furthermore, as the dye has to be reduced on the surface of the cathode, the area of the cathode must be large to promote the reduction. This reduction via radical process successfully reduces the sulfur dyes; however, the concentration of the dye required to get the desired shade is higher than that for a conventional reduction process due to the limited cathode–dye particle interaction. The reduction rate has to be increased significantly for industrial application of direct electrochemical reduction to vat dyes, and the experimental data clearly indicates that the diffusion-controlled reduction of the intermediate radical anion is the rate-limiting step of this electrochemical process.

This limiting reaction rate critically depends on the thickness of the diffusion layer at the electrode surface, and an increase of the flow in the cathode zone could enhance the reduction rate significantly. Therefore, further investigations were focused on enhancing the radical concentration. Organic solvents can increase the radical concentration and, consequently, the reaction rates, but only concentrations higher than 20% (v/v) of these organic solvents (e.g., methanol, isopropanol, ethanol, Dimethylformamide) are capable of stabilizing the radical species. The use of such high amounts of solvents in the dye bath is not realistic at an industrial level, and the precipitation of leuco dye with cationic surfactants also creates several problems during the process (Bond et al., 1997; Roessler, Dossenbach and Rys 2002b).

These investigations conceded that it is not industrially feasible to reduce indigo electrochemically on a planar electrode as the current efficiency in all experiments was below 20%. The most limiting factor seems to be the poor contact between indigo particles and the electrode. Also, due to well-defined reduction states in vat dyes, no significant self-mediating effect like sulfur dyes was observed. Thus, the direct cathodic reduction is not so productive with the vat and indigo dyes. Therefore, to implement the direct cathodic reduction, two different approaches were suggested further; one must be a change in cathode design to improve dye particle interaction with the electrode, and the other is the electrocatalytic hydrogenation process (Miled et al. 2014).

4.3.1.4 Direct Electrochemical Reduction on Graphite Electrode

The limiting factor in employing direct electrochemical reduction is the poor contact between the insoluble dye particles and the electrode. The indigo microcrystals can be immobilized on the surface of electrode materials for the reduction, but the results

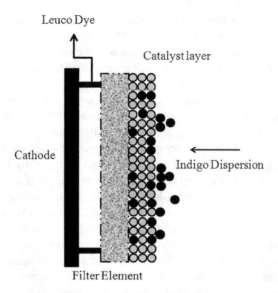

FIGURE 4.5 Concept of precoated catalyst layer electrode.

obtained after immobilization were very similar to the results obtained by indigo dissolved in solvents(Bond et al. 1997; Roessler et al. 2002a, b). Therefore, the precoat-layer-cell was proposed as a remedy to bring the indigo into contact with the cathode. This reactor works on the filtration concept, which is made by an electrically conductive filter fabric support and a cathodically polarized layer of Raney-nickel formed in situ by precoating filtration (Figure 4.5). However, due to significant drawbacks connected with this technology, such as a large pressure drop built up during the filtration process and the constant risk of clogging the reactor, this research was abandoned. Furthermore, the reactor's performance was poor (Bechtold and Brunner 2005; Roessler et al. 2003).

Ray and Rossler also investigated the use of graphite granules as a high surface-area material in a fixed and fluidized bed reactor to improve the industrial practicality of direct electrochemical reduction. Roessler and Crettenand investigated the application of electrochemical reduction of several vat dyes and even mixtures of them on a fixed bed cathode consisting of graphite granules. Because graphite is an inexpensive and stable high-surface-area material, it can be used as an electrode for direct electrochemical reduction. In contrast to electron transfer from the cathode, direct electron transfer between the dye and graphite is used to make leuco dye directly from the dye suspension. Furthermore, the pressure loss induced by granular material is substantially lower than that caused by fine Raney nickel powder in the precoat-layer-cell. Moreover, for this objective, an electrochemical fixed or fluidized bed reactor is a cost-effective reactor design. Chemisorption is likewise impossible or very weak under the given conditions due to the large hydrogen overvoltage on graphite. As a result, it appears that conventional electron transport is the most appropriate route for dye reduction. (Roessler et al. 2003; Roessler and Crettenand 2004).

Further, the work was aligned to improve the graphite surface's chemical activity to increase the reduction rate. The chemical activity of the graphite (carbon) surface is linked to oxygen functionalities (i.e., quinone and hydroquinone groups), which can be found directly on the carbon black surface or as part of more complex compounds. Hence by selectively generating quinine-like functionalities on the graphite electrode surface, it is feasible to accelerate electron transfer and also the adsorption rate on graphite electrode (Bechtold and Brunner 2005; Roessler and Jin 2003).

The oxidative pretreatment of the graphite (i.e., soaking in hydrogen peroxide or preanodization) can activate the graphite surface and increase the reduction rate due to the development of quinone and hydroquinone functionalities. Due to the production of quinoid functional groups, treating graphite with acidified H_2O_2 could practically double the reduction rate (Bechtold and Brunner 2005). Therefore, electrochemical preanodization appears to be a highly effective method for altering the electrochemical behavior of graphite electrodes. Another intriguing method for improving electrocatalytic capabilities is to covalently bond quinoid molecules to the graphite surface. The electron transfer mediators are immobilized on the electrode, which can undergo quick electron transfer with the electrode as well as with the indigo. The carboxy, hydroxyl functionalities obtained from the oxidation of graphite could be functionalized and hence utilized for the immobilization of suitable electron transfer mediators. Anthraquinones and quinones were utilized as redox-active molecules, particularly in the reduction of indigo quinones. Figure 4.6 shows the immobilization with (a) an amine or hydroxyl functional group via the carboxylic groups (b) modified with quinone.

Various immobilization concepts exist, and the most appropriate one is determined by the graphite and mediator molecules. Here are some of the anthraquinone derivatives that are used to reduce vat and indigo colors (Figure 4.7). Quinoid compounds such as 1,8-dihydroxyanthraquinone and 5-amino-acenaphthenequinone have been discovered to be the most active catalysts for increasing the reduction rate. In this way, the process works as a hybrid of a pure surface-based reduction and a mediator-catalyzed reduction (Bechtold and Brunner 2005). Using unmodified graphite electrodes, the maximum indigo concentration of 10 g/L was reduced successfully, while now it is feasible to reduce an extended variety of indanthrene vat dyes and indigo suspensions up to 100 g/L after modifying the graphite surface.

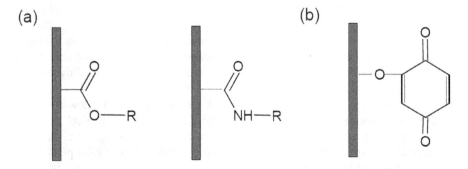

FIGURE 4.6 Covalent modification of graphite with (a) Carboxylic group (b) Quinone.

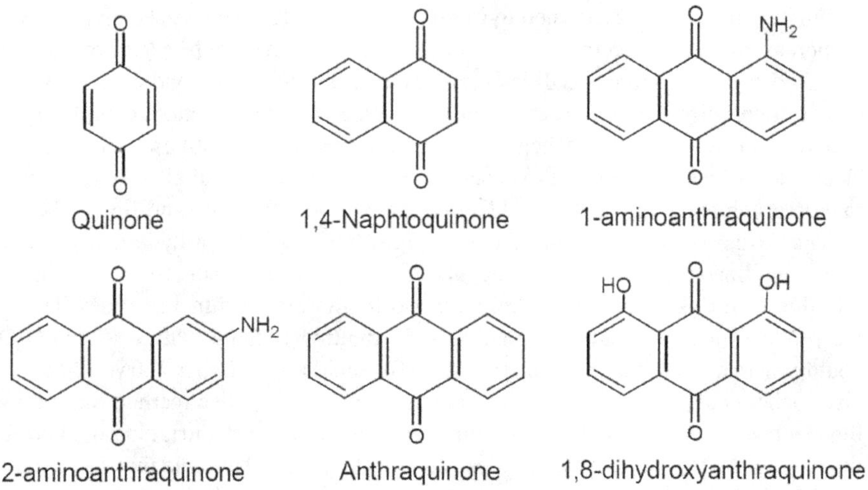

FIGURE 4.7 Anthraquinone derivatives for immobilization on graphite.

Current yields up to 86% in the presence of low current densities can be achieved without the blocking of the reactor (Roessler and Crettenand 2004).

Kulandainathan et al., in their work, employed the iron-deposited graphite surfaces as cathode materials for the electrochemical vat-dyeing process using a very low concentration of sodium dithionite. The electrodeposited graphite employed high-surface-area cathodes and can be a better alternative to employing multiple cathodes in the electrochemical cell. Also, the adherence and reusability of the electrodeposit can be improved significantly by employing organic (additives like gelatin and animal glue) and inorganic (like thiocyanate) adsorbents. The surface-bound Fe^{3+}/Fe^{2+} redox sites on the electrode determine the electrochemical activity, and the electrodes could be reused for several dyeing processes. Various Vat dyes like CI Vat Violet-1, C.I. Vat Green-1, and CI Vat Blue-4 could be efficiently dyed employing these above electrode materials. The color yield and wash fastness rating of the dyed samples were found to be equivalent to conventionally dyed samples (Kulandainathan et al. 2008).

Carbon felt was recently used as a cathode for the direct electrochemical reduction of CI Vat Yellow-1. Using cyclic voltammetry and electrolytic investigations, it was confirmed that the carbon felt is an appropriate electrode material for the direct electrochemical reduction of the dye. After electrochemical dyeing, the color shade and washing fastness of the dyed fabric were comparable to traditional dyeing processes (Yang et al. 2021).

These findings clearly lay the groundwork for the creation of a low-cost, constantly operating, and environmentally friendly direct electrochemical reduction of dispersed vat pigments. The introduction of surface functionality via chemical modification is an interesting study area. However, establishing a pilot plant is difficult, and optimizing the cell structure will be necessary to achieve this (Roessler and Jin, 2003).

Electrochemical Technology for Dyeing

4.3.2 Indirect Method or Mediator-Enhanced Electrochemical Reduction

The indirect electrochemical reduction technique does not require the direct contact of the dyestuff with the cathode, as in direct electrochemical reduction. Here the reduction takes place between the separated surfaces of the electrode and insoluble dye pigment through the redox mediator or an electron-carrier (Figure 4.8). As a result, this process is referred to as indirect electrochemical reduction. Employing a soluble redox mediator can greatly enhance the rate of the electron transfer between dyes and cathode to reduce the dye (Ahmed and El-Shishtawy 2010; Božič and Kokol 2008; Kulandainathan et al. 2007a; Holme 2002).

The indirect electrochemical reduction process was patented by Bechtold in 1993 and has since published a number of related research papers. Mediators, also known as reversible redox systems, are used as reducing agents in this approach, and they go through both reduction and oxidation cycles. The mediators oxidize themselves after traditionally reducing the dye. The oxidized mediators are then reduced at the cathode surface, making them available for dye reduction, and this cycle is continued throughout the dyeing process. Thus, the mediator behaves like a conventional reducing agent, which continuously is regenerated cathodically (Bechtold et al. 1994).

The primary objective of the reversible redox system is to generate a constant reduction potential in the dye liquor. After dyeing, the unexhausted color gets precipitated by air oxidation and can be removed by filtering. Therefore, the liquor containing the mediator can be recycled for additional dyeing processes after the dye has been removed. This appears to be the most important feature of this process in terms of cost

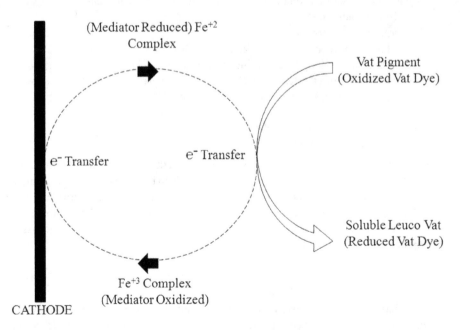

FIGURE 4.8 Mechanism of indirect electrochemical reduction.

and environmental friendliness. As a result, indirect electrochemical reduction not only lowers the cost of reducing chemicals, but it also allows for the first time the dye liquid to be closed cycled. The fundamental benefit of this technology is that the state of dye reduction can be judged by the redox potential measurement and can be controlled by adjustment of the cell current. While in the conventional reducing system, often a surplus of reducing agents has to be applied to guarantee stable dye bath conditions as it cannot be monitored (Bechtold et al. 1994; Kulandainathan et al. 2007a).

On an electrochemical dyeing experiment, DyStar Textilfarben GmbH collaborated with the University of Innsbruck's Institute for Textile Chemistry and Textile Physics (TID) in Dornbirn, Austria, and Getzner Textil AG. In this patented technique, an electric current is paired with a recyclable mediator. It eliminates the need for nonregenerative reducing agents in the application of vat and sulfur dyes to textiles, which prevent dye bath recycling and pollute industrial effluent. The project marks a turning point in the use of electrochemical dyeing in the industry (DyStar 2004).

4.3.2.1 Mediator Systems

The redox mediators are chemicals with electrochemical activity; they assist the transfer of electrons between the electrode surface and the dye. This process is repeated, and the mediator functions as an electron shuttle between the dye and electrode. Different redox mediators can be used for the indirect electrochemical reduction of dyes. The following sections discuss some of the important mediator systems that can be employed as reversible redox systems for indirect electrochemical reduction.

1. **Organic mediator systems:** Substituted anthraquinones: e.g., 9,10-anthraquinone-2-sulfonate, 9,10-anthraquinone-2,5-disulfonate, dihydroxy-9–10-anthraquinone.
2. Inorganic Mediator Systems
   a. **Iron-amine complexes:** Fe(II)/Fe(III) with Triethanolamine (TEA), Hydroxyethylethylenediaminetriacetic acid (HETDA), Bicine as ligands
   b. **Iron-carbohydrates complexes:** Fe(II)/Fe(III) with D-gluconate (DGL), hepta-D-gluconate (HDGL) ligands

4.3.2.2 Organic Mediator Systems

4.3.2.2.1 Substituted Anthraquinones

Substituted anthraquinones can undergo reversible reduction or oxidation processes, which makes them an interesting class of compounds to be of potential value as mediators for indirect cathodic reduction of sulfur, indigo, and vat dyes. Anthraquininoid compounds such as anthraquinone sulfonic acids and hydroxyl anthraquinones and are of interest, especially in the case of sulfur dyes where the dyes have the risk of precipitation in presence of iron-sulfide (iron-complex mediator).

4.3.2.2.2 Mechanism of Indirect Electrochemical Reduction Using Anthraquinones

The mechanism of indirect dye reduction using anthraquinone as a mediator system is shown in Scheme 4.2. The reduction of an anthraquinone derivative (AQD) occurs under highly alkaline conditions by a series of two quick single-electron transfer

$$\text{AQD} + e^- \rightleftharpoons \text{AQD}^{\bullet-} \quad \text{---------- (1)}$$

$$\text{AQD}^{\bullet-} + e^- \rightleftharpoons \text{AQD}^{2-} \quad \text{---------- (2)}$$

$$2\,\text{AQD}^{\bullet-} \rightleftharpoons \text{AQD} + \text{AQD}^{2-} \quad \text{---------- (3)}$$

$$2\,\text{AQD}^{2-} \rightleftharpoons \text{AQD} + \text{Dye}^{2-} \quad \text{---------- (4)}$$

SCHEME 4.2 Mechanism of indirect reduction using anthraquinone mediator system.

processes, as indicated in (1) and (2). In most aqueous solutions, the radical anion does not occur; hence the total reaction is a two-electron transfer with the creation of the anthraquinone dianion. Both the transfer of a second electron to the radial anion (2) and the disproportionation of the radical anion according to equation (3) can lead to the formation of AQD^{2-}. If the redox couple $\text{AQD}/\text{AQD}^{2-}$ can produce a sufficiently negative redox potential in solution, the dispersed dye will be reduced according to equation (4), and the oxidized form of the anthraquinone is regenerated (Bechtold, Fitz-Binder and Turcanu 2010).

Bechtold and his coworkers performed a series of experiments with different anthraquinones mono- and disulfonic acids, hydroxyanthraquinones, and various substituted derivatives for the reduction of sulfur and vat dyes below the cathodic potential of the mediator system. The cathode peak potential obtained with anthraquinoid systems depends upon the anthraquinone derivatives, and the maximum cathodic potential of −850 mV vs. Ag/AgCl/3M KCl was obtained with 1,2,5,8-tetrahydroxyanthraquinone among the various studied anthraquinone derivatives. Anthraquinoid concentration ranges from 0.5×10^{-3} to 3×10^{-3} mol/L and is good along with alkali (NaOH) for dye reduction, while by using a closed apparatus (avoid the input of oxygen from the air), the concentration of mediator required can be minimized (Kulandainathan et al. 2007b).

However, the relatively low rate of conversion at the electrodes is the main disadvantage with these organic mediator systems. Also, the limited solubility of anthraquinone-based mediators at lower pH results in their deposition on the cellulosic fibers, which lower the concentration of the mediator system in recovery. Furthermore, because many of the anthraquinones studied are colored compounds, absorption on the fiber must be prevented to avoid undesirable dyeing color changes. Moreover, the anthraquinone must be chemically stable under dye bath conditions in order to be effective. The concentration of electrochemically active mediator will decrease as it decomposes or rearranges in alkaline solution.

In continuation to the above work, further, a series of anthraquinones were studied by cyclic voltammetry for their suitability as a mediator in indirect dye reduction. Cyclic voltammetry is a powerful tool for determining a compound's capacity to act as a mediator in indirect electrolysis. Among the different groups of anthraquinones studied, the hydroxy-9,10-anthraquinones showed promising negative cathodic current peak potentials (Ep) and mid-point potentials ($E_{1/2}$), but their lack of chemical stability and tendency to stain cotton samples during the dyeing process make their application questionable. The 1,2-dihydroxy-9,10-anthraquinone-2-sulfonic acid (Alizarin Red S) was found to be the most promising substituted anthraquinones among the studied representatives, with a negative mid-point potential of −848 mV, good chemical stability under dye bath conditions, and low staining of the dyed material. However, the majority of these anthraquinones' mid-point potentials were not found to be sufficiently negative for cathodic reduction of indigo or other vat dyes (Bechtold, Fitz-Binder and Turcanu 2010).

Further, the efficiency of different dihydroxy-9,10-anthraquinones was studied for the indirect cathodic reduction of dispersed CI Vat Blue-1. The 1,8-Dihydroxy-9,10-anthraquinone had the highest efficiency, followed by the 1,2- and 1,4-substituted dihydroxy-anthraquinones, and 1,2-Dihydroxy-9,10-anthraquinone-3-sulfonic acid (Alizarine Red S) (Turcanu and Bechtold 2011; Turcanu, Fitz-Binder and Bechtold 2011).

A broader range of anthraquinones has been investigated and found to have promising electrochemical characteristics for indirect cathodic reduction; however, their reducing power is not sufficiently negative to achieve the reduction of the majority of vat dyes now in use. The limited maximum cathodic potential attained with anthraquinones confined their application to the reduction of sulfur dyes.

4.3.2.3 Inorganic Mediator Systems

In inorganic mediators, primarily the reducing effect of low-valency metal salts, such as those used in the conventional copperas and zinc dust methods, was utilized for the application of vat dye on cotton. However, due to the enormous quantities used, these reducing agents have various disadvantages, including precipitation in the dye bath and serious metal concentration in the effluent. Heavy metals like Ni and Co complexes have also been used to accelerate reduction operations using traditional reducing agents, but these systems have significant issues in practice. Bechtold et al. studied a wide range of complex salts in order to find redox systems with a large negative reduction potential under homogeneous conditions, i.e., without precipitation (Kulandainathan et al. 2007a).

4.3.2.3.1 Iron-Amine Complexes

The most suitable inorganic redox systems are the salts of Fe (II)/Fe(III) complexed with ligands. Ligands form a complex with iron by coordination to act as a mediator system. Ligands play a significant role in cathodic peak potential in the indirect system. The stability of the coordination molecule in an alkaline medium and the maximal possible redox potential are the major requirements for finding acceptable ligands for Fe(II/III) complexes. Only HEDTA (Hydroxyethylethylenediaminetriacetic acid), bicine, diethanolamine, and TEA (Triethanolamine) have exhibited good Fe(II) complex stability in alkaline solution among the different ligands investigated. The reduced form of these complexes, particularly the more stable one containing TEA as a ligand, is ideally suited to reducing vat dyestuffs. Fe(II) complexes with TEA as a ligand in an alkaline solution have potentials up to −1050 mV vs Ag/AgCl, allowing the reduction of all contemporary vat, indigo, and sulfur dyes (Bechtold and Turcanu 2006; Kulandainathan et al. 2007b; Saïd et al. 2008).

4.3.2.3.2 Iron-Carbohydrates Complexes

In another category of carbohydrate-based ligands: Gluconic acid complexes (D-gluconate, Hepta-D-gluconate) exhibit greater stability at lowered pH, but the current density achieved with such carbohydrate complexes is lowered compared to the iron-amino complexes.

4.3.2.3.3 Indirect Electrochemical Reduction Mechanism Using Iron Complex

The following is a general reaction for the indirect electrochemical dyeing method employing the Fe(II/III)–Ligand complex (Scheme 4.3). In the first reduction step, the iron complex is cathodically reduced, which is a reversible electrochemical reaction (1). Reduced complexes diffuse from the electrode to the dyestuff particles, and at the surface of the dispersed particle, one electron is transferred, resulting in the

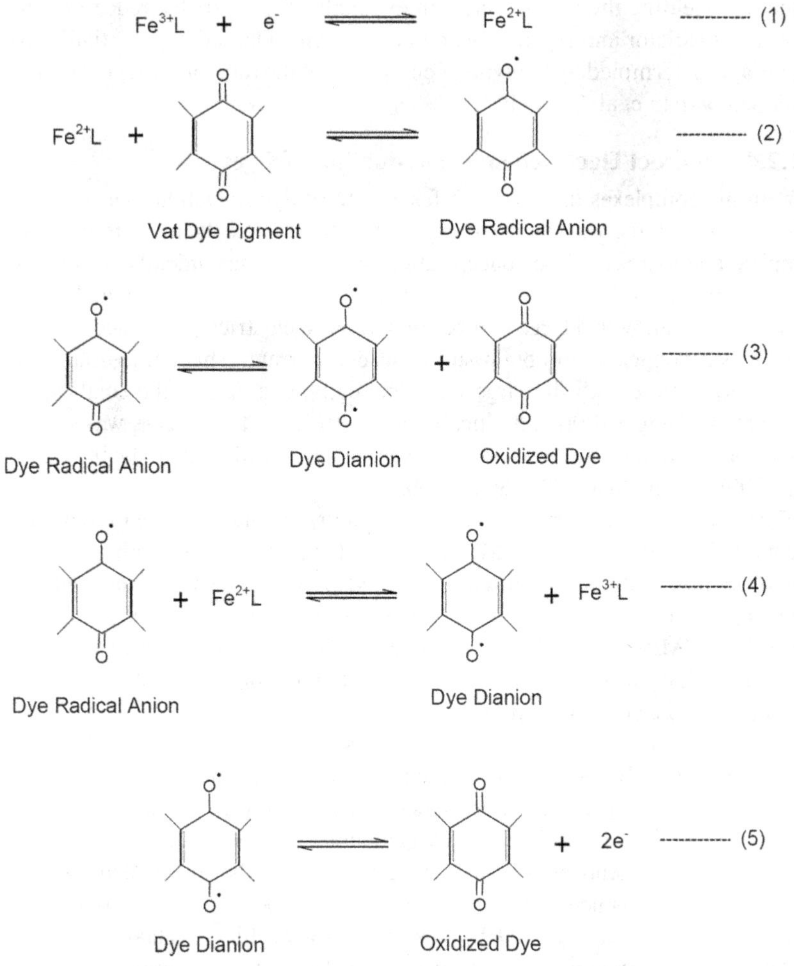

$$Fe^{3+}L + e^- \rightleftharpoons Fe^{2+}L \quad (1)$$

$$Fe^{2+}L + \text{Vat Dye Pigment} \rightleftharpoons \text{Dye Radical Anion} \quad (2)$$

$$\text{Dye Radical Anion} \rightleftharpoons \text{Dye Dianion} + \text{Oxidized Dye} \quad (3)$$

$$\text{Dye Radical Anion} + Fe^{2+}L \rightleftharpoons \text{Dye Dianion} + Fe^{3+}L \quad (4)$$

$$\text{Dye Dianion} \rightleftharpoons \text{Oxidized Dye} + 2e^- \quad (5)$$

SCHEME 4.3 Indirect electrochemical reduction using Fe(II/III)–ligand complex mediator system.

creation of a dye radical anion, which is an intermediate stage in the formation of a completely reduced dye dianion (2).

There are two ways to make a dye dianion or completely reduce the dye. In fact, most dyes require two electrons for complete reduction; the two dye radical anions may generate one fully reduced dye dianion molecule and regenerate one dye molecule (3). In another proposed reaction mechanism, one-electron transfer to generate a reduced mediator to a dye radical anion and construct a fully reduced dye dianion molecule as a reaction product (4) one-electron transfer to generate a reduced mediator from a dye radical anion and construct a fully reduced dye dianion molecule as a reaction product is another proposed reaction process for the creation of a reduced dye molecule (4). In a separate reaction, the equilibrium is established between the fully reduced dye dianion molecule and the dye's insoluble oxidized state (5).

The reversible redox system (Fe(III)-complex) is continually reduced by cathodic reduction, yielding the ferrous form. In dye baths, the reversible redox system functions as a mediator and regenerable reducing agent. The redox potential in dye bath potential is determined by the type of complex and the ratio of Fe(II) to Fe(III) forms (Kulandainathan et al. 2007a,b).

4.3.2.4 Indirect Electrochemical Reduction of Dyes

Iron-amine complexes have a much faster rate of dye reduction than sodium dithionate. However, the rate-determining step is the electrochemical reduction of the complex, and increasing the concentration of the electrochemically active iron complex is limited, resulting in low current density for cathodic reduction. As a result, a multicathode cell with a large number of cathodes, electrically coupled to one or two anodes, was suggested to obtain sufficient cell current. This configuration permits the cell to operate with the largest feasible cathode area and the smallest possible anode area. Using a 1000 A multicathode electrolyzer, the process was successfully tested for continuous dyeing of cotton yarn on a full-scale indigo dyeing range (Saïd et al. 2008; Bechtold and Turcanu 2009).

Earlier experiments were performed with a low mediator to dye ratio with the use of a multielectrode system, but Miled proposed that the vat dyes with different chemical structures (indigoid, anthraquinone, and polycyclic) can be reduced using Fe (III)-TEA mediator system at high mediator to dye ratio (10:1) in a simply designed electrolyzer (Miled et al. 2008). However, higher concentrations of mediators can raise the toxicity load, and the electrochemical reduction will not remain a cleaner alternative to the conventional one.

The improvement of current efficiency was studied and optimized with Indanthrene Brilliant Green FFB since it is one of the most widely used vat dyes in the textile industry. The impacts of operating parameters such as reaction temperature, current density, NaOH concentration, and Fe(III)-TEA mediator were compared utilizing orthogonal electrolytic experiments with a stainless steel cathode. The current density was found to be the main influence factor in the indirect electrochemical reduction of the dye, and high current efficiency of 49.9% was successfully achieved (Huanhuan et al. 2011).

The possibility of continuous electrochemical reduction of indigo dye using a simple electrolytic cell was also investigated at room temperature employing Fe(III)-TEA mediator. After completion of dye reduction, the dye bath was circulated through the electrochemical reactor by a peristaltic pump to allow the continuous renewal of the reducing capacity. CIELab values of the dyed samples show only small differences in color depth and shade, indicating, in general, an unchanged dyeing behavior of indigo. Also, the fastness properties with respect to light, washing, and rubbing of dyed samples were appeared to be equivalent to conventional dyeing (Miled et al. 2018).

In another attempt mixture of triethanolamine and D-gluconate was investigated for mediator purposes. The use of a mixture of ligands permits the successful combination of advantages of both ligand systems (high current density of Fe(TEA), high stability of Fe(DGL). The addition of Fe(DGL) and Fe(HDGL) with Fe(TEA) make ligand exchange reactions and give efficient vat dye reduction(Bechtold and Brunner 2005; Kulandainathan et al. 2007a). For electrochemical vat dyeing, an alternate iron complex, Fe(III)+ oxalate system in the presence of gluconate, can work as redox complexes. According to voltammetric studies, the Fe(II)+TEA+gluconate system, as well as the Fe(II)+oxalate+gluconate system, catalyze the electrochemical vat dyeing system via the Fe(II)-gluconate intermediate and can be used to dye a variety of vat dyes such as Vat Green-1, Vat Violet-1, Vat Brown-1, Vat Yellow-2, and Vat Black-27 (Firoz Babu et al. 2009).

Further, the binuclear Ca^{2+}-Fe^{2+}/Fe^{3+}+D-gluconate complexes (Abdelileh et al. 2020) or Ca^{2+}-Fe^{2+}/Fe^{3+}+Triethanolamine complexes were investigated with analytical techniques for dispersed vat dyes reduction. The addition of calcium ions into Fe(II/III) complexes causes a shift of formal redox potential of these complexes toward more negative values due to the formation of the binuclear complex (Bechtold and Turcanu 2004).

It was discovered that adding Calcium+gluconate to the Fe (III)+TEA mediator system increased the effectiveness of electrochemical vat dyeing. Fe (III)+gluconate complex itself did not exhibit any distinct volumetric redox response, but the presence of gluconate exhibit a competitive ligand phenomenon, and the content of TEA in the mediator system may be reduced to some extent. Studies were also carried out with other hydroxyethyl group-based ligands in place of TEA (Bechtold and Turcanu 2006). In another study, the recipe for the (Ca^{2+}+Fe^{3+}+TEA) redox system was adjusted in terms of reduction efficiency (RE) and current efficiency (CE) for the indirect electrochemical reduction of indigo. With a CE of 75.9% at 50°C, the best optimization recipe was 20 g/L NaOH, 5 g/L $Fe_2(SO4)3$, 30 g/L TEA, 5 g/L calcium gluconate, and 3 g/L indigo (Tan et al. 2018).

A device for electrochemical reduction and dyeing was recently developed to combine electrolysis and dyeing in one device. In this indirect electrochemical reduction of indigo, the iron-triethanolamine-calcium gluconate (Fe-TEA-Ca) combination served as a crucial intermediary. The use of ultrasonic waves not only improves current efficiency, but it also cleans the cathodes. After dyeing, washing with an oxalic acid solution has been proven to be an efficient approach to remove iron salt from the fabric. As a result, the dyed fabric produced by this procedure could meet industrial color fastness standards (Yi et al. 2020).

These mediators, on the other hand, are costly and not completely safe from a toxicological perspective. In addition, the mediator must be isolated from the soluble leuco dye by ultra-filtration prior to the dyeing process, and the concentration of the mediator in the filtrate must be raised by nanofiltration. This allows the recovery of the mediator (with 15% losses) as well as the recycling of water and textile auxiliaries. Filtration, on the other hand, significantly raises the expenses of this reduction process and creates various technical issues, such as a large pressure drop built up during filtration and the constant risk of blocking the reactor (electrochemical cell). As a result, despite the passage of time, this mediator-technique is still in development; the production trials are currently limited to only pilot plant level. As a result, innovative alternatives for enhancing the eco-efficiency of vat dye application are still needed.

4.3.2.5 Requirements for a Mediator System

Mediators play a crucial role in oxidative and reductive mechanisms. Therefore, various experimental aspects need to be taken into consideration for the mediator system.

- Sufficient negative reduction potential to reduce dye, i.e., the reduction potential of the mediator system has to be somewhat higher than the reduction potential of the dye.
- Reduction potentials should be within the range of the thermodynamic potentials of the anode and cathode, i.e., the conversion from oxidized form to the reduced form of the mediator should take place at the electrode without a marked over-voltage (hydrogen liberation as side reaction).
- High current density and high current efficiency of the redox mediator-system
- High reaction rate between the reduced mediator and oxidized dyestuff
- The mediator should be chemically stable and soluble in both its reduced and oxidized forms
- A high number of electrochemical cycles without losses in activity
- Low concentration of chemicals employed to achieve effective reduction potential
- Minimum side reactions, i.e., no dyestuff destruction due to the formation of radicals
- Reproducible and similar dyeing results compared with conventional
- Easy and economical techniques for recycling of mediator system and waste water

Some important groups of mediator systems for indirect cathodic reductions with suitable examples are with their suitability:

Mediator System	Suitable Example	Redox Potential (mV vs Ag/AgCl)	Temperature (°C)	pH	Suitable Dyestuff
Anthraquinones	1,2-Dihydroxyanthraquinone; Anthraquinone-1,5-sulfonate	−500 to −700	95	11–13	Sulfur
Iron-Amine Complex	Fe (II/III) + TEA (Triethanolamine)	−1000	80	12–14	Vat, Indigo, Sulfur
Iron-Carbohydrate Complex	Fe (II/III) + DGL (D-Gluconate)	−750	40	11–12	Indigo
Binuclear Iron-Complex	Ca$^+$ + Fe (II/III) + TEA Ca$^+$ + Fe (II/III) + DGL	−1250	40	11–14	Vat, Indigo, Sulfur

4.3.2.6 Reducing Power of Mediator Systems

The reducing action of different redox systems can be characterized by its half-wave potential, which basically defines the characteristics of the electroactive material. According to the Nernst equation, the desired potential (reduction/oxidation) can be achieved by combining the appropriate proportion of reduced and oxidized species. In an electrochemical system, both the oxidized and reduced forms of the redox pair are present in the solution at the same time. Therefore, the potential prevailing in the solution can be varied by altering the concentration of these oxidized or reduced species using the voltage provided. However, for commercial applicable systems, the maximum cathodic potential should be achieved, and it can be equivalent to the standard potential of reduced and oxidized species to get a stable potential (Kulandainathan et al. 2007a).

In practice, this means that when reduced and oxidized species are present in the same quantity, then maximum cathodic potential (E) is equivalent to standard potential ($E^{o\prime}$) and half-wave potential ($E_{1/2}$), i.e., $E = E^{o\prime} \approx E_{1/2}$ in Scheme 4.4, where C_{ox} and C_{re} are the concentrations of oxidized and reduced species, and z is the number of electrons transferred in half cell reaction.

$$E = E^{o\prime} + 2.3026 \frac{RT}{nF} \log_{10} \frac{C_{ox}}{C_{rd}}$$

The half-wave potential can be easily seen in a cyclic voltammetry plot. When scanning the potential applied to an electrochemical cell, the current from the reaction will follow two different curves, depending on the scan direction. As the applied potential is scanned, the current will reach a peak value and then fall back to a low saturation value at extreme potentials. The half-wave potential is located right between the two peaks potential values in a back-and-forth cyclic voltammetry sweep curve (Figure 4.9) and provides a straightforward way to estimate the $E^{o\prime}$ for a reversible electron transfer (Elgrishi et al. 2018).

4.3.2.7 Indirect Electrochemical Dyeing Process

Bechtold and his colleagues (Bechtold et al. 1994) developed an appropriate indirect electrochemical dyeing technique. They suggested a dyeing system that includes an electrochemical cell for dye reduction from where the reduced dye liquor is transferred to the dyeing vessel. Additional advantages of this technique include the ability to recycle the redox-mediator system for several dye reduction cycles. In the electrochemical cell, a reduction system consisting of triethanolamine (TEA), ferric sulfate, and alkali (NaOH) in a predetermined proportion was utilized. A magnetic stirrer is included in the electrochemical cell vessel to ensure that the ferric sulfate is distributed evenly throughout the system.

$$E = E^{o\prime} + 2.3026 \frac{RT}{nF} \log_{10} \frac{C_{ox}}{C_{rd}}$$

SCHEME 4.4 Nernst equation for cathodic potential.

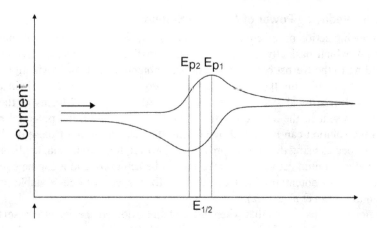

FIGURE 4.9 Cyclic voltammetry sweep curve.

A working copper cathode served as an electron source at the bottom of the vessel. With the use of a voltmeter, the amount of electron production liberated at the cathode (known as cathode potential) was measured with respect to a reference electrode (Ag/AgCl/3M KCl). The dyeing system's actual reduction potential was measured between the copper cathode and the reference electrode. In order to avoid re-oxidation at the anode, it is critical that the functioning cathode be covered by a membrane.

Prior to adding the dye to the reduction system, the ferric ion must first be converted to ferrous ion, and the dissolved oxygen must be removed from the system. So, initially, for the first 15–20 minutes, voltage is applied to the cathode to stabilize the reduction potential in the system. After that, the dye is introduced to the system, and the voltage applied to the cathode is maintained for electron liberation. The electrons liberated at the cathode are expected to convert the ferric ion to ferrous ion. The ferrous ions then react with sodium hydroxide to generate ferrous hydroxide, which can be used as a reducing agent. The TEA in the system acts as a ligand, keeping ferrous hydroxide soluble so that its reducing property can be used. The concept of mediator circulation with the electrochemical dyeing process in the flowchart (Figure 4.10) and a line diagram for electrochemical dyeing arrangement is shown in Figure 4.11.

The dye is added to the system after the liberation of ferrous ions from the ferric salt has begun. The dye is reduced quickly since the reduction potential has already been created in an aqueous solution containing ferric sulfate, TEA, and NaOH. The actual dying is done in a dyeing compartment, which also holds the fabric sample. The compartment has a perforated bottom from where the reduced dye solution is circulated with the help of a peristaltic pump. The colored fabric's oxidation and posttreatment are done separately.

Without generating hydrogen, a working potential of up to −1200 mV versus Ag/AgCl/3M KCl can be achieved at the cathode using a power supply unit. In conventional reducing system the dye bath potential depends on the type and concentration of the reducing agent used and the dye bath temperature. While the indirect

Electrochemical Technology for Dyeing

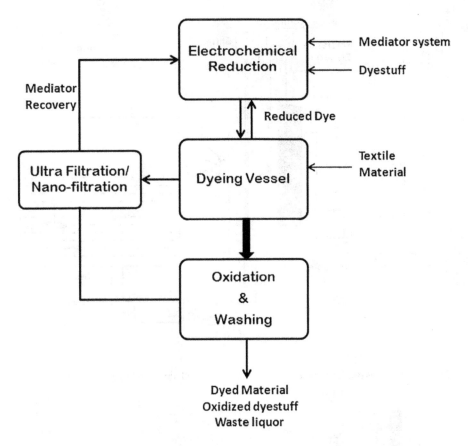

FIGURE 4.10 Indirect chemical process flowchart.

electrochemical reduction technique using mediator permits direct control of the dye bath potential without the undesirable connection between temperature and reduction potential (Bechtold et al. 1997).

4.3.3 Electrocatalytic Hydrogenation

Electrocatalytic hydrogenation differs significantly from the previously reported indirect and direct electrochemical dye reduction procedures. In the above processes, the dyes are reduced by transfer of electrons directly from the cathode or either a mediator or the leuco radical anion of the dye, while in electrochemical hydrogenation, the in-situ hydrogen produced by electrolysis of water reacts chemically with dye at the electrode surface. Electrochemical hydrogenation involves a sequence of reduction steps; firstly, the hydrogen is produced in situ by electrolysis of water and reacts with vat dyes at the electrode surface to reduce to its leuco form (Figure 4.12). The leuco dye prepared by catalytic hydrogenation is converted back into the pigment form in a conventional manner by air oxidation (Ahmed and El-Shishtawy 2010). The reaction mechanism of electrocatalytic hydrogenation is shown in Scheme 4.5.

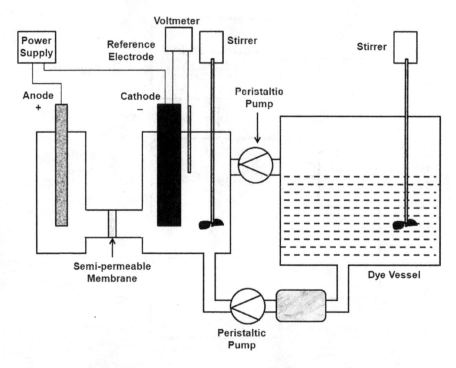

FIGURE 4.11 Basic instrument for electrochemical reduction and dyeing process.

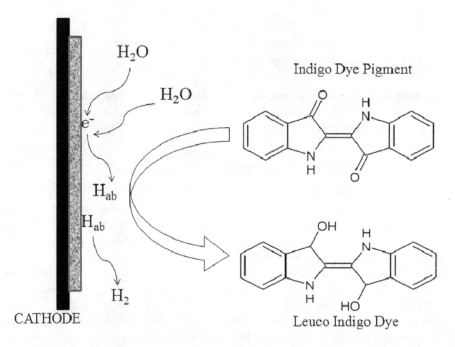

FIGURE 4.12 Mechanism of dye reduction with electrocatalytic hydrogenation.

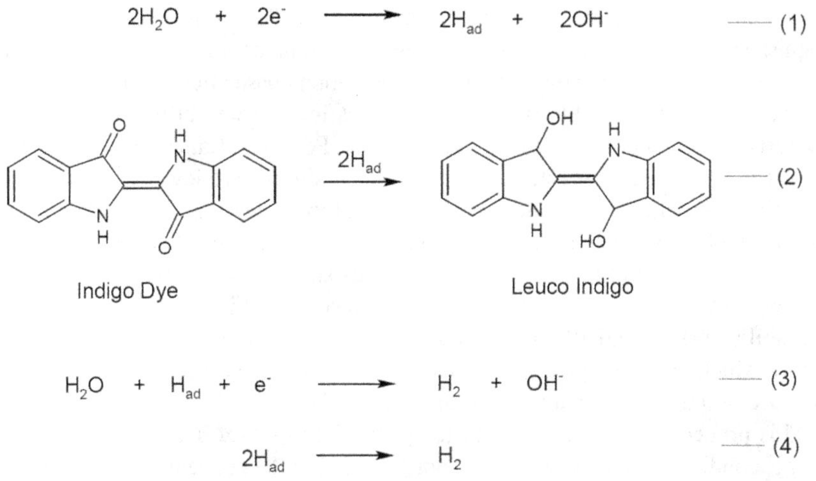

Electrochemical Technology for Dyeing

$$2H_2O + 2e^- \longrightarrow 2H_{ad} + 2OH^- \quad\quad (1)$$

Indigo Dye $\xrightarrow{2H_{ad}}$ Leuco Indigo (2)

$$H_2O + H_{ad} + e^- \longrightarrow H_2 + OH^- \quad\quad (3)$$

$$2H_{ad} \longrightarrow H_2 \quad\quad (4)$$

SCHEME 4.5 Reaction mechanism of electrolytic hydrogenation.

Electrochemical hydrogenation has been successfully applied to a variety of substrates, and it has a number of benefits over chemical hydrogenation. In chemical hydrogenation, the kinetic barrier formed by hydrogen molecule splitting is completely circumvented, and high temperatures and pressures are avoided. Furthermore, electrochemical hydrogen generation directly on the catalyst electrode surface eliminates the need for chemical hydrogenation's hydrogen compression, transportation, and storage (Roessler et al. 2002c; Roessler and Jin 2003).

The major portion of the electrochemically produced hydrogen does not react with the vat dyes; instead, molecular hydrogen gas is formed, which is then emitted from the cathode. The hydrogenation process competes with this hydrogen evolution reaction, and this competition determines the efficiency of electrocatalytic hydrogenation. This problem of hydrogen ion to liberate hydrogen gas instead to act as a reducing agent was resolved by utilizing a hydrogenation metal catalyst, which causes homolytic or heterolytic bond splitting of the H – H bond as expected (Scheme 4.6). Metal catalysts, which adsorb hydrogen and generate a bulk hydride phase, also eliminated the risk of fire and explosion caused by gaseous hydrogen. Even when hydrogen is diluted within an inert gas carrier and loses its inflammability, the bulk hydride ensures the constant surface activity of hydrogen and hence the maintenance of catalytic activity.

As a result, more research on electrolytic hydrogenation of vat dyes was conducted using electrodes composed of a thin grid coated with a metal layer embedded with fine particles of catalyst. The surface of a metal powder catalyst (Raney-nickel or platinum black) that is electrically conductive and has a low hydrogen-overvoltage, indicating that hydrogen production is less likely. The catalytic material serves as both an electrode (for

$$H_2 \rightarrow 2H^\cdot \text{ (homolytic)} \quad\quad H_2 \rightarrow H^+ + H^+ \text{ (hetrolytic)}$$

SCHEME 4.6 Homolytic and heterolytic bond splitting of H_2.

producing hydrogen) and a catalyst for hydrogenation. Thus, this process is clearly distinguished from an electron transfer process (electronation – protonation mechanism), in which hydrogen is only involved in the protonation phase after the electron transfer.

The industrial feasibility of this process was investigated utilizing several metal catalyst electrode materials (Pb-black, Rh-black, Pd-black, Raney-nickel, and Raney-cobalt) in a divided flow cell. Palladium and rhodium have lower catalytic activity than platinum. Nickel is the most active catalyst in Raney-type electrodes, although its current efficiency is just 3.5%. Nickel has an extremely low electro hydrogenation rate, which is unusual for a metal that demands such harsh conditions. Nonetheless, Raney-nickel was chosen as the electrode material since it is intriguing in terms of availability, prices, and alkaline media stability. The stability of platinum-black electrodes, which are among the most active, has been demonstrated to be poor, making their use in industry difficult (Roessler et al. 2002c).

This procedure could be used to reduce varieties of vat dyes. In the instance of indigo, conditions in the system were optimized, and the concentration of indigo was scaled up to 10 g/L. Unfortunately, with 95% conversion, only a poor current efficiency of 12.7% could be achieved using these optimal circumstances. To achieve an industrially practicable reduction rate for stock solutions, a massive electrode surface of several hundreds of square meters would be required. Most likely, the approach will only be effective for dye bath stabilization.

The precoat-layer cell developed to improve the contact between the dye particles, and the electrode was also employed for the electrocatalytic hydrogenation. Noble metal particles supported on graphite granules have recently been researched as an electrode material in a fixed- and fluidized bed reactor in order to address the aforementioned issues. Although the pressure drop over the granular material was substantially lower than that over fine Raney nickel powder, reasonable and good electrogenesis efficiencies could still be achieved. Noble metals, on the other hand, are highly expensive, and the catalyst's long-term performance has been demonstrated to be poor (Roessler et al. 2003).

The mechanical strength of electrodes manufactured with metallic powder particles is often low, without the use of binding material. As a result, two different types of Raney nickel electrodes were investigated: RVC carbon and lanthanum phosphate linked Raney nickel electrodes. With these new sorts of electrodes, no genuine breakthrough could be made. The electrode stability could be improved in all cases, but the catalytic activity was significantly reduced. Raney nickel electrodes manufactured by electro-deposition are widely recognized for their short lifetime. Because of friction forces, the active time of fixed and fluidized bed electrodes is significantly shorter (Roessler and Jin 2003).

Furthermore, it appears that electrochemical procedures can also be employed to stabilize or regenerate sodium dithionite. It is feasible to synthesize a powerful reducing species from sodium hydrosulfite via cathodic reduction with a redox potential higher than that of hydrosulfite itself, which should lower dithionite consumption by 30% (Roessler and Jin 2003).

Sulfur dyes are thus directly reduced, whereas vat dyes require a mediator system for reduction. Alternatively the dyes having higher reduction potential (vat) can also be reduced using in-situ generated H_2 at cathode surface. This approach for reducing vat and sulfur dyes shows to be a viable and green alternative in terms of economic and environmental factors because it does not require any reducing agent, so direct utilization of vat bath is possible. The electrocatalytic hydrogenation might be a promising method for vat dye reduction. The electrode stability, on the other hand, is critical in an industrial application. Unfortunately, an electrode material with both high mechanical strength and high catalytic activity has not been found till now. As a result, more research in this area is required (Comisso and Mengoli 2003).

4.4 ADVANTAGES OF ELECTROCHEMICAL DYEING PROCESS

4.4.1 Environment

Traditional reduction techniques for vat and sulfur dyes use hazardous chemicals (e.g., sodium-dithionite, sodium-sulfide, etc.) that not only pollute the air by creating H_2S but also taint sewage systems and cause disposal issues. The electric potential for dye reduction is involved in electrochemical reduction. It lowered the chemical load in wastewater, improving the dyeing process' ecological profile and extending its long-term viability. It will be possible to reduce the sulfate concentration in wastewater from current levels of 640–1600 mg/L to levels substantially below the legal limit of 200 mg/L (Abdelileh et al. 2020). Control through redox-potential measurement, mediator system concentration, total charge flow, and cell current measurement allowed savings in chemical costs arid in the effluent loading of 50%–75% (Holme 2002).

4.4.2 Economy

A high number of dyeing cycles of the mediator without losses in the activity, recovery of water, and textile auxiliaries is possible. Chemical, water, and energy savings will result in financial gains. The entire textile sector is impacted because chemical prices are lowered by 80%, wastewater recycling expenses are decreased by 85%, electrode material for cathode and anode is inexpensive, and the cell is simple to construct and maintain, resulting in a cost-effective process (Holme 2002).

4.4.3 Health

The indirect electrochemical reduction allows for multiple baths recycling using the redox-mediator system, making it an environmentally friendly process. Minimum side reactions, e.g., dyestuff destruction due to the formation of radicals. Sulfates and sulfites, which are harmful, are not present in effluent; hence there is no negative impact on aquatic life. Compounds in low concentrations and nontoxic chemicals are used. A cost-effective method for recycling chemicals and water used in washing.

4.4.4 Dyeing Quality

Fully controlled dyeing parameters and maximum process reliability through control of redox potential as needed, ranging from −0 to −1200 mV, just by varying the cell current, results in a very good dye reduction rate (10 mg dye/min) and dye pick up may go up to 85%–90%. Stable concentrations of chemicals in the dye bath result in constant dye bath exhaustion. Also, the controlled and defined dyeing conditions result in good reproducibility and dyeing results. Better overall fastness property compared with the technique is already in use.

4.5 FUTURE PROSPECTS

Recently, electrochemical technology was employed in a novel approach for the application of electroactive dye to textiles, in which the coloration was carried out without the need for external heating and solely by the application of electrical potential.

By applying a potential to electroactive synthetic dyes, it was discovered that they might be dyed electrochemically on cotton, silk, and polyester. The fabric was embedded on the anode in this electrochemical process, and voltage was applied for coloration. Without the use of heat, the dye molecules are anticipated to be adsorbed on the fabric's surface via electrochemical attraction. When compared to conventional dyeing, electrochemical coloring with fabric embedded on anode resulted in enhanced dye uptake on the fabrics, and the dye uptake increased as the applied potential increased. Here, the dye-ability of traditional and electrochemical coloration was compared on cotton with direct, vat, and reactive dyes, silk with acid dye, and polyester with disperse dye. The dye-ability results revealed that the depth of shade was independent of the fabric substrate, and color strength (K/S) obtained on samples was comparable in both conventional and embedded systems (Kalapriya, Prabhu and Nithya 2017).

In another study, the dye-ability of polyester fabric was improved by embedding it on graphite electrodes. A better dye-ability was observed in electrochemical than conventional dyeing. The electrochemical method of dyeing polyester fabric using disperse dyes can be considered a greener one as no heat its used, and it saves energy when dyed at a lower potential with better uptake (Kalapriya, Prabhu and Ramakrishnan 2019).

4.6 CONCLUSION

Several ecological alternatives have been developed to reduce the toxicity and wastewater associated with traditional vat and sulfur dye reduction. Eco-friendly reducing agents (thiourea–dioxide, sodium borohydride, hydroxyketones, glucose-NaOH, Fe(II)-ligand complex, etc.), assistance with ultrasound, magnetic fields, or UV to reduce the need for chemicals and electrochemical reduction using electric energy. Electrochemical reduction methods are a suitable alternative since they employ electric energy instead of a reducing agent and don't require any tertiary treatment to treat the final effluents.

The electrochemical reduction can be done by direct and indirect reduction mechanism. Although the direct electrochemical reduction is a simple technique to reduce dyes, the effectiveness is poor due to low contact of dispersed dye pigments with the cathode surface. The reduction potential achieved with the direct method can reduce the sulfur dye, but for vat and indigo, it is insufficient to meet the industry's needs. Direct electrochemical reduction of sulfur dyes and indigo dyes can be achieved via radical mechanism with muticathode electrolyzer cell or Raney-nickel precoat layer electrodes. Direct electrochemical reduction on graphite electrodes with modification or immobilization of quinine compounds can improve the reduction efficiency with increased surface area.

The indirect electrochemical reduction by using a mediator system improves the reduction efficiency by transferring electron. Substituted anthraquinones, iron-amine and iron-carbohydrates are reversible redox systems for the indirect electrochemical reduction. The mediator system can be recycled after dyeing and this appears to be the most important feature of this process in terms of cost. Measurement of current efficiency and electrochemical parameters were used to optimize the mediator-system, and iron salt–complexes were shown to be effective mediators. Electrolytic hydrogenation is based on the generation of hydrogen in situ through water electrolysis and the dyes adsorbed on electrode surface react with this hydrogen to reduce. Various electrodes and catalysts have been investigated for the electrolytic hydrogenation process.

A lot of investigations have been done the laboratory as well as industrial-scale applications of this technology. In comparison to the conventional method, various attempts have been made to achieve higher reduction rates with increased color values and fastness ratings. Taking all of these researches into account, we can conclude that electrochemical techniques constitute a promising field for the dyeing of vat and sulfur dyes as it avoids the use of toxic reducing chemicals. However, in practice, the dyestuff is partially reduced with a conventional reducing agent and before the electrochemical process completes the dye reduction. This facilitates the complete dye reduction and improved stability to the reduced dye bath. A high current efficiency and enough rate of reduction are the most challenging engineering tasks to make electrochemical reduction industrially feasible.

REFERENCES

Abdelileh, M., Manian, A. P., Rhomberg, D., Ticha, M. B., Meksi, N., Aguiló-Aguayo, N. and Bechtold, T. 2020. "Calcium-iron-D-gluconate complexes for the indirect cathodic reduction of indigo in denim dyeing: A greener alternative to non-regenerable chemicals". *Journal of Cleaner Production*. 266 (1): 121753. https://doi.org/10.1016/j.jclepro. 2020.121753.

Ahmed, N. S. E. and El-Shishtawy, R. M. 2010. "The use of new technologies in coloration of textile fibers". *Journal of Materials Science*. 45: 1143–1153. https://doi.org/10.1007/s10853-009-4111-6

Bechtold, T. and Brunner, H. 2005. "Electrochemical Processes in Textile Processing". In: *New Developments in Electrochemical Research*. Ed: Nunez, M., 113–166, Nova Science Publishers Inc, New York.

Bechtold, T., Burtscher, E., Kühnel, G. and Bobleter, O. 1997. "Electrochemical reduction processes in indigo dyeing". 113 (4): 135–144. https://doi.org/10.1111/j.1478-4408.1997.tb01886.x

Bechtold, T., Burtscher, E., Turcanu, A. and Bobleter, O. 1994. "The reduction of vat dyes by indirect electrolysis". *Journal of the Society of Dyers and Colourists*. 110 (1): 14–19. https://doi.org/10.1111/j.1478-4408.1994.tb01586.x

Bechtold, T., Fitz-Binder, C. and Turcanu, A. 2010. "Electrochemical characteristics and dyeing properties of selected 9,10-anthraquinones as mediators for the indirect cathodic reduction of dyes". *Dyes and Pigments*. 87 (3): 194–203. https://doi.org/10.1016/j.dyepig.2010.03.026.

Bechtold, T. and Turcanu, A. 2004. "Fe^{3+}--gluconate and Ca^{2+}-Fe^{3+}--gluconate complexes as mediators for indirect cathodic reduction of vat dyes – Cyclic voltammetry and batch electrolysis experiments". *Journal of Applied Electrochemistry*. 34: 1221–1227. https://doi.org/10.1007/s10800-004-1707-z

Bechtold, T. and Turcanu, A. 2006. "Iron-complexes of bis(2-hydroxyethyl)-amino-compounds as mediators for the indirect reduction of dispersed vat dyes – Cyclic voltammetry and spectroelectrochemical experiments". *Journal of Electroanalytical Chemistry*. 591 (1): 118–126. https://doi.org/10.1016/j.jelechem.2006.03.040.

Bechtold, T. and Turcanu, A. 2009. "Electrochemical reduction in vat dyeing: greener chemistry replaces traditional processes". *Journal of Cleaner Production*. 17 (18): 1669–1679, https://doi.org/10.1016/j.jclepro.2009.08.004.

Bechtold, T., Turcanu, A., Brunner, H. and Schrott, W. 2009. "Model calculations to optimise multi-cathode flow through electrolysers: direct cathodic reduction of CI Sulphur Black 1". *Journal of Applied Electrochemistry*. 39: 1963–1973: https://doi.org/10.1007/s10800-009-9906-2

Bechtold, T., Turcanu, A. and Schrott, W. 2008. "Electrochemical reduction of CI sulphur black 1—correlation between electrochemical parameters and colour depth in exhaust dyeing". *Journal of Applied Electrochemistry*. 38: 25–30. https://doi.org/10.1007/s10800-007-9390-5

Bond, A. M., Marken, F., Hill, E., Compton, R. G. and Hugel, H. 1997. "The electrochemical reduction of indigo dissolved in organic solvents and as a solid mechanically attached to a basal plane pyrolytic graphite electrode immersed in aqueous electrolyte solution". *Journal of the Chemical Society, Perkin Transactions*. 2 (9): 1735–1742. https://doi.org/10.1039/A701003F.

Božič, M. and Kokol, V. 2008. "Ecological alternatives to the reduction and oxidation processes in dyeing with vat and sulphur dyes". *Dyes and Pigments*. 76 (2):299–309. https://doi.org/10.1016/j.dyepig.2006.05.041.

Chakraborty, J. N. 2014. *Fundamentals and Practices in Colouration of Textiles*. Woodhead Publishing, New Delhi, India.

Comisso, N. and Mengoli, G. 2003. "Catalytic reduction of vat and sulfur dyes with hydrogen". *Environmental Chemistry Letters* 1: 229–232. https://doi.org/10.1007/s10311-003-0056-1.

DyStar. 2004. "Electrochemical dyeing process from dystar: first pilot plant for dyeing cheeses with Indigo". *Pigment & Resin Technology*. 33 (3). https://doi.org/10.1108/prt.2004.12933cab.003.

Elgrishi, N., Rountree, K. J., McCarthy, B. D., Rountree, E. S., Eisenhart, T. T. and Dempsey, J. L. 2018. "A practical beginner's guide to cyclic voltammetry". *Journal of Chemical Education*. 95 (2). 197–206. https://doi.org/10.1021/acs.jchemed.7b00361.

Firoz Babu, K., Senthil, R. K., Kulandainathan, A. M. and Noel, M. 2009. "Ferric-oxalate-gluconate based redox mediated electrochemical system for vat dyeing'. *Journal of Applied Electrochemistry*. 39: 1025–1031. https://doi.org/10.1007/s10800-008-9750-9.

Holme, I. 2002. "Recent developments in colorants for textile applications". *Surface Coatings International Part B: Coatings Transactions*. 85: 243–264. https://doi.org/10.1007/BF02699548.

Huanhuan J., Xiaofang, G. E., Yinghua, X. U., Wenwen, Z., Chunan, M. A. 2011. "Indirect electrochemical reduction of indanthrene brilliant green FFB". *Chinese Journal of Chemical Engineering*. 19 (2); 199–204. https://doi.org/10.1016/S1004-9541(11)60154-7.

Kalapriya, K., Prabhu., H. G. and Nithya, S. 2017. "Comparative study on the dyeability of cotton, silk and polyester fabrics by conventional and electrochemical methods". *Rasayan Journal of Chemistry*. 10 (4): 1330–1333. http://dx.doi.org/10.7324/RJC.2017.1041923.

Kalapriya, K., Prabhu., H. G. and Ramakrishnan, G. 2019. "Electrochemical dyeing of polyester fabric using disperse dyes". *Rasayan Journal of Chemistry*. 12 (1):319–323. http://dx.doi.org/10.31788/RJC.2019.1215004.

Kulandainathan, M. A., Muthukumaran, A., Patil, K. and Chavan, R. B. 2007a. "Review of the process development aspects of electrochemical dyeing: its impact and commercial applications". *Coloration Technology*. 123 (3). 143–151. https://doi.org/10.1111/j.1478-4408.2007.00082.x.

Kulandainathan, M. A., Muthukumaran, A., Patil, K. and Chavan, R. B. 2007b. "Potentiostatic studies on indirect electrochemical reduction of vat dyes". *Dyes and Pigments*. 73 (1):47–54. https://doi.org/10.1016/j.dyepig.2005.10.007.

Kulandainathan, M. A., Kiruthika, K., Christopher, G., Firoz Babu, K., Muthukumaran, A. and Noel, M. 2008. "Preparation of iron-deposited graphite surface for application as cathode material during electrochemical vat-dyeing process". *Materials Chemistry and Physics*. 112 (2): 478–484. https://doi.org/10.1016/j.matchemphys.2008.06.010.

Miled, W., Ladhari, N., Saïd, A. H. and Roudesli, S. 2008. "Indirect electrochemical reduction of vat dyes: process optimization for indigoid, anthraquinonic and polycyclic vat dyes". *International Review of Chemical Engineering*. 20 (10): 352–358.

Miled, W., Ladhari, N., Saïd, A. H. and Roudesli, S. 2018. "Electrochemical reduction of indigo: dyeing and fastness properties". DOI: 10.13140/RG.2.2.19012.32648.

Miled W., Said A. H. and Roudesli, S. 2014. "Electrochemical reduction behavior of vat and sulfur dyes: a comparative study". *International Journal of Applied Research on Textile*. 2 (2): 1–10.

Roessler, A. and Crettenand. D. 2004. "Direct electrochemical reduction of vat dyes in a fixed bed of graphite granules". *Dyes and Pigments*. 63 (1): 29–37. https://doi.org/10.1016/j.dyepig.2004.01.005.

Roessler, A., Crettenand, D., Dossenbach, O., Marte, W. and Rys, P. 2002a. "Direct electrochemical reduction of indigo". *Electrochimica Acta*. 47 (12): 1989–1995. https://doi.org/10.1016/S0013-4686(02)00028-2.

Roessler, A., Crettenand, D., Dossenbach, O. and Rys, P. 2003. "Electrochemical reduction of indigo in fixed and fluidized beds of graphite granules". *Journal of Applied Electrochemistry*. 33: 901–908. https://doi.org/10.1023/A:1025876114390.

Roessler, A., Dossenbach, O., Marte, W. and Rys, P. 2002c. "Electrocatalytic hydrogenation of vat dyes". *Dyes and Pigments*. 54 (2): 141–146. https://doi.org/10.1016/S0143-7208(02)00035-9.

Roessler, A., Dossenbach, O., Meyer, U., Marte, W. and Rys, P. 2001. "Direct electrochemical reduction of indigo". CHIMIA. 55 (10): 879–882.

Roessler, A., Dossenbach, O. and Rys, P. 2002b. "Direct electrochemical reduction of indigo: process optimization and scale-up in a flow cell". *Journal of Applied Electrochemistry* 32: 647–651. https://doi.org/10.1023/A:1020198116170.

Roessler, A. and Jin, X. 2003. "State of the art technologies and new electrochemical methods for the reduction of vat dyes". *Dyes and Pigments*. 59 (3); 223–235. https://doi.org/10.1016/S0143-7208(03)00108-6.

Saïd, A. H., Miled, W., Ladhari, N. and Roudesli, S. 2008. "New prospects in the electrochemical dyeing of indigo". *Journal of Applied Sciences*. 8 (13): 2456–2461.

Tan, X., Xiong, W., Zhang, J., Xu, J. and Yi, H. 2018. "Indirect electrochemical reduction of indigo and dyeing." *Key Engineering Materials*. 773 (July): 379–383. https://doi.org/10.4028/www.scientific.net/kem.773.379.

Turcanu, A. and Bechtold, T. 2011. "Indirect cathodic reduction of dispersed indigo by 1,2-dihydroxy-9,10-anthraquinone-3-sulphonate (Alizarin Red S)". *Journal of Solid State Electrochemistry*. 15: 1875–1884. https://doi.org/10.1007/s10008-010-1204-8.

Turcanu, A., Fitz-Binder, C. and Bechtold, T. 2011. "Indirect cathodic reduction of dispersed CI Vat Blue 1 (indigo) by dihydroxy-9,10-anthraquinones in cyclic voltammetry experiments". *Journal of Electroanalytical Chemistry*. 654 (1–2): 29–37. https://doi.org/10.1016/j.jelechem.2011.01.040.

Yang, Z., Shen, W., Chen, Q. and Wang, W. 2021. "Direct electrochemical reduction and dyeing properties of CI Vat Yellow 1 using carbon felt electrode". *Dyes and Pigments*. 184: Article No. 108835. https://doi.org/10.1016/j.dyepig.2020.108835.

Yi, C., Tan, X., Bie, B., Ma, H. and Yi, H. 2020. "Practical and environment-friendly indirect electrochemical reduction of indigo and dyeing". *Scientific Reports*. 10: 4927. https://doi.org/10.1038/s41598-020-61795-5.

5 Application of Microwave Irradiation in Coloration of Textiles

Aminoddin Haji
Yazd University

CONTENTS

5.1 Introduction 89
5.2 Dyeing of Polyester...90
5.3 Dyeing of Polypropylene 90
5.4 Dyeing of Cotton 92
5.5 Dyeing of Wool and Silk 93
5.6 Dyeing of Acrylic and Nylon..96
5.7 Conclusion and Future Outlook..96
References..97

5.1 INTRODUCTION

Microwave is a low-energy wave with a frequency in the range of 1000–300,000 MHz. When ions and polar molecules such as water are exposed to the microwave, the increased agitation of the molecules (due to dipolar rotation and ionic conduction mechanisms) results in the conversion of electromagnetic energy to heat. Microwave has the ability of volumetric and selective heating, which allows more uniform and quick heating of the solutions compared with the conventional heating methods, which reduce the energy consumption and processing times (Chen et al. 2019; Elshemy and Haggag 2019). Furthermore, another advantage of microwave heating is the speed of switching on and off and pollution-free environment due to using no fuel for heating (Metaxas 1991).

Dyeing is one of the most water and energy-consuming processes in the textile industry. Heat is applied in dyeing processes to increase the dyeing rate and exhaustion of the dyes. The application of more uniform and efficient methods of heating can improve the efficiency and uniformity of dyeing processes. In traditional heating, which is done by conduction, heat is generated by an external source and passes through the walls of the dyeing vessel first. This method is inefficient and slow and depends on the thermal conductivity of the vessel walls and the solvent. On the other hand, in microwave heating, microwaves couple directly with the materials inside the dyeing vessel, leading to a rapid and uniform rise in the temperature (Kim et al. 2003; Hayes 2002). Microwave heating is a promising candidate for the replacement of the traditional method of heating in textile dyeing and finishing. The efficiency of the dye extraction from natural sources such as plant parts is significantly increased by the use of microwave heating, and the required time for efficient extraction is shortened (Kiran et al. 2020; Yusuf, Shabbir, and Mohammad 2017).

5.2 DYEING OF POLYESTER

Polyethylene terephthalate (PET) is the most common polyester fiber used in the textile industry. This fiber is hydrophobic and has a high degree of crystallinity and orientation. Furthermore, the glass transition temperature (T_g) of PET fiber is around 110°C, which makes its dyeing difficult at boil or temperatures below it. Several methods such as surface modification, ultrasonic dyeing, microwave heating, and supercritical carbon dioxide ($SCCO_2$) dyeing have been used for the improvement of polyester dyeing (Rehman et al. 2020; Abou Elmaaty, El-Taweel, and Elsisi 2018). Kim et al. used microwave heating for pad-steam dyeing of polyester fabric with disperse dyes. The polyester fabric was successfully dyed by steaming for 7 minutes. The addition of urea and sodium chloride in the padding solution improved the dye uptake and color strength (Kim et al. 2003). Another study revealed that the pretreatment of polyester with dimethylformamide (DMF) under microwave heating significantly improved the uptake of disperse dyes in dyeing by conventional method at 100°C and 130°C (Kale and Bhat 2011). Microwave treatment of polyester fiber and disperse dyeing bath for 1–6 minutes improved the dyeability at boil (Adeel, Khan et al. 2018).

Al-Mousawi et al. synthesized five thienobenzochromene disperse dyes and compared the dyeability of polyester fabrics with these dyes under conventional and microwave heating at 130°C. As shown in Figure 5.1, the color strength of the samples dyed with all synthesized disperse dyes using microwave heating was higher compared with the samples dyed with the same dyes using conventional heating. The chemical structure of the synthesized dyes is shown in Figure 5.2. The fastness of the samples dyed using both methods against washing and perspiration was excellent (Al-Mousawi, El-Apasery, and Elnagdi 2013). The studies of Oner et al. revealed that microwave heating enhanced the exhaustion and fastness properties of six disperse dyes on poly(butylene terephthalate) fabric (Öner, Büyükakinci, and Sökmen 2013). In natural dyeing of polyester fabric with henna, microwave heating enhanced the efficiency of the extraction and reduced the dyeing and mordanting time up to 65%, and enhanced the color strength and fastness properties of the dyed fabrics (Arain et al. 2019; Rabia et al. 2019).

5.3 DYEING OF POLYPROPYLENE

Polypropylene fiber (PP) is highly hydrophobic and lacks any functional groups in its chemical structure. PP fibers are usually dyed by the mass pigmentation method in the melt spinning process. However, it can be dyed at light shades by disperse dyes and shows lower fastness properties compared with polyester fibers. Several

Microwave Irradiation in Textile Coloration

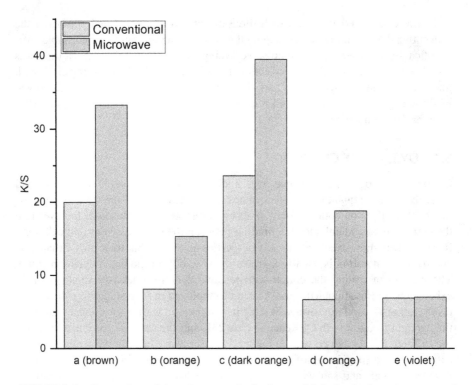

FIGURE 5.1 Comparison of the color strength of polyester fabric dyed using microwave and conventional heating (Al-Mousawi, El-Apasery, and Elnagdi 2013).

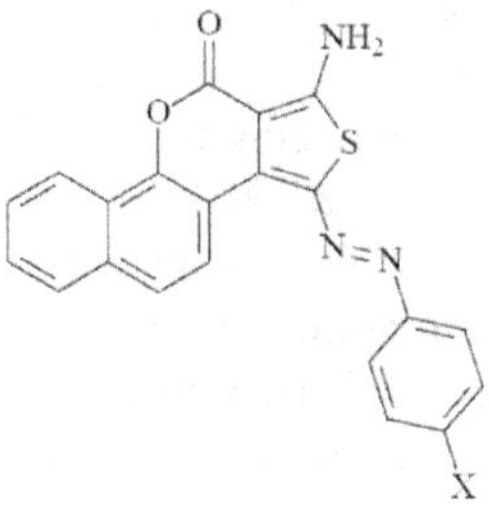

FIGURE 5.2 Chemical structure of the synthesized disperse dyes (Al-Mousawi, El-Apasery, and Elnagdi 2013).

approaches, such as blending with dyeable polymers and nanoparticles, copolymerization, and surface modification methods, have been employed to improve the dyeability of PP fibers. In microwave dyeing of PP fibers with CI Disperse Blue 56, increasing the microwave power level, dyeing time, and temperature increased the color strength. The optimum dyeing time was 15–20 minutes. The diffusion coefficient was higher in microwave dyeing compared with the conventional and ultrasonic dyeing methods (Yiğit and Teker 2011).

Kocak et al. dyed PP fibers with three disperse dyes at boil and compared the conventional and microwave heating methods. The color strength of the samples was significantly improved using microwave dyeing, while the whole process time was 90% lower compared with the conventional heating method. The blending of PP with linear low-density polyethylene and elastomer of ethylene-vinyl acetate improved its dyeability by conventional and microwave heating methods (Koçak et al. 2015; Sahinbaskan et al. 2017).

5.4 DYEING OF COTTON

Cotton can be dyed with anionic dyes, including direct, reactive, sulfur, vat, and azoic dyes. Large quantities of electrolytes are usually used to improve the exhaustion of these dyes on cotton. Microwaves can promote the diffusion of the dyes into the cotton fibers and enhance the binding between the dye molecules and cellulose. Reactive dyes show excellent fastness properties on cotton, but their poor exhaustion and fixation on cotton is usually a problem, especially in the batch dyeing method. The process of dyeing the cotton with reactive dyes includes two stages, namely exhaustion, and fixation. Microwaves can be employed in both stages to enhance the result. Lei et al. used microwave heating in comparison with conventional heating in the dyeing of cotton with CI Reactive blue 231, and the samples were dyed according to Table 5.1. As can be seen in Figure 5.3, microwave heating improved both dye exhaustion and fixation significantly. Also, the dyeing time was shortened, and the dosage of alkali and salt was lowered in microwave dyeing. XRD results showed that microwave irradiation did not change the inner structure of cotton, and the tensile strength of the cotton fabric was unchanged. Microwave heating improved the kinetic energy of dye molecules leading to better penetration and interaction with cellulose (Lei et al. 2013). Haggag et al. showed that the application of microwave heating (150 W for 5–20 minutes) in the dyeing of cotton with CI Reactive Red 195 and CI Reactive Blue 221 saves 90 minutes in dyeing time, 75% salt used, and 20% alkali used (Haggag, El-Molla, and Mahmoued 2014).

Microwave heating can be useful in the natural dyeing of cotton as well. Both dye extraction and application are improved using microwave heating. It has been shown that microwave heating can improve the color strength and fastness properties of cotton when dyed with harmal seeds and neem bark. The mordanting process using

TABLE 5.1
Abbreviations Used in Reactive Dyeing and Fixing of Cotton Samples

Method of Dyeing and Fixing	Abbreviation
Conventional Dyeing and Fixing	CC
Microwave Dyeing and Fixing	MM
Microwave Dyeing and Conventional Fixing	MC
Conventional Dyeing and Microwave Fixing	CM

Source: Lei et al. 2013.

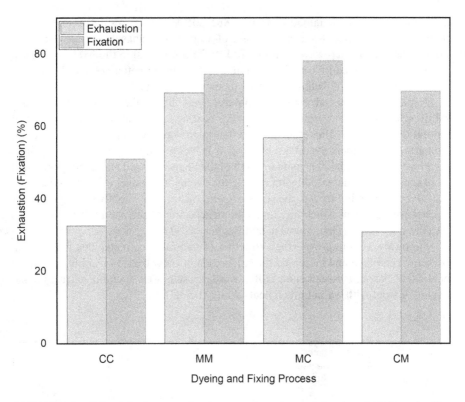

FIGURE 5.3 Effect of microwave heating on exhaustion and fixation of CI Reactive blue 231 on cotton (Lei et al. 2013).

alum and acacia (as bio-mordant) was also improved (Adeel, Zuber et al. 2018, Adeel, Rehman et al. 2020).

5.5 DYEING OF WOOL AND SILK

Microwave treatment influences the physical and morphological properties of wool fibers. Microwave treatment of wool fibers in wet state decreases the tensile strength of the fibers, and this effect increases with increasing the duration and power of the microwave treatment. FTIR investigations showed that microwave treatment has no significant effect on the chemical structure of wool. XRD results confirmed the decrease in crystallinity of wool fibers when treated with microwave under wet conditions. SEM images revealed the damage of surface scales of wool fibers. The concentration of S-S bonds in the microwave-treated wool is lower than the untreated fibers, as confirmed by Raman spectroscopy. Microwave treatment slightly decreases the whiteness of wool fibers (Xue and Jin-Xin 2011). These changes can influence the dyeability of wool fibers significantly. Studies by Xue showed that the pretreatment of wool fibers with microwave for 1–3 minutes improves the dyeability with reactive and metal complex 1:1 dyes. The dye uptake and diffusion coefficient of the microwave-treated wool fibers were higher than the untreated sample, while the adsorption behavior remained constant (Xue 2016). Similar results were found in microwave pretreatment and pad dyeing of wool with Lanasol reactive dye (Xue 2017). The use of microwave heating (700 W for 3 minutes) in the fixation stage of wool dyeing by the pad-fix method resulted in the same or better color yield as compared with 24 hours of batching at room temperature (Zhao 2017).

Despite the various advantages of natural dyes, the uptake and fixation of natural dyes on wool fibers are low. Microwave heating can enhance the exhaustion of natural dyes on wool. Haji et al. showed that microwave heating improved the color strength of wool fibers dyed with eucalyptus leaves and decreased the dyeing time. As shown in Figure 5.4, the color strength of the nonmordanted and alum-mordanted wool samples was increased by increasing the dyeing time under microwave heating. As can be seen in Figure 5.5, the surface scales of wool are slightly destroyed under microwave treatment which facilitates the penetration of the natural dye molecules into the wool fibers (Haji, Mehrizi, and Baghery 2017).

Microwave also improved the extraction of natural dyes molecules from Arjun bark and cochineal and increased the exhaustion of these dyes on wool fibers (Adeel et al. 2019, Adeel, Hussaan et al. 2018). Similar results were found in extraction and dyeing of wool with walnut green peel (Wang et al. 2021).

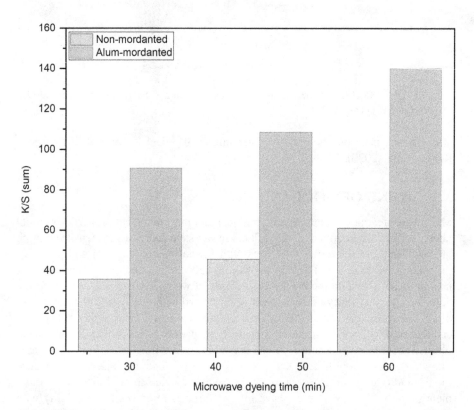

FIGURE 5.4 Effect of treatment time on color strength of wool in dyeing with eucalyptus leaves under microwave heating (Haji, Mehrizi, and Baghery 2017).

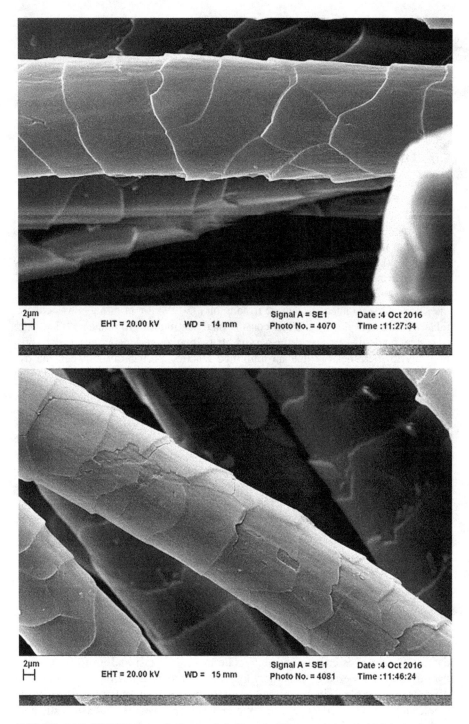

FIGURE 5.5 SEM images of untreated (left) and microwave-treated (right) wool fibers (Haji, Mehrizi, and Baghery 2017).

Microwave has been employed in natural dyeing of silk fibers as well. Application of microwave in the extraction and dyeing of silk with saffron influenced the process efficiency up to 50% and increased the dye exhaustion and color strength (El-Khatib, Ali, and Ramadan 2014). Microwave heating also enhanced the dyeability of bio-mordanted silk with safflower, neem, and cinnamon barks as well as tannins extracted from coconut coir (Kiran et al. 2020, Adeel, Naseer et al. 2020, Adeel, Habib et al. 2020, Zuber et al. 2019).

5.6 DYEING OF ACRYLIC AND NYLON

Popescu et al. dyed acrylic knit fabric with curcumin extracted from turmeric under microwave irradiation. The process was done without any additive and resulted in uniform and florescent shade at shorter duration compared with the conventional dyeing method. The highest color strength was obtained when water/ethanol was used as the solvent for the dye extraction for 15 minutes. The influence of microwave heating on the dyeing of acrylic fibers with CI Basic Blue 28 was investigated by Nourmohammadian and Gholami. The results showed that the dyeability of acrylic fibers was significantly improved under microwave irradiation. The color strength of the samples dyed under microwave irradiation (720 W) for 14 minutes was comparable with the samples dyed under conventional heating for 110 minutes. Microwave treatment made the surface of the acrylic fibers rougher and increased the dye sorption due to local heating and an amplified reaction probability between the dye and acrylic fibers. The tensile strength of the fibers was unchanged (Davoudzadeh Gholami and Nourmohammadian 2008).

Microwave heating is effective in dyeing of nylon fibers and saves time and energy. Reactive dyeing of nylon 6 under microwave irradiation (without salt) was much faster than the conventional heating and saves 40 minutes when obtaining the same color strength. Salt-free dyeing is another advantage of microwave heating as an eco-friendly method which saves money, time, and energy, without changing the tensile strength and surface morphology of nylon fibers (Ghazal 2020). Disperse dyeing of nylon fibers can be done using microwave irradiation at shorter times (5 minutes), without carrier and dispersing agent, and very good fastness properties can be obtained (Haggag, El-Molla, and Ahmed 2015).

5.7 CONCLUSION AND FUTURE OUTLOOK

Textile industry consumes high amounts of energy in heating and cooling processes in different stages of the production such as dyeing, finishing, etc. The conventional method of heating is highly energy- and time-consuming. Microwave heating is a promising alternative for conventional heating in textile dyeing which saves energy and time. The heating is fast and uniform, and the dyeing time is lowered considerably by using microwave heating. The dyeing of different textile fibers including polyester, polypropylene, nylon, acrylic, cotton, wool, and silk with different synthetic and natural dyes has been investigated, and the results confirmed the effectiveness of microwave heating in improvement of these dyeing processes. It has no inverse effect on the physical properties of the textiles, and the fastness properties of the dyed goods remain unchanged as well and thus could be a significant tool in coloring and finishing of textile substrates in future as well.

REFERENCES

Abou Elmaaty, Tarek M., Fathy M. El-Taweel, and Hanan G. Elsisi. 2018. "Water-free dyeing of polyester and nylon 6 fabrics with novel 2-oxoacetohydrazonoyl cyanide derivatives under a supercritical carbon dioxide medium." *Fibers and Polymers* 19 (4):887–893. doi: 10.1007/s12221-018-7766-2.

Adeel, Shahid, Noman Habib, Saba Arif, Fazal ur Rehman, Muhammad Azeem, Fatima Batool, and Nimra Amin. 2020. "Microwave-assisted eco-dyeing of bio mordanted silk fabric using cinnamon bark (Cinnamomum Verum) based yellow natural dye." *Sustainable Chemistry and Pharmacy* 17:100306. doi: 10.1016/j.scp.2020.100306.

Adeel, Shahid, Muhammad Hussaan, Fazal-ur Rehman, Noman Habib, Mahwish Salman, Saba Naz, Nimra Amin, and Nasim Akhtar. 2018. "Microwave-assisted sustainable dyeing of wool fabric using cochineal-based carminic acid as natural colorant." *Journal of Natural Fibers* 16 (7):1026–1034. doi: 10.1080/15440478.2018.1448317.

Adeel, Shahid, Samreen Gul Khan, Sania Shahid, Muhammad Saeed, Shumaila Kiran, Muhammad Zuber, Muhammad Suleman, and Nasim Akhtar. 2018. "Sustainable dyeing of microwave treated polyester fabric using disperse yellow 211 dye." *Journal of the Mexican Chemical Society* 62, 1–10.

Adeel, Shahid, Khadija Naseer, Sadia javed, Saqib Mahmmod, Ren-Cheng Tang, Nimra Amin, and Saba Naz. 2020. "Microwave-assisted improvement in dyeing behavior of chemical and bio-mordanted silk fabric using safflower (Carthamus tinctorius L) extract." *Journal of Natural Fibers* 17 (1):55–65. doi: 10.1080/15440478.2018.1465877.

Adeel, Shahid, Fazal-Ur Rehman, Muhammad Kaleem Khosa, Tahira Anum, Muhammad Shahid, Khalid Mahmood Zia, and Mohammad Zuber. 2020. "Microwave assisted appraisal of neem bark based tannin natural dye and its application onto bio-mordanted cotton fabric." *Iranian Journal of Chemistry and Chemical Engineering (IJCCE)* 39 (2):159–170. doi: 10.30492/ijcce.2020.34225.

Adeel, Shahid, Fazal-Ur Rehman, Khalid Mahmood Zia, Muhammad Azeem, Shumaila Kiran, Mohammad Zuber, Muhammad Irfan, and Muhammad Abdul Qayyum. 2019. "Microwave-supported green dyeing of mordanted wool fabric with Arjun bark extracts." *Journal of Natural Fibers* 18 (1): 136–150. doi: 10.1080/15440478.2019.1612810.

Adeel, S., M. Zuber, Rehman Fazal Ur, and K. M. Zia. 2018. "Microwave-assisted extraction and dyeing of chemical and bio-mordanted cotton fabric using harmal seeds as a source of natural dye." *Environmental Science and Pollution Research* 25 (11):11100–11110. doi: 10.1007/s11356-018-1301-2.

Al-Mousawi, Saleh Mohammed, Morsy Ahmed El-Apasery, and Mohamed Hilmy Elnagdi. 2013. "Microwave assisted dyeing of polyester fabrics with disperse dyes." *Molecules* 18 (9):11033–11043.

Arain, Rabia Almas, Farooq Ahmad, Zeeshan khatri, and Mazhar Hussain Peerzada. 2019. "Microwave assisted henna organic dyeing of polyester fabric: a green, economical and energy proficient substitute." *Natural Product Research* 35 (2): 327–330. doi: 10.1080/14786419.2019.1619721.

Chen, Xiaoping, Lihua Zhan, Yongwei Pu, Minghui Huang, Xiwen Chen, Chenglong Guan, Xintong Wu, and Jiayang He. 2019. "Variation of voids and inter-layer shear strength of advanced polymer-matrix composites at different pressures with high-pressure microwave." *Journal of Engineered Fibers and Fabrics* 14:1558925019863958. doi: 10.1177/1558925019863958.

Davoudzadeh Gholami, M. and F. Nourmohammadian. 2008. "An investigation of the dyeability of acrylic fiber via microwave irradiation." *Progress in Color, Colorants and Coatings* 1 (1):57–63. doi: 10.30509/pccc.2008.75708.

El-Khatib, E. M., N. F. Ali, and M.A. Ramadan. 2014. "Environmentally friendly dyeing of Silk fabrics using microwave heating." *International Journal of Current Micribiology and Applied Sciences* 3 (10):757–764.

Elshemy, Nagla and Karima Haggag. 2019. "New trend in textile coloration using microwave irradiation." *Journal of Textiles, Coloration and Polymer Science* 16 (1):33–48. doi: 10.21608/jtcps.2019.9928.1019.

Ghazal, Heba. 2020. "Microwave irradiation as a new novel dyeing of polyamide 6 fabrics by reactive dyes." *Egyptian Journal of Chemistry* 63 (6):12–13.

Haggag, K, MM El-Molla, and KA Ahmed. 2015. "Dyeing of nylon 66 fabrics using disperse dyes by microwave irradiation technology." *International Research Journal of Pure and Applied Chemistry* 8 (2): 103–111.

Haggag, K, MM El-Molla, and ZM Mahmoued. 2014. "Dyeing of cotton fabrics using reactive dyes by microwave irradiation technique." *Indian Journal of Fibre & Textile Research* 39 (4):406–410.

Haji, Aminoddin, Mohammad Khajeh Mehrizi, and Zahra Baghery. 2017. "Eco-friendly dyeing of wool with eucalyptus leaves enhanced by plasma treatment and microwave heating." *8th Texteh International Conference*, Bucharest, Romania.

Hayes, Brittany L. 2002. *Microwave Synthesis: Chemistry at the Speed of Light*. Matthews, USA: CEM Publishing.

Kale, Manik J. and Narendra V Bhat. 2011. "Effect of microwave pretreatment on the dyeing behaviour of polyester fabric." *Coloration Technology* 127 (6):365–371. doi: 10.1111/j.1478–4408.2011.00332.x.

Kim, Sam Soo, Su Gyung Leem, Han Do Ghim, Joon Ho Kim, and Won Seok Lyoo. 2003. "Microwave heat dyeing of polyester fabric." *Fibers and Polymers* 4 (4):204–209. doi: 10.1007/bf02908280.

Kiran, Shumaila, Atya Hassan, Shahid Adeel, Muhammad Abdul Qayyum, Muhammad Sajjad Yousaf, Muhammad Abdullah, and Noman Habib. 2020. "Green dyeing of microwave treated silk using coconut coir based tannin natural dye." *Industria Textila* 71 (3):227–234.

Koçak, Dilara, Mehmet Akalin, Nigar Merdan, and Burcu Yılmaz Şahinbaşkan. 2015. "Effect of microwave energy on disperse dyeability of polypropylene fibers." *Marmara Journal of Pure and Applied Sciences* 27 (1):27–31.

Lei, Ning Ning, De Li Gong, Xiao Rui Ling, and Yi Dong Shi. 2013. "Researches on microwave dyeing cotton fabrics." *Advanced Materials Research* 627:343–347 doi: 10.4028/www.scientific.net/AMR.627.343.

Metaxas, A.C. 1991. "Microwave heating." *Power Engineering Journal* 5 (5): 237–247.

Öner, Erhan, Yeşim Büyükakinci, and Nihal Sökmen. 2013. "Microwave-assisted dyeing of poly(butylene terephthalate) fabrics with disperse dyes." *Coloration Technology* 129 (2):125–130. doi: 10.1111/cote.12014.

Rabia, Samad Arain, Basheer Arain Samad, Hussain Peerzada Mazhar, and Ayoub Arbab Alvira. 2019. "An efficient ultrasonic and microwave assisted extraction of organic Henna dye for dyeing of synthetic polyester fabric for superior color strength properties." *Industria Textila* 70 (4):303–308. doi: 10.35530/IT.070.04.1551.

Rehman, Fazal ur, Shahid Adeel, Muhammad Jawwad Saif, Muhammad Kaleem Khosa, Muhammad Naveed Anjum, Muhammad Kamran, Muhammad Zuber, and Muhammad Asif. 2020. "Ultrasonic assisted improvement in dyeing behaviour of polyester fabric using disperse red 343." *Polish Journal of Environmental Studies* 29 (1): 261–265.

Sahinbaskan, Burcu, Emine Dilara Koçak, Nigar Merdan, and Mehmet Akalın. 2017. "Dyeing of polypropylene blends by using microwave energy." *Journal of Engineered Fibers and Fabrics* 12 (2):20–27.

Wang, Xuemei, Fan Yi, Wentai Zhang, and Xiaohan Guo. 2021. "Optimization of the application of walnut green peel pigment in wool fiber dying and fixing process under microwave-assisted condition." *Journal of Natural Fibers*. doi: 10.1080/15440478.2020.1870632.

Xue, Zhao. 2016. "Study of dyeing properties of wool fabrics treated with microwave." *The Journal of the Textile Institute* 107 (2):258–263. doi: 10.1080/00405000.2015.1024974.

Xue, Zhao. 2017. "Effect of microwave pretreatment on dyeing performance of wool fabric." *Journal of Textile Engineering & Fashion Technology* 1 (6):217–222.

Xue, Zhao and He Jin-Xin. 2011. "Effect of microwave irradiation on the physical properties and structures of wool fabric." *Journal of Applied Polymer Science* 119 (2):944–952. doi: 10.1002/app.32792.

Yiğit, Elif Atabek and Murat Teker. 2011. "Disperse dyeability of polypropylene fibres via microwave and ultrasonic energy." *Polymers and Polymer Composites* 19 (8):711–716. doi: 10.1177/096739111101900812.

Yusuf, Mohd, Mohd Shabbir, and Faqeer Mohammad. 2017. "Natural colorants: historical, processing and sustainable prospects." *Natural Products and Bioprospecting* 7 (1):123–145. doi: 10.1007/s13659-017-0119-9.

Zhao, Xue. 2017. "Research on microwave pad dyeing process for wool fabric." *Research Journal of Textile and Apparel* 21 (4):263–275.

Zuber, Mohammad, Shahid Adeel, Fazal-Ur Rehman, Fozia Anjum, Majid Muneer, Muhammad Abdullah, and Khalid Mahmood Zia. 2019. "Influence of microwave radiation on dyeing of bio-mordanted silk fabric using neem bark (Azadirachta indica)-based tannin natural dye." *Journal of Natural Fibers* 17 (10): 1410–1422. doi: 10.1080/15440478.2019.1576569.

6 Functional Value-Added Finishing of Textile Substrates Using Nanotechnology
A Review

Ahmad Faraz
Glocal University

Mohammad Faizan
Nanjing Forestry University

Shamsul Hayat
Aligarh Muslim University

CONTENTS

6.1 Introduction 101
6.2 Antimicrobial Properties 102
6.3 Antistatic Finishes 104
6.4 UV-Protective Finishes 104
6.5 Antioxidant Finishes 105
6.6 Nanomaterials for Coloring Textile Materials 105
6.7 Environmental Considerations 106
6.8 Conclusion and Future Prospect 106
References 107

6.1 INTRODUCTION

Nanotechnology has become one of the most important interdisciplinary research areas during the last decades because of its valuable impact on almost all the things used by us in our daily lives. It has wide applications in several different fields. Nanotechnology deals with the materials that come under the nanoscale. Nanoparticles have a large number of applications and are utilized by all of us in some way or another way. Going to the definition, particles with at least one dimension and size between 1 and 100 nm are called nanoparticles (Faraz et al. 2019). Nanoparticles have unique properties than their bulk particles and become more important as they can be applied in various fields. The unique feature present in nanoparticles is due to their small size, large surface area, high reactivity, and efficient optical and magnetic properties. This unique feature makes them a valuable material used in various fields like medicine, industries, agriculture, and electronics (Al-Huqail et al. 2018).

The unique properties of nanomaterials also bring a bright future and new energy for textile industries. Enormous economic potential processes by nanomaterials attracted researchers or scientists, and businessmen from all over the world. Nanotechnology is utilized to develop several fabric treatments to increase certain qualities of the fabric, such as their durability, smoothness, slit strength, resistance to abrasion, and wrinkle resistance (Joshi and Bhattacharyya 2011). The difference between nano-treated fabric is that it provides high strength, permanent antistatic behavior, and durability, and on the other hand, traditionally treated wrap component of the fabric provides necessary softness, ease, and aesthetic appearances (Patra and Gouda 2013). Textile industries move toward nanotechnology because it provides beneficial, cheap, eco-friendly, durable, and high tensile strength fibers. Recently, it has been reported that reliance on the conventional textile business of global textile industries is very difficult under present circumstances, although there is enormous capital investment. The reason behind the fall-off of conventional textile businesses is the increase in the number of competitors. Subsequently, these products are available in the market but at higher prices, which reduces the profit percentage to the minimum level (Mahmud et al. 2017). Moreover, fabrics coated with nanoparticles will not affect their other qualities. Therefore, the preference of nanotechnology over conventional methods in textile industries is increasing day by day. A US-based Burlington industry name Nano-Tex was the first to utilize nanoparticles in textiles (Russell 2002). Use of nanotechnology in textile proved very supportive as it improves the different qualities like water repellence, soil resistance, wrinkle resistance, antibacterial, antistatic and UV-protection, flame retardation, improvement of dyeability, and the number goes on (Samal et al. 2010). Coating is a common technique used to apply nanoparticles to textiles. Different methods are available through which nanoparticles can be coated onto fabrics such as spraying, printing, washing, rinsing, and padding. Among these, the most frequently used one is padding coating (Ward 2003; Yeo et al. 2003). Finishing of final fabrics is the most important criterion to improve the quality of fabrics. While finishing the fabrics, it was kept in mind that the other quality of fabric should not be affected. Nanotechnology provides all these things without affecting the inherent property of fabrics. In finishing textile fabric with nanomaterials, the colloidal solution or ultrafine dispersion of nanomaterials is coated on the fabrics to enhance some of the desired quality (Ghosh et al. 2018; Joshi and Adak 2018; Haji et al. 2015).

6.2 ANTIMICROBIAL PROPERTIES

Our main aim when we use clothes is to protect ourselves from contamination of harmful organisms, but textile fabrics themselves provide a very efficient ground for the growth of microorganism especially those utilized in hospitals, for infants, underwear, and sportswear as microorganisms are easily attracted toward them because of their large surface area. A slight modification in the textile fabric surface can bring new properties in them like antibacterial activity, self-decontamination, hydrophilicity, hydrophobicity, and biocompatibility without affecting the mechanical strength of textile (Li et al. 2017). These conventional methods of textile finishing may prove very costly and also not good for the environment. Nanotechnology provides a way to combat these problems as it is cheap, easy to use, and eco-friendly with sustainable use.

Recent development in nanotechnology and interdisciplinary studies provided a great platform in the antibacterial finishing of textile. In the last few years, it has been reported that various nanomaterials have antimicrobial properties including copper, zinc, gold, silver magnesium, and titanium (Vidic et al. 2013; Hajipour et al. 2012; Shrivastava et al. 2007). Use of nanoparticles in antibacterial finishing in textile was cited in many kinds of literature. Out of different nanoparticles used for antimicrobial properties, silver nanoparticles (AgNPs) are most commonly used as an antimicrobial agent, and they are very effective in controlling both gram-positive and gram-negative bacteria (Shabbir and Mohammad 2018). A very good example of antibacterial finishing in textile is provided by Zahran et al. (2014). During his work, he prepared AgNPs-alginate composite and coated them onto cotton fabric. Results demonstrated that cotton fabric coated with the AgNPs-alginate composite has excellent antibacterial activity against the tested bacteria, *Escherichia coli*, *Staphylococcus aureus*, and *Pseudomonas aeruginosa*. However, with successive washing, there is a slight decrease in the cotton fabrics' antibacterial feature, but fabrics still have efficient antibacterial properties.

There are some reports on titanium oxide nanoparticles (TiO_2 NPs) used as antimicrobial agents in textile industries. TiO_2 NPs applied on cotton fabrics show antimicrobial efficiency even after ten washes, as Khurana and Adivarekar (2013) reported. Moreover, broad-spectrum antibacterial activity was exhibited by ZnO NPs when applied on cotton, polyamide, and bamboo fabric (Zhang et al. 2013). Better results can also be obtained by mixing the ZnO NPs with some synthetic materials. When ZnO NPs are mixed with chitosan and applied on fabrics, they protect them from both gram-positive and gram-negative bacteria (Petkova et al. 2014). Self-cleaning, UV protective, and antibacterial activities of synthesized ZnO NPs in textile industries have been reported by Çakir et al. (2012). ZnO NPs are synthesized in reverse micelle core and coated on fabric to enhance UV protection, self-cleanness, and antibacterial activity. The result shows that fabric has antibacterial activity against gram-negative bacteria (*E. coli*) and gram-positive bacteria (*Staphylococcus aureus*) due to ZnO NPs coating (Çakir et al. 2012). Moreover, cotton fabrics coated with ZnO NPs have antibacterial activity against human pathogen *Staphylococcus epidermidis* and *Klebsiella pneumoniae* as reported by Rajamani et al. (2013). ZnO NPs were green-synthesized using the extract of the Areca nut plant. So it is clear that green synthesis can also be used in textile industries to produce antimicrobial fabrics which are eco-friendly and cost-friendly. Green-synthesized AgNPs reported a similar result of antibacterial activity in fabrics against pathogen *E. coli* and *S. aureus* (Shahid-ul-Islam et al. 2019).

6.3 ANTISTATIC FINISHES

Static charges develop easily in synthetic fibers such as nylon and polyester, reducing water absorbing capacity. It is very often noticed that due to the triboelectric effect in fabric, static charges are accumulated. To overcome these static build-up of charges, textile industrialists used antistatic substances to reduce the accumulation of static charges on fabric. The antistatic substance has both hydrophilic and hydrophobic characteristics; the hydrophobic side chain interacts with the surface of the material, and the hydrophilic side chain interacts with the air moisture and binds the water molecules (Weiguo 2002). Nanotechnology gives very promising results in this area by reducing poor antistatic properties exhibited by synthetic fibers. Synthetic fibers coated with TiO_2 NPs, ZnO whiskers, and nanoantimony-doped tin oxide (ATO) usually have very good antistatic properties. The possible reason may be that TiO_2, ZnO, and ATO are efficient in electric conductivity, which keeps static charges away from synthetic fibers (Asif and Hasan 2018; Wong et al. 2006). Antistatic textile can be developed by incorporating a conducting nanofiller, which can provide a very good appearance without any further reduction in other quality (Saleem and Zaidi 2020).

6.4 UV-PROTECTIVE FINISHES

Clothes protect us from dirt or cold or microorganism and play a very important role in protecting us from the harmful effect of UV radiations if they are designed accordingly. We know that depletion of the ozone layer increased the entry of UV rays toward the earth surface. These UV rays are hazardous for animals and humans as we protect ourselves using cream/sunscreen lotion or UV protective clothes (Shabbir and Mohammad 2018). Thus, protection against UV radiation has become an important property of clothes and textiles (Dhineshbabu and Bose 2019). Reports suggest that nanotechnology has the potential to produce fabrics that have UV-blocking properties. Sol-gel methods are used while preparing UV-resistant cotton fabric by NPs. UV transmittance and UV protection factor (UPF) were measured to find out the ability of a textile to have UV protective properties (Feng et al. 2007). Metal oxide NPs such as TiO_2, ZnO, Fe_3O_4, and CeO_2 show promising results as they improve the UV protective properties in fabric. The nanosized TiO_2 and ZnO are better placed than their bulk counterpart as the latter is more efficient in UV absorptions (Wong et al. 2006). Fei et al. (2006) found excellent results while working with nanorod of TiO_2 on fabrics. Fabric treated with nanorod of TiO_2 has good antibacterial activity along with better UV protective properties. IN FABRICS, the UV protective quality is generated by NPs due to their large surface area, which effectually blocks the UV rays (Pant et al. 2011). UV blocking properties in fabrics can also be changed by changing other qualities like thickness, fabric color, fabric opening, and types of fabrics (Tsuzuki and Wang 2010). Transmittance rate decides the protective nature of fabrics against UV. It has been found that fabrics having 5% or lower transmittance show UV protective capacity (Teng and Yu 2003). With the help of polyurethane-based MnO_2-$FeTiO_3$ nanocomposites, Dhineshbabu and Bose (2019) produce cotton fabrics with excellent UV blocking properties and fire resistance ability. This quality is maintained even after ten cycles of washing, which suggests sustainable use of nanotechnology in textile industries. Antibacterial activity and UV protectivity with green-synthesized AgNPs have also been reported in cotton fabrics (Filipič et al. 2020).

6.5 ANTIOXIDANT FINISHES

Another important area in textile industries that needed improvement is producing hygienic clothes with antioxidant properties. Although antioxidant is not a new concept in biology, it surely is a new thing when we talk about it in clothing. In the last few years, the production of fabrics with good antioxidant quality has got much attention as this type of fabrics can easily neutralize the harmful oxidative radicles (Koh and Hong 2014; Shahid-ul-Islam et al. 2019). The demand for hygienic clothes in medical fields increases and clothes with antioxidative properties can be a good choice in this. Without any exception, nanotechnology proved handy in this field too as results show that NPs have the ability for this. Wool fabric treated with biosynthesized AgNPs shows a great tendency of having antioxidant quality. The NPs were synthesized with plant extract and then coated on wool fabrics, which shows the antioxidant properties with other qualities (Shabbir and Mohammad 2018). It may deduce that plant extract and NPs both mutually play a role in this. Chitosan NPs embedded with iodine are incorporated in viscose fabrics to examine the antioxidative properties. Results show that viscose fabrics treated with the aforementioned chemical had antioxidative properties. This type of material can be used to make medical textiles, especially in wound dressings, tampons, gauzes, plasters, etc. (Zemljič et al. 2018). Moreover, Shahid-ul-Islam et al. (2019) also found similar results while working on linen fabric with chitosan-AgNPs.

Chitosan-AgNPs are synthesized using pineapple extract and coated onto linen fabrics that have antibacterial and antioxidant activity. From the above results, it can be concluded that nanotechnology can also be employed for the production of fabrics with antioxidant activity. A study reported by Shahid-ul-Islam et al. (2019) also provides new insights into textile fields and suggests that pomegranate peel extract which is used in the syntheses of AgNPs can be a valuable functional agent for chemical processing of cotton and might be a potential candidate in medical and healthcare textile segment.

6.6 NANOMATERIALS FOR COLORING TEXTILE MATERIALS

Color is one of the attractive parts of clothes. New and unique color combination increases the clothes' importance and their marketing. The old conventional method used in textile for coloring the fibers is not eco-friendly. The hour needs to improve the coloring properties, hydrophilicity, and adsorption properties, and all these can be achieved by modifying the fabric/textile surface. The surface properties of fibers can be changed by chemical methods like alkaline hydrolysis and aminolysis. Due to their small size, NPs have increased surface area and activity which can easily modify the surface of fibers (Afshari et al. 2019). Nanoparticles impart different colors on textile fibers depending on their synthesis methods and other characterization. Shabbir and Mohammad (2018) showed that green-synthesized AgNPs give blue to yellow color in wool fabrics depending on their size. AgNPs were synthesized using plant extract containing naphthoquinones, phenolics/flavonoids, and polyphenols as a reducing agent and coated onto the wool fabric for color development, antibacterial activity, and UV protection at the same time. Different techniques were used to confirm the presence of AgNPs on wool fibers. For coloration characteristics, color strength (K/S) and CEIL were measured. On the basis of characterization and value plotted on graph, AgNPs impart yellow to blue color onto wool fibers. Similar results were also reported by AgNPs manufactured from honeysuckle on silk fibers (Zhou and Tang 2018). Silk fibers treated with AgNPs have faint yellow color as compared to untreated silk fibers that are white. Moreover, AuNPs were synthesized to provide high-quality color onto wool fibers. Different techniques of NP characterization like SEM and EDS confirm the adsorption of AuNPs onto wool fibers, and results also suggest that there is a strong correlation between Au and sulfur of amino acid cysteine which strengthens the bond between the wool fibers and makes the color fading tough (Johnston et al. 2009). Moreover, various application of AuNPs in wool fibers and silk fibers are reported by different authors (Ahmed et al. 2017; Tang et al. 2017).

6.7 ENVIRONMENTAL CONSIDERATIONS

Environmental consideration should always be in mind when doing experiments because the outcome or end products may interfere with the environment. Huge uses of nanoparticles in various fields always raise a concern about their release into the environment and consequence effects on living things. With unauthorized and large-scale uses of NPs, it becomes necessary to realize the fate and behavior of engineered nanomaterials (ENMs) in the environment (Ray et al. 2009). A very informative and comprehensive review about nanoparticles and their effects on the environment and outdoor workplace was provided by Taghavi et al. (2013). NPs used in fabrics are intentionally or unintentionally released into the environment and deposit in water, soil and, the atmosphere and affects other organisms positively or negatively according to their concentration, type, and size. The toxicity of manufactured NMs on living cells and their fate into the environment was reported by Ray et al. (2009). A positive aspect of NMs is that they are now using to clean the atmosphere as they are used in the phytoremediation process (Song et al. 2019; Gong et al. 2019; Souza et al. 2020).

6.8 CONCLUSION AND FUTURE PROSPECT

Nontechnology is one of the most emerging fields of research, and its application in various products makes it more interesting for upcoming research techniques. As we see above, NPs play a huge role in textile industries, agriculture systems, electronics, and many more commercial sectors; we can say without any doubt that in coming years, nanotechnology would be the leading research industry, and more unimagine products will be made with the help of ENMs. In textile industries, NPs are used to make fabrics with antimicrobial, antistatic, UV protective, and antioxidative properties without affecting their intrinsic quality. Moreover, functionalities imparted to nanofinished fabrics with the help of NPs are more durable, cheap and long-lasting and can withstand many cycles of washing. In the future, nanotechnology can be used in many more industries, which can have novel properties other than their commercial production. Nanotechnology can be used in the agriculture system to enhance productivity and stress-tolerant plants. In textile, nanotechnology can be used to make fabrics lighter and water-resistant and which can be maintained after a number of washings.

REFERENCES

Afshari, Sepideh, Majid Montazer, Tina Harifi, and Mahnaz Mahmoudi Rad. 2019. "A coloured polyester fabric with antimicrobial properties conferred by copper nanoparticles." *Coloration Technology* 135 (6): 427–438. doi:10.1111/cote.12430.

Ahmed, Hanan B., Nancy S. El-Hawary, and Hossam E. Emam. 2017. "Self-assembled aunps for ingrain pigmentation of silk fabrics with antibacterial potency." *International Journal of Biological Macromolecules* 105: 720–729. doi:10.1016/j.ijbiomac.2017.07.096.

Al-Huqail, Asma A., Maysa M. Hatata, Arwa A. AL-Huqail, and Mohamed M. Ibrahim. 2018. "Preparation, characterization of silver phyto nanoparticles and their impact on growth potential of lupinus termis L. Seedlings." *Saudi Journal of Biological Sciences* 25 (2): 313–319. doi:10.1016/j.sjbs.2017.08.013.

Asif, Akmah, and Md Zayedul Hasan. 2018. "Application of nanotechnology in modern textiles: a review." *International Journal of Current Engineering and Technology* 8(2): 227–231.

Çakir, Burçin Acar, Leyla Budama, Önder Topel, and Numan Hoda. 2012. "Synthesis of ZnO nanoparticles using PS-b-PAA reverse micelle cores for UV protective, self-cleaning and antibacterial textile applications." *Colloids and Surfaces A: Physicochemical and Engineering Aspects* 414: 132–139. doi:10.1016/j.colsurfa.2012.08.015.

Dhineshbabu, N R, and Suryasarathi Bose. 2019. "UV resistant and fire retardant properties in fabrics coated with polymer based nanocomposites derived from sustainable and natural resources for protective clothing application." *Composites Part B: Engineering* 172: 555–563.

Faraz, Ahmad, Mohammad Faizan, Fareen Sami, Husna Siddiqui, John Pichtel, and Shamsul Hayat. 2019. "Nanoparticles: biosynthesis, translocation and role in plant metabolism." *IET Nanobiotechnology* 13 (4): 345–352. doi:10.1049/iet-nbt.2018.5251.

Feng, X X, L L Zhang, J Y Chen, and J C Zhang. 2007. "New insights into solar UV-protective properties of natural dye." *Journal of Cleaner Production* 15 (4): 366–372.

Filipič, Jana, Dominika Glažar, Špela Jerebic, Daša Kenda, Anja Modic, Barbara Roškar, Iris Vrhovski, et al. 2020. "Tailoring of antibacterial and uv-protective cotton fabric by an in situ synthesis of silver particles in the presence of a sol-gel matrix and sumac leaf extract." *Tekstilec* 63 (1): 4–13. doi:10.14502/Tekstilec2020.63.4–13.

Ghosh, Gourhari, Ajay Sidpara, and P P Bandyopadhyay. 2018. "High efficiency chemical assisted nanofinishing of HVOF sprayed WC-Co coating." *Surface and Coatings Technology* 334: 204–214.

Gong, Xiaomin, Danlian Huang, Yunguo Liu, Guangming Zeng, Rongzhong Wang, Piao Xu, Chen Zhang, Min Cheng, Wenjing Xue, and Sha Chen. 2019. "Roles of multiwall carbon nanotubes in phytoremediation: cadmium uptake and oxidative burst in: boehmeria nivea (L.) Gaudich." *Environmental Science: Nano* 6 (3): 851–862. doi:10.1039/c8en00723c.

Haji, Aminoddin, Ahmad Mousavi Shoushtari, Firoozmehr Mazaheri, and Sayyedeh Ensieh Tabatabaeyan. 2015. "RSM optimized self-cleaning nano-finishing on polyester/wool fabric pretreated with oxygen plasma." *Journal of the Textile Institute* 107 (8): 985–994. doi:10.1080/00405000.2015.1077023.

Hajipour, Mohammad J., Katharina M. Fromm, Ali Akbar Ashkarran, Dorleta Jimenez de Aberasturi, Idoia Ruiz de Larramendi, Teofilo Rojo, Vahid Serpooshan, Wolfgang J. Parak, and Morteza Mahmoudi. 2012. "Antibacterial properties of nanoparticles." *Trends in Biotechnology* 30 (10): 499–511. doi:10.1016/j.tibtech.2012.06.004.

Johnston, James H, Fern M Kelly, Kerstin A Burridge, and Thomas Borrmann. 2009. "Hybrid materials of conducting polymers with natural fibres and silicates." *International Journal of Nanotechnology* 6 (3–4): 312–328.

Joshi, M, and B Adak. 2018. "Sector 6. Nanotechnology-based textiles: a solution for emerging automotive sector." In *Rubber Nanocomposites and Nanotextiles*, Ed. B. Banerjee, 207–266. De Gruyter: Berlin, Germany.

Joshi, Mangala, and Anita Bhattacharyya. 2011. "Nanotechnology–a new route to high-performance functional textiles." *Textile Progress* 43 (3): 155–233.

Khurana, N, and R V Adivarekar. 2013. "effect of dispersing agents on synthesis of nano titanium oxide and its application for antimicrobial property." *Fibers and Polymers* 14 (7): 1094–1100.

Koh, Eunmi, and Kyung Hwa Hong. 2014. "Gallnut extract-treated wool and cotton for developing green functional textiles." *Dyes and Pigments* 103: 222–227.

Li, Jiwei, Jinmei He, and Yudong Huang. 2017. "Role of alginate in antibacterial finishing of textiles." *International Journal of Biological Macromolecules* 94: 466–473. doi:10.1016/j.ijbiomac.2016.10.054.

Mahmud, Sakil, Md Nahid Pervez, Mst Zakia Sultana, Md Ahsan Habib, and Hui Hong Liu. 2017. "Wool functionalization by using green synthesized silver nanoparticles." *Oriental Journal of Chemistry* 33 (5): 2198–2208. doi:10.13005/ojc/330507.

Pant, Hem Raj, Madhab Prasad Bajgai, Ki Taek Nam, Yun A. Seo, Dipendra Raj Pandeya, Seong Tshool Hong, and Hak Yong Kim. 2011 "Electrospun nylon-6 spider-net like nanofiber mat containing TiO2 nanoparticles: a multifunctional nanocomposite textile material." *Journal of Hazardous Materials* 185: 124-130.Patra, Jayanta Kumar, and Sushanto Gouda. 2013. "Application of nanotechnology in textile engineering: an overview." *Journal of Engineering and Technology Research* 5 (5): 104–111.

Petkova, Petya, Antonio Francesko, Margarida M. Fernandes, Ernest Mendoza, Ilana Perelshtein, Aharon Gedanken, and Tzanko Tzanov. 2014. "Sonochemical coating of textiles with hybrid ZnO/chitosan antimicrobial nanoparticles." *ACS Applied Materials and Interfaces* 6 (2): 1164–1172. doi:10.1021/am404852d.

Rajamani, Ranjithkumar, Selvam Kuppusamy, and M Shanmugavadivu. 2013. "Antibacterial textile finishing via green synthesized silver nanoparticles." *Journal of Green Science and Technology* 1 (2): 111–113.

Ray, Paresh Chandra, Hongtao Yu, and Peter P. Fu. 2009. Toxicity and environmental risks of nanomaterials: challenges and future needs. *Journal of Environmental Science and Health - Part C Environmental Carcinogenesis and Ecotoxicology Reviews* 27. doi:10.1080/10590500802708267.

Russell, E. 2002. "Nanotechnologies and the shrinking world of textiles." *Textile Horizons* 9 (10): 7–9.

Saleem, Haleema, and Syed Javaid Zaidi. 2020. "Sustainable use of nanomaterials in textiles and their environmental impact." *Materials* 13 (22): 1–28. doi:10.3390/ma13225134.

Samal, Subhranshu Sekhar, P. Jeyaraman, and Vinita Vishwakarma. 2010. "Sonochemical coating of Ag-TiO$_2$ nanoparticles on textile fabrics for stain repellency and self-cleaning-the indian scenario: a review." *Journal of Minerals and Materials Characterization and Engineering* 9 (6): 519–525. doi:10.4236/jmmce.2010.96036.

Shabbir, Mohd, and Faqeer Mohammad. 2018. "Multifunctional AgNPs@Wool: colored, UV-protective and antioxidant functional textiles." *Applied Nanoscience* 8 (3): 545–555. doi:10.1007/s13204-018-0668-1.

Shahid-ul-Islam, B. S. Butola, Abhishek Gupta, and Anasuya Roy. 2019. "Multifunctional finishing of cellulosic fabric via facile, rapid in-situ green synthesis of AgNPs using pomegranate peel extract biomolecules." *Sustainable Chemistry and Pharmacy* 12 (February): 100135. doi:10.1016/j.scp.2019.100135.

Shrivastava, Siddhartha, Tanmay Bera, Arnab Roy, Gajendra Singh, P. Ramachandrarao, and Debabrata Dash. 2007. "Characterization of enhanced antibacterial effects of novel silver nanoparticles." *Nanotechnology*. doi:10.1088/0957-4484/18/22/225103.

Song, Biao, Piao Xu, Ming Chen, Wangwang Tang, Guangming Zeng, Jilai Gong, Peng Zhang, and Shujing Ye. 2019. "Using nanomaterials to facilitate the phytoremediation of contaminated Soil." *Critical Reviews in Environmental Science and Technology* 49 (9): 791–824. doi:10.1080/10643389.2018.1558891.

Souza, Lilian Rodrigues Rosa, Luiza Carolina Pomarolli, and Márcia Andreia Mesquita Silva da Veiga. 2020. "From classic methodologies to application of nanomaterials for soil remediation: an integrated view of methods for decontamination of toxic metal(Oid) S." *Environmental Science and Pollution Research* 27 (10): 10205–10227. doi:10.1007/s11356-020-08032-8.

Taghavi, Sayed Mohammad, Mahdiye Momenpour, Maryam Azarian, Mohammad Ahmadian, Faramarz Souri, Sayed Ali Taghavi, Marzieh Sadeghain, and Mohsen Karchani. 2013. "Effects of nanoparticles on the environment and outdoor workplaces." *Electronic Physician* 5 (4): 706–70612. doi:10.14661/2013.706–712.

Tang, Bin, Xu Zhou, Tian Zeng, Xia Lin, Ji Zhou, Yong Ye, and Xungai Wang. 2017. "In situ synthesis of gold nanoparticles on wool powder and their catalytic application." *Materials* 10 (3): 2–13. doi:10.3390/ma10030295.

Teng, Cuiqing, and Muhuo Yu. 2003. "Preparation and property of poly(ethylene terephthalate) fibers providing ultraviolet radiation protection." *Journal of Applied Polymer Science* 88 (5): 1180–1185. doi:10.1002/app.11773.

Tsuzuki, Takuya, and Xungai Wang. 2010. "Nanoparticle coatings for uv protective textiles." *Research Journal of Textile and Apparel* 14 (2):9–20.

Vidic, Jasmina, Slavica Stankic, Francia Haque, Danica Ciric, Ronan Le Goffic, Aurore Vidy, Jacques Jupille, and Bernard Delmas. 2013. "Selective antibacterial effects of mixed ZnMgO nanoparticles." *Journal of Nanoparticle Research* 15 (5). doi:10.1007/s11051-013-1595-4.

Ward, D. 2003. "Small-scale technology with the promise of big rewards." *Technical Textiles International* 12 (2).

Weiguo, Dong. 2002. "Research on properties of nano polypropylene/TiO$_2$ composite fiber." *Journal of Textile Research* 23 (1): 22–23.

Wong, Y. W.H., C. W.M. Yuen, M. Y.S. Leung, S. K.A. Ku, and H. L.I. Lam. 2006. "Selected applications of nanotechnology in textiles." *Autex Research Journal* 6 (1): 1–8.

Yeo, Sang Young, Hoon Joo Lee, and Sung Hoon Jeong. 2003. "Preparation of nanocomposite fibers for permanent antibacterial effect." *Journal of Materials Science* 38 (10): 2143–2147.

Zahran, M K, Hanan B Ahmed, and M H El-Rafie. 2014. "Facile size-regulated synthesis of silver nanoparticles using pectin." *Carbohydrate Polymers* 111: 971–978.

Zemljič, Lidija Fras, Zdenka Peršin, Olivera Šauperl, Andreja Rudolf, and Mirjana Kostić. 2018. "Medical textiles based on viscose rayon fabrics coated with chitosan-encapsulated iodine: antibacterial and antioxidant properties." *Textile Research Journal* 88 (22): 2519–2531. doi:10.1177/0040517517725117.

Zhang, Guangyu, Yan Liu, Hideaki Morikawa, and Yuyue Chen. 2013. "Application of ZnO nanoparticles to enhance the antimicrobial activity and ultraviolet protective property of bamboo pulp fabric." *Cellulose* 20 (4): 1877–1884.

Zhou, Yuyang, and Ren Cheng Tang. 2018. "Facile and eco-friendly fabrication of AgNPs coated silk for antibacterial and antioxidant textiles using honeysuckle extract." *Journal of Photochemistry and Photobiology B: Biology* 178: 463–471. doi:10.1016/j.jphotobiol.2017.12.003.

7 Advanced Use of Nanomaterials in Coloration and Discoloration Processes

*Faten Hassan Hassan Abdellatif
and Hend M. Ahmed*
National Research Centre

CONTENTS

7.1 Introduction 112
7.2 Classification of Dyes 112
    7.2.1 Synthetic Dyes 112
        7.2.1.1 Reactive Dyes 112
        7.2.1.2 Direct Dyes 113
        7.2.1.3 Vat Dyes 113
        7.2.1.4 Sulfur Dyes 113
        7.2.1.5 Disperse Dyes 114
        7.2.1.6 Acid Dyes 114
        7.2.1.7 Basic Dyes 114
    7.2.2 Natural Dyes 114
7.3 Advanced Use of Nanotechnology in the Textile Industry 115
    7.3.1 Utilization of Nanodisperse Dyes as Alternative to Bulk Dyes 116
    7.3.2 Utilization of Nanomaterials for Textile Coloration 116
7.4 Advanced Use of Nanomaterials in Discoloration Processes 122
    7.4.1 Utilization of Carbonaceous Nanomaterials for Dye Removal 123
    7.4.2 Utilization of Metallic Nanomaterials for Dye Removal 124
    7.4.3 Utilization of Magnetic Nanomaterials for Dye Removal 125
    7.4.4 Utilization of Nanomaterial for Dye Degradation 125
7.5 Conclusion and Future Prospects 127
References 127

DOI: 10.1201/9781003140467-7

7.1 INTRODUCTION

Dyes are materials used to confer color to different substrates, e.g., textiles, plastics, paper, cosmetics, etc. Huge amounts of dyes are used annually in different industries (Carneiro, Nogueira, and Zanoni 2007; Chequer et al. 2013; Zollinger 1987). These materials consist of chromophore groups based on anthraquinone, azo, nitro, methine, aryl methane, carbonyl, etc. which are responsible for their coloring ability. Dyes can cling to the substrate physically, chemically, or by mechanical retention (Bafana, Devi, and Chakrabarti 2011).

Dyeing and printing processes are important steps of wet processing of textiles. These processes are considered value-adding processes since they provide the fabric with beauty. Dyeing can take place by batch, continuous, or semi-continuous methods. Depending on the fabric type and the desired quality, the dyeing method can be chosen (Shang 2013).

Generally, dyeing occurred by the interaction of dye and fabric, which adsorbs the dye from its aqueous solution, and the diffusion of dye in the interior parts of the fibers. Moreover, dyeing can take place by precipitation of dye inside textile fibers such as vat dye or by proceeding chemical reaction with the fiber such as reactive dye (Shang 2013).

Printing is localized coloring of parts of the fabric. However, dyeing and printing are complex processes since the used dye is strongly dependent on the type of the fibers, structure of fabric, and dyeing technology (Yeung and Shang 1999).

7.2 CLASSIFICATION OF DYES

7.2.1 Synthetic Dyes

The great efforts made for dye synthesis led to the presence of a remarkable number of dyes in markets. Synthetic dyes can be classified depending on their chemical structure into anthraquinone, azo, phthalocyanine, sulfur, indigo, nitro, and nitroso. Industrially, dyes can be categorized into reactive, direct, vat, dispersed, acid, basic, etc. (Benkhaya, El Harfi, and El Harfi 2017).

7.2.1.1 Reactive Dyes

Reactive dyes are mainly used for coloring cellulosic textiles, but they can also dye wool and nylon. Their chemical structure usually consists of reactive functional groups, typically pyrimidine, triazine, or sulfone groups, which form a covalent bond with the functional groups of the fiber, i.e., hydroxyl group in cellulosic fibers. Reactive dyes consist of chromogen which is responsible for color display. The reactive functional groups are usually linked with the chromogen through bridging links (Ahmed 2005). The chemical attachment of reactive dye with the fabric led to outstanding fasting and durable coloring properties even after washing with hot water. Nevertheless, the fixations of the conventional reactive dyes do not exceed 70% even in printing. The rest of the dyestuff subjects to hydrolysis rather than binding to the fiber, which increases the pollutants in the effluent (Morris, Lewis, and Broadbent 2008). Furthermore, the high water solubility of reactive dyes leads to great difficulty in their removal from wastewater. Reactive dyes are not biodegradable and accumulate in the environment leading to serious environmental problems (Cooper 1993).

7.2.1.2 Direct Dyes

Direct dyes are complex sulfonic acids containing diazo groups, used for dyeing of different types of fibers, e.g., cellulosic, wool, or silk fibers. Mordant or fixations are not required in case of using this class of dyes, but they need auxiliaries similar to what is used with reactive dyes (Bomgardner 2018). The dyes can attach physically to the fiber by hydrogen bonding or van der Waals forces. Direct dyes are soluble salts, and their aqueous solutions are usually used for dyeing cotton at a temperature near the boiling point. Electrolytes such as sodium chloride or sodium sulfate are essential to facilitate the aggregation of the dye on the fabric. Moreover, using of a nonionic wetting agent can improve dyeing homogeneity and dye uptake. To enhance the fastness properties of direct dyes, compounds like quaternary ammonium can be used. However, these compounds are toxic and have an adverse effect on aquatic organisms (Bomgardner 2018).

7.2.1.3 Vat Dyes

Vat dyes are an insoluble class of dyes, used particularly with cellulosic fibers. Their structure is usually based on anthraquinone or indigo. Vat dyes are expensive and lead to excellent fastness properties and UV stability. They are mainly used with textiles usually exposed to harsh washing such as military clothing or with textiles that are daily exposed to UV such as furnishings.

Vat dyes cannot be used directly for dyeing fabrics. They should be converted to water-soluble form using redox chemistry principles by utilizing sodium hydrosulfite as a reducing agent. After absorption of dyes into the fabric, the fabric is treated with oxidizing agent such as hydrogen peroxide to convert the dye again to its insoluble form. Since vat dyes are water-insoluble, production of wastewater is reduced. However, the discharge of heavy metal during the production processing leads to dangerous environmental problems (Chequer et al. 2013).

7.2.1.4 Sulfur Dyes

This class of dyes is used for dyeing cellulosic and cellulosic blend and in some cases dyeing of silk fabrics. They are mainly used for obtaining the dark shades with excellent wash and moderate lightfastness properties. Sulfur dyes are cheap and insoluble in water. The reaction of sulfur compound with phenol and amino compound leads to sulfur dye with high molecular weight. Sulfur dyes can be synthesized from aromatic compounds such as benzene, naphthalene, diphenyl, diphenylamine, and azobenzene using different reactions routes, e.g., substitution, reduction, oxidation, and ring formation. To covert the dyes into water-soluble compounds, sodium hydrosulfide is used, which has great environmental concern. Other systems can be used as an alternative to sodium hydrosulfide such as glucose alone or with hydroxyl acetone (Clark 2011).

7.2.1.5 Disperse Dyes

Disperse dyes are mainly used with hydrophobic fabrics, e.g., polyester, nylon, and acrylic fabrics. This type of dyes is usually insoluble or slightly soluble in water. As a result, it is used for dyeing fabric in disperse form. Subsequently, disperse dyes are usually used in large quantities. The amount of remaining dye in dyeing bath is huge and leads to big environmental concerns (Clark 2011).

7.2.1.6 Acid Dyes

The structure of acid dye usually includes anthraquinone, triphenylmethane, or metal complexes such as cupper complexes. Acid dye becomes soluble in water by interaction with one to four sulfonate groups. Acid dye is used for dyeing wool by partial interaction between sulfonate groups of the dye and ammonium groups on the wool fiber in addition to the interaction provided by van der Waals forces.

7.2.1.7 Basic Dyes

Basic dyes are used for dyeing of acrylic, nylon, and modified polyester fabrics. Due to the poor migration of basic dyes at high temperature, a retarder is usually added to overcome the basic dyes' sensitivity problem. In aqueous solutions, basic dyes form positive charge which is attracted to the fabric bearing negative charge by electrostatic charges (Benkhaya, El Harfi, and El Harfi 2017).

7.2.2 NATURAL DYES

Natural dyes (Yusuf, Shabbir, and Mohammad 2017) can be classified into three categories depending on their origin:

- from plants
- from animals
- from minerals

Based on the chemical structure, natural dyes can be classified as followed:

- indigoids give blue color
- pyridine give yellow color
- carotenoids give yellow, red and oranges colors
- quinonoid give color range from yellow to red
- flavonoids give color range of pale to dark yellow and blue
- dihydropyran
- betalains
- tannins

In addition to textile coloration, natural dyes have other functions such as antibacterial, UV protection, anti-insect, and anti-inflammatory. Natural dyes were widely used before the emerging of synthetic dyes. Although natural dyes are considered sustainable, they need a lot of water and place for plantation. For the dyeing process, mordant is usually used to bind the natural dye with the fabric to increase the dye resistance to wash and sunlight. Mordant usually contains hazard materials such as heavy metal. Moreover, high concentration of mordant such as iron or alum is required to have a successful dyeing process, which is dangerous for the environment and humans (Elsahida et al. 2019).

7.3 ADVANCED USE OF NANOTECHNOLOGY IN THE TEXTILE INDUSTRY

Recently, nanotechnologies have emerged and find reasonable applications in numerous industries, e.g., textiles, information technology, and electronic and wastewater treatments. Materials at nanoscale exhibit valuable properties compared to that of bulk materials. The size of nanomaterials ranged from 1 to 100 nm.

Nanomaterials can be synthesized using different synthetic methods as followed:

1. **Chemical vapor condensation:** Where the precursor is heated in a tube until the molecules decompose. The resulted gas steam expands quickly and reduces the particle growth, followed by condensation of the particles on a cold substrate which can be collected after their scraping off. In this method, the composition of the nanomaterials can be controlled with high purity. Furthermore, no agglomeration can take place. However, it is difficult to control the size distribution and the size of the particles (Chang et al. 1994).
2. **Plasma:** By applying plasma on the metal precursor in the presence of hydrogen and argon gases, nanosized materials such as Ag, Cu, Fe, etc. could be obtained. This method is cheap and produces nanomaterials in high yield (Tavakoli, Sohrabi, and Kargari 2007).
3. **Pyrolysis by laser:** In this method, the gaseous state of metal precursor is interacted with the resonant of laser. This eco-friendly method produces nanoparticles with uniform shape and size. Nevertheless, this method is expensive and consumes energy intensively (Grimes et al. 2000).
4. **W/O microemulsion:** In this method, nanomaterials are formed by dispersion of water droplet at the oil phase interface. The water droplets are stabilized by using surfactant. This method is simple and produces nanoparticles with controlled size and shape. Nevertheless, it is an expensive method, produces low nanoparticles yield, and uses a high amount of liquids (Ganachari et al. 2017).
5. **Sol-gel:** This method produces nanoparticles with excellent control of composition and uniform and ultrafine particles. Sol-gel method consists of five main steps: hydrolysis, polycondensation, aging, drying, and thermal decomposition (Parashar, Shukla, and Singh 2020).
6. **Microbial synthesis of nanoparticles:** In this method, microbes act as bio-factories for nanoparticle production by either intracellular or extracellular method. The used approach depends on the location of nanoparticles. In the intracellular approach, the formation of the nanoparticles occurs in the cell of the organism leading to nanoparticle formation with less size than that obtained by extracellular approach (Narayanan and Sakthivel 2010).

The utilization of nanomaterials in the textile industry has been spread due to the incredible functions offered by nanomaterials. These functionalities can be antibacterial, flame retardant, UV protection, electric conductivity, superhydrophobicity, and drug delivery.

7.3.1 Utilization of Nanodisperse Dyes as Alternative to Bulk Dyes

The main problem associated with the dyeing process is the use of huge amounts of dyes especially disperse dyes. This class of dyes is water insoluble and suitable for dyeing hydrophobic fabrics. The immense quantities of used dyes are very dangerous for the environment as well as cause dye degradation and fabric staining.

The innovation brought about by nanotechnology is the reason for the emergence of new class of dyes suitable for the innovated micro- or nanopolyester fibers. The surface area of the fibers at the nano level extremely increased leading to the need for more dyes for obtaining the desired color shades. Disperse dyes with nano size can play a great role in decreasing the used dyes' quantities, enhancing the quality of dyeing and thus reducing the environmental detriments. Many methods have been developed for nanodisperse dye synthesis including (Yin et al. 2020):

- milling
- homogenization by high pressure
- emulsion polymerization
- supercritical fluid method
- antisolvent precipitation

Choi et al. have synthesized six different nanodisperse dyes using the O/W nanoemulsion method in the presence of sodium lauryl sulfate and caprylic triglyceride as surfactant. The size of nanodisperse dyes ranged from 110 to 130 nm. The dye exhaustion rate reached only 40%, which can be attributed to the solubility of the nanodyes into the micelles of the surfactant. Moreover, the nanoemulsion was very stable and led to the decrease of the dye uptake. However, the color properties of the ultra-microfiber sites are better than that of the regular polyester site (Choi and Kang 2006).

Yin et al. have created an easy method for preparing a stable aqueous nanodispersion of C.I. disperse yellow 54 disperse dye by the combination of antisolvent precipitation and spray drying methods. The size of the nanodye was about 120 nm. This nanodye dispersibility in water and wettability were better than those of the raw dye. Moreover, the color properties associated with nanodye are better than those of the raw dye (Yin et al. 2020).

7.3.2 Utilization of Nanomaterials for Textile Coloration

Nanomaterials can produce color by surface plasmon resonance property (Figure 7.1). This property resulted from the interaction between the surface of the nanomaterials and the light radiation. The color produced from the nanomaterials is differing from that produced from dyes. The color of nanomaterials depends on their shape and

Advanced Use of Nanomaterials 117

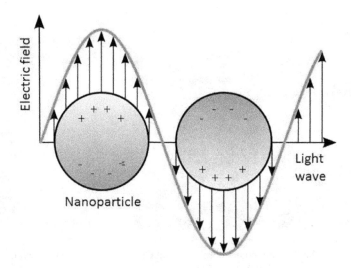

FIGURE 7.1 Localized surface plasmon presented on a nanoparticle surface (Unser et al. 2015).

size. Metals like Au, Ag, and Cu and alkali metals possess free electrons and have plasmon resonances in the visible region (Dang et al. 2011), which is considered the main source of their intense color. The electron behavior of the bulk particles differs from that of the nanoparticles. The light scattering and absorption of nanomaterials exhibit amazing characteristics. The optical properties of the nanoparticles are also influenced by their surrounding environment, its refractive index, and the distance between the nanoparticles. The color of gold NPs change from ruby to violet to blue (Figure 7.2). The Ag NPs have yellow color (Figure 7.2), and the color of Cu NP ranged from red to black to violet and changed to green upon oxidation (Figure 7.3) (Slepicka et al. 2019; Dang et al. 2011).

Recently, nanoparticles have been explored as a promising eco-friendly dyeing procedure. In this method, the textile substrates are colored without using the usual low molecular weight organic dyeing materials. Using nanoparticles in this context

FIGURE 7.2 Colloidal gold (a, b) and silver (C) (Slepicka et al. 2019).

FIGURE 7.3 The different colors obtained from Cu NP, red Cu sol (1), black (2), violet (3) upon onset oxidation (4) (Dang et al. 2011).

will reduce the generation of refractory wastewater and decrease the environmental pollution caused by dyeing processes. In this method, the nanoparticles can be dispersed in water where they can be bonded to the textile substrates. The remaining nanoparticles in water can be easily removed by filtration or precipitation. Moreover, the chemical oxygen demand (COD) of nanoparticles in wastewater is less than that of the conventional dyeing processes. Thus, the purification of wastewater in this method is easier and cheaper (Ding et al. 2019). Furthermore, coloration of textiles with the different nanoparticles led to high intensity, high fad resistance, facile color tunability, and producing structure coloring rather than surface coloring with the conventional dyeing materials (Zhao et al. 2012).

Among nanoparticles, Ag NPs found great interest for innovating new sustainable textile products. Ag NPs are used for developing medical and protective clothing. According to the size and shape of nanoparticles, they could confer textile substrate different functionalities such as antimicrobial, self-cleaning, UV-protection, coloration, etc.

Wu et al. used Ag NPs for obtaining multifunctional cotton fabrics, i.e., durable antibacterial and self-healing superhydrophobic properties as well as coloring the fabric with satisfactory color fastness. The fabric was treated using the solution dipping method. The finishing formula consisted of branched poly(ethylenimine) (PEI), AgNPs, and fluorinated decyl polyhedral oligomeric silsesquioxane. AgNPs are used as an alternative to the organic dyes for dyeing cotton fabric with providing abundant colors (Figure 7.4) (Wu et al. 2016).

Tang et al. assembled anisotropic Ag NPs onto cotton fibers. Poly (diallyldimethylammonium chloride) was used to link the nanoparticles with the cotton fibers at room temperature. The localized surface plasmon resonance property of Ag NPs was responsible for the coloration of the cotton fibers with different colors. The electrostatic force between the ingredients may be the driven force for the coloration process. The resulted color was dependent on the solution pH. The color properties

Advanced Use of Nanomaterials

FIGURE 7.4 Ag NPs used for coloring cotton fabrics with tunable colors. (Reprinted from (Wu et al. 2016), copyright 2015 WILEY-VCH Verlag GmbH & Co. KGaA, Weinheim with permission.)

and the fastness to washing were satisfactorily good which open new opportunities for developing the coloration process with Ag NPs (Figure 7.5) (Tang et al. 2012).

Imam et al. used two solvent-less procedures: sorption and padding, for instant deposition of AgNPs onto cotton fabric. The AgNP content of cotton fabric using the sorption procedure was extremely better than the contents obtained by the padding procedure. The deposition of AgNPs onto cotton fabric with increasing the concentration led to changing its color to gray and reddish yellow as well as enhancing the color strength values (Emam et al. 2016).

Hasan et al. reported a novel method for dyeing polyester fabric with green synthsized AgNPs by the dip coating technique. Chitosan was used as a natural eco-friendly reductant and stabilizing agent because of its nontoxicity and bio-compatibility. The amino group of chitosan led to the formation of yellow, blue, and red colors of green synthesized AgNPs by interaction with silver nitrate. For coloration of polyester, the three colors of green synthesized AgNPs were added to the fabric in one step separately. During this process, the Ag^+ aggregate, by raising the temperature to 135°C the reducing end group of polyester convert the Ag^+ to Ag^0 which color the polyester fabric (Hasan et al. 2019).

Different nanoparticles have promising characteristics, which enable them to replace the conventional dyeing materials. For example, carbon black can confer textile substrates with black color with the ability to control the color darkness.

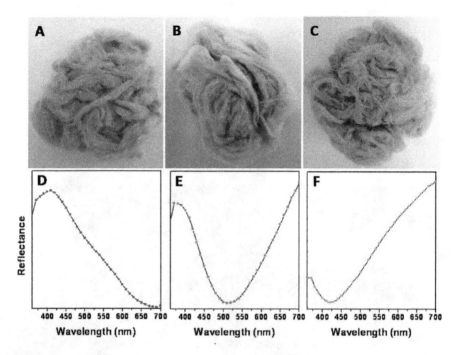

FIGURE 7.5 A to C are cotton fibers colored with Ag NPs. D to F are the corresponding reflectance spectra of the silver nanoparticle-treated cotton fibers. (Reprinted from (Tang et al. 2012), Copyright © 2012, American Chemical Society with permission.)

Moreover, nanoparticles such as cobalt green, cobalt blue, and iron oxide red can provide RGB color system.

Carbon black CB is used widely in the painting industry as an inorganic pigment. CB NPs have good absorption and give excellent color properties and refraction of light. The characteristics of CB NPs provide great opportunities for their application in the textile dyeing process. CB NPs with 8 nm can diffuse into the polyester and acrylic fibers at temperature above their glass transition temperature. Dyeing process using CB NPs occurs in short time with saving energy and reducing the used water. Nevertheless, dyeing of cotton by CB NPs in the exhaustion process is difficult since CB NPs are hydrophobic, and their dispersion in aqueous solution is very difficult.

Li et al. succeeded in directly dyeing some natural and synthetic fabrics such as cotton, wool, acrylic, and nylon fabrics by modified CB NPs using exhaustion method. The CB NPs underwent the oxidation process in nitric acid to acquire carboxylic groups on their surfaces. The modified CB NPs acted as direct or acid dyes in dyeing cotton, wool, acrylic, and nylon fabrics. The dyeing process occurred without using dispersing agent or binder. The exhaustion rate was very low because a small part of nanoparticles is dispersed in single particle form (Li and Sun 2007).

Wang et al. used another procedure for dyeing cotton by CB NPs using the exhaustion method. The cotton fabric was pretreated by introducing cationic group, e.g., quaternized amine group. The quaternized cotton fabric could increase the attraction of cotton fabric to CB NPs, thus enhancing its color properties (Wang et al. 2012)

Advanced Use of Nanomaterials

Ding et al. have innovated a new clean method by using cobalt blue, cobalt green, red iron, and carbon black as alternative to the conventional dyeing materials to avoid the environmental problems from its origin. The nanoparticles were reacted with silane coupling agent to gain C=C groups. These groups permit the nanoparticles to participate in the radiation-initiated cograft polymerization reaction of vinyl monomers like trimethylolpropane trimethacrylate and hydroxyl ethyl acrylate. The nanoparticles were strongly immobilized onto cotton fabric by covalent bond network which led to the coloration of the fabric with a series of different colors, e.g., black, red, green, and blue (Figure 7.6). The covalent bonding networks prevent the release of the nanoparticles even after 100 laundering cycles. Furthermore, the resulting color was easily controlled by changing the used amount of the nanoparticles. This new method has extremely low COD, chroma, TSS, and dangerous ion concentration (Figure 7.7) (Ding et al. 2019).

Gao et al. produced structure coloration of textile substrates using uniform silica nanoparticles. Natural sedimentation self-assembly of silica nanoparticle suspension was used for coating the textile substrates. The size of the nanoparticles ranged from 207 to 350 nm, and this particular size range led to the structure coloring of textile substrates. The color resulted from Bragg diffraction of white light with the ordered structure of silica nanoparticles. The resulted color was dependent on

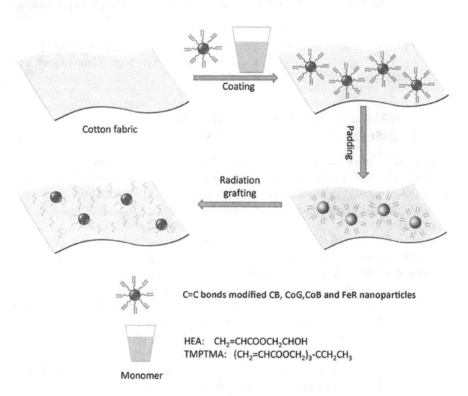

FIGURE 7.6 Immobilization of carbon black NPs onto cotton fabric. (Reprinted from (Ding et al. 2019), Copyright Royal Society of Chemistry with permission.)

FIGURE 7.7 Colored cotton with different nanoparticles and the resulted wastewater. (Reprinted from (Ding et al. 2019), Copyright Royal Society of Chemistry with permission.)

the silica nanoparticle diameter which varied from red, yellow, green, and cyan to blue. The light fastness of the treated fabric was extremely high (Gao, Rigout, and Owens 2017).

7.4 ADVANCED USE OF NANOMATERIALS IN DISCOLORATION PROCESSES

The textile industry is considered the main chemical exhaustion sector, and it is the most global polluting industry for running water. The size and the composition of the chemical substances in effluents of textile processing vary from the enormous presence of dyes and finishing agents to the presence of salts, surfactants, and additives (Cao et al. 2014).

Most types of dyes such as reactive, direct, basic, and acid dyes are soluble in water. Dyes are toxic and carcinogenic materials, form gene mutations, disturb the balance of the entire food chain, and cause serious diseases for animals and humans. Their high solubility increases the difficulty of their removal by conventional methods (Hassan and Carr 2018). The basic use of dyes in the textile industry is to color the textile substrates and make them attractive for the consumer. The chromophoric groups present in the dye compositions are responsible for the color characteristic. Unfortunately, the chromophoric groups make aesthetic damage in the body of water. Furthermore, the presence of color in water prevents the penetration of sunlight inside water causing reduction in the photosynthesis reaction and decrease in the dissolved oxygen in water, which has an adverse impact on the marine life (Wardman 2017).

Various techniques have been utilized for remediation of organic dyes from water, e.g., photocatalytic degradation, membrane separation (Abdellatif et al. 2016, 2018; Abdellatif et al. 2015; Abdellatif et al. 2017), chemical oxidation, flocculation biological treatment, coagulation, aerobic or anaerobic treatment, and adsorption. The adsorption technique is widely used for organic dye remediation due to its simplicity, low price, and great efficiency. The biomass material obtained from agriculture, polymers, by-product of many industries, and clays are examples for adsorbent used for wastewater remediation (Emam, Abdellatif, and Abdelhameed 2018). Nevertheless, these adsorbents have low capacities, which limit their industrial application (Abdellatif and Abdellatif 2020; Abdellatif et al. 2020; Zhang et al. 2013).

Nanomaterials have been exploited as a new adsorbent because of their distinguished properties, e.g., high reactive surface area, quantum size effects, and electron conduction. These unique properties offer nanomaterials as promising strong adsorbents for wastewater treatments. Various nanomaterials such as nanometal oxides, polymer-based nanomaterials, and graphene-based nanomaterials have been reported for removal of organic dyes.

7.4.1 Utilization of Carbonaceous Nanomaterials for Dye Removal

Carbonaceous nanomaterials including graphene, carbon nanofibers (CNFs), and carbon nanotubes (CNTs) have been widely used as a superior adsorbent for dye removal from wastewater due to their high capacity and affinity. These nanomaterials consist of carbon molecules linked together covalently. They adsorb the organic dyes using their hydrophobic character, π-π stacking, covalent interaction, and electrostatic or hydrogen bonding. Graphene is an allotrope of 2 D carbon networks with honeycomb lattice; CNFs are stacked cylindrical nanographene layers in a cone structure, while CNTs are single wall of rolled up graphene sheets (Cai et al. 2017). The different carbonaceous nanomaterials have different adsorption abilities due to the different space offered by each of them. Among carbonaceous nanomaterials, graphene oxide exhibits excellent dye remediation from wastewater due to the presence of numerous carboxyl, carbonyl, and epoxy hydroxyl groups on its surface which increase the efficiency of graphene oxide as the adsorbent material (Tan et al. 2015).

CNTs include two classes: single-walled and multiwalled. Maleki et al. modified the multiwalled CNT by free radical polymerization of unsaturated monomers functionalized with -NH and -NH$_2$ groups on the surface of CNT. The introduction of −NH and −NH$_2$ groups on the surface of CNT increases the ability of CNT to chelate dyes and metal ions. The modified CNT was investigated for adsorption of some anionic dyes such as Acid Blue 45 and Acid Black 1separatly and in binary system. The adsorption capacities of modified CNT for Acid Blue 45 and Acid Black 1 were 714 and 666 mg/g, respectively. The results indicated that modified CNT is a very good adsorbent with very high adsorption capacity for remediation anionic dyes single or binary systems (Maleki et al. 2017).

Li et al. synthesized a new magnetic hybrid from polypyrrole-multiwall CNT and CoFe$_2$O$_4$; CNTs-CoFe$_2$O$_4$@PPy; by chemical oxidative polymerization of pyrrole in the presence of CNTs-CoFe$_2$O$_4$. This composite was investigated for the adsorption of both anionic and cationic dyes. The CNTs-CoFe$_2$O$_4$@PPy exhibited high adsorption capacities with fast adsorption rate in the studied pH range toward anionic dyes better than cationic dyes. Furthermore, CNTs-CoFe$_2$O$_4$@PPy showed efficient catalytic activation for generation of reactive radical, e.g., OH and SO$_4$ from peroxymonosulfate. These reactive radicals are efficient in 100% methylene blue degradation after 30 minutes using 1.0 g/L catalyst, 4.0 mM peroxymonosulfate and 50 mg/L methylene blue solution. This indicated that CNTs-CoFe$_2$O$_4$@PPy was an effective adsorbent and catalyst for the remediation of wastewater containing both cationic and anionic dyes (Li et al. 2017).

Dastgerdi et al. synthesized pure CNT and CNT-functionalized COOH by the chemical vapor deposition method at 1000°C and using methane gas as a carbon source in a quartz tube. The synthesized adsorbents were investigated for the removal of indigo carmine dye. The results exhibited that the adsorption efficiency of CNT-COOH (98.7%) is better than that of pure CNT (84%). The adsorption of indigo carmine dye onto pure CNT was attained by π–π interaction between them. In addition to π–π interaction, CNT-COOH adsorbed indigo carmine dye by hydrogen bonding interaction that led to a quick adsorption process (Dastgerdi, Meshkat, and Esrafili 2019).

Saharan et al. fabricated Sn-MnO$_2$/CNT nanocomposites by coprecipitation. This nanocomposite was efficient in adsorption of anionic dyes from aqueous solutions (Saharan et al. 2019). Another nanomagnetic composite CNT/MgO/CuFe$_2$O$_4$ was synthesized by Foroutan et al. using the coprecipitation method for removal of two cationic dyes, i.e., methyl violet dye and Nile blue dye. BET analysis of the magnetic composite exhibited its meso-porous structure with specific surface area of 127.58 m^2/g. The adsorption capacity for methyl violet dye and Nile blue dye was 36.46 and 35.60 mg/g, respectively (Foroutan et al. 2020).

Mohamed et al. created a superior adsorbent for azo dye removal by impregnation of anthracite with a small amount of multiwalled CNT at 950°C for 2 Hours. Azo dye such as methyl orange was used to investigate the adsorption efficiency of the synthesized adsorbent. The driving mechanism of azo dye adsorption by An/CNT at pH < pH of the point of zero charge was the electrostatic interaction between the function groups containing oxygen and nitrogen present on the composite surface and the azo dye, while at pH > pH of the point of zero charge, the hydrogen bonding is the driving mechanism for adsorption process (Mohamed, Li, and Zayed 2020).

7.4.2 Utilization of Metallic Nanomaterials for Dye Removal

Metallic nanomaterials consist of at least one metal in their composition. TiO$_2$ NPs, ZnO NPs, MnO NPs, and MgO NPs are reported for dye removal from aqueous solution. The characteristics of these materials and their dye removal mechanisms are different since they do not belong to the same group in the periodic table.

Manganese oxides and their composites have been reported as a very good adsorbent for a wide range of organic dyes from wastewater, e.g., methylene blue and methyl orange. The commonly used manganese adsorbents are manganese dioxide or birnessite (δ-MnO$_2$), pyrolusite (β-MnO$_2$), hausmannite (Mn$_3$O$_4$), manganite (MnOOH), cryptomelane (α-MnO$_2$), and Mn$_2$O$_3$. Manganese oxides and their composites showed great adsorption capacity and quick rate for different dyes due to the large surface area and low surface charge. In case of cationic dyes, higher adsorption was obtained at higher pH, while in case of anionic dyes, better adsorption was obtained at lower pH (Islam et al. 2019).

MgO is widely used in many fields. It is used as a catalyst for many organic reactions and adsorbents for dyes and heavy metals from aqueous solutions, electrochemical biosensor, and antimicrobial agent. MgO NPs can be synthesized using the conventional method of nanoparticle synthesis. MgO NPs were reported for Remazol Red RB-133 remediation from aqueous solution. Complete removal of the dye could be reached using MgO NPs (Mahmoud, Ibrahim, and El-Molla 2016).

7.4.3 Utilization of Magnetic Nanomaterials for Dye Removal

The main problem with using the nanomaterials in wastewater treatment is their separation from water using the conventional separation methods, i.e., centrifugal separation that is less effective and increasing the cost of the process. Combining the nanomaterials with magnetic materials could be a solution for this problem. In this case, the nanomaterials could be separated by the action of external magnetic field (Wang et al. 2018).

Wang et al. prepared magnetic carbon nanomaterials by the one-pot hydrothermal method. Ferric ammonium citrate was the iron source, and biologically regenerated glucose was the source of carbon. Upon increasing the temperature during the synthetic process, the iron transformed to alpha structure and carbon transformed to graphite structure. This transforms the sample's magnetism from supermagnetism into ferromagnetism. The carbon graphite structure helps in adsorbing organic dyes such as rhodamine B, while the amorphous iron causes photocatalytic degradation of the dye. The magnetism of the synthesized samples facilitates their recovery and reuse (Wang et al. 2018).

Abdellatif et al. created three new physically cross-linked Iota-carrageenan-poly(amidoamine)-based magnetic aerogels. The large surface area, the high nitrogen content, the presence of sulfonate groups and the magnetic nanoparticles led to a synergetic effect in removal of cationic dye, i.e., Alphanol fast blue dye (Abdellatif et al. 2020).

7.4.4 Utilization of Nanomaterial for Dye Degradation

Recently, nanomaterials have found a wide range of applications in remediation of organic dyes by photocatalytic degradation. This process is very essential, since some organic dyes such as textile dyes are not degraded by the conventional biological degradation processes. Nanoparticles such as TiO_2 NPs and ZnO NPs upon exposure to visible light or UV generate free radicals such as OH at the surface of the nanoparticles. The generated free radicals initiate the oxidation process of the organic dyes leading to its degradation to simple and nontoxic compounds (Figure 7.8). Another degradation mechanism can be occurred by the adsorption of the organic dyes on the surface of the nanoparticles. Upon reacting of the dyes with the generated free radicals, the dyes degraded to the simplest form of nontoxic compound (Hu et al. 2003).

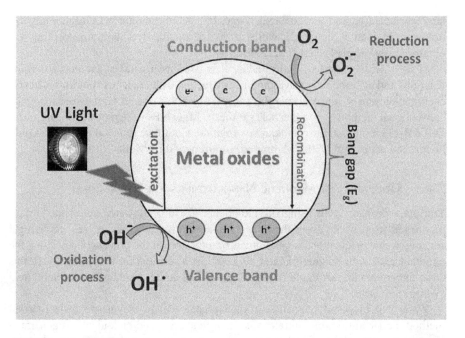

FIGURE 7.8 Photocatalytic mechanism of metal oxide materials. (Reprinted from (Gnanasekaran et al. 2017), copyright 2017 Elsevier.)

Both TiO$_2$NPs and ZnO NPs have high photo-catalytic activity under UV due to their wide band gab. The degradation efficiency of these nanometal oxides is mainly governed by several factors, i.e., dye concentration, pH of the solution, nanometal oxide dose, particle size, and the energy of the band gab.

The main drawbacks of using TiO$_2$ NPs in practical application are the short lifetime of photo-induced electron-hole pairs and their wide band gab. Furthermore, photocatalysis can be used with low dye concentration. Recently, materials with both characteristics – high adsorption efficiency and suitable photocatalytic activity – are synthesized. These materials can improve the dye remediation by adsorbing dyes from massive volume of wastewater, and subsequently, the adsorbed dyes are removed by photocatalysis. This synergetic effect can be attained by titanate nanomaterials, e.g., titanate nanosheets, titanate nanotubes, titanate nanorods, and titanate nanofibers (Chen et al. 2012; Pan et al. 2012).

Nguyen et al. synthesized titanate nanosheets and titanate nanotubes by the hydrothermal method. 0.5 wt% Pt was doped onto titanate nanosheets and titanate nanotubes by the photodeposition method. The adsorbability and photocatalytic performance of the prepared materials were investigated by using two cationic dyes: methylene blue and rhodamine B and two anionic dyes: methyl orange and naphthol blue black. The doping process of Pt onto titanate nanomaterials improves their catalytic activity compared to undoped nanomaterials. The changing in the morphology of TiO$_2$ nanoparticle improves the adsorption efficiency and photocatalytic activity for the remediation of cationic dyes than the anionic dyes. Pt-doped titanate nanosheets attained 100% removal of methylene blue and rhodamine upon exposure for UV irradiation while the photocatalytic activity of these materials was much slower in case of anionic dyes (Nguyen and Juang 2019).

Gnanasekaran et al. synthesized different nanomatal oxides, i.e., CeO_2, CuO, NiO, Mn_3O_4, SnO_2, and ZnO by the precipitation method. The photocatalytic ability of these metal oxides was investigating for methyl orange and methylene blue degradation. Under the exposure of the metal oxide surfaces to UV light, the valence electrons of the metal oxides distribute and transfer to the conduction band. Subsequently, the resulting holes in the valence band react with water molecules and lead to hydroxyl radical generation. ZnO was superior in both dyes' degradation due to the spherical shape and crystalline structure of this oxide (Gnanasekaran et al. 2017).

7.5 CONCLUSION AND FUTURE PROSPECTS

Nanotechnology is continuously growing, and it has been introduced in many industries, i.e., textile industries. Nanotechnology offers a great emerging of new smart materials which can penetrate every stage of the textile industry. Recently, various nanomaterials have been used as an alternative to conventional organic dyes. The nanomaterials could color different textile fabrics by their localized surface plasmon characteristic. The colored fabrics exhibited moderate to very good color properties which prove the nanomaterials as new promising coloring materials.

Textile industries usually generate effluents containing high levels of hazardous compounds, like dyes. The presence of organic dyes in the wastewater causes serious environmental problems. These organic compounds are carcinogenic and cause a mutagenic effect on living creatures. Therefore, the remediation of wastewater and its reuse is an important practice to control the safety of the environment to protect human health and ecological system. In this chapter, nanomaterials were introduced as a promising solution for wastewater treatments. Various nanomaterials can adsorb organic dyes as well as degrading dyes by their photocatalytic activity. Nevertheless, the knowledge about using nanomaterials in textile coloring or in wastewater treatments is not enough, and more studies should be conducted to have further information about the environmental impact of nanomaterials and the influence of wide spreading their applications. The release of nanomaterials during synthesis and treatment should be studied to address the drawbacks of their accumulation in the environment and to reduce the expected health problems. Moreover, further studies should evaluate the cytotoxicity of existing and modified nanomaterials. Nanomaterials should be adopted for use in mass processes and creating cost-competitive synthetic techniques.

REFERENCES

Abdellatif, F. H. H. and M. M. Abdellatif. 2020. "Bio-based i-carrageenan aerogels as efficient adsorbents for heavy metal ions and acid dye from aqueous solution." *Cellulose* 27 (1):441–453.

Abdellatif, F. H. H., J. Babin, C. Arnal-Herault, and A. Jonquieres. 2015. "Grafting of cellulose and cellulose derivatives by cuaac click chemistry." In: *Cellulose-Based Graft Copolymers: Structure and Chemistry*, edited by V. K. Thakur, 568–597. Boca Raton, FL: CRC Press.

Abdellatif, F. H. H., J. Babin, C. Arnal-Herault, L. David, and A. Jonquieres. 2016. "Grafting of cellulose acetate with ionic liquids for biofuel purification by a membrane process: influence of the cation." *Carbohydrate Polymers* 147:313–322 doi: 10.1016/j.carbpol.2016.04.008.

Abdellatif, F. H. H., J. Babin, C. Arnal-Herault, L. David, and A. Jonquieres. 2018. "Grafting cellulose acetate with ionic liquids for biofuel purification membranes: influence of the anion." *Carbohydrate Polymers* 196:176–186. doi: 10.1016/j.carbpol.2018.05.008.

Abdellatif, F. H. H., J. Babin, C. Arnal-Herault, C. Nouvel, J. L. Six, and A. Jonquieres. 2017. "Bio-based membranes for ethyl tert-butyl ether (ETBE) bio-fuel purification by pervaporation." *Journal of Membrane Science* 524:449–459 doi: 10.1016/j.memsci.2016.11.078.

Abdellatif, M. M., S. M. M. A. Soliman, N. H. El-Sayed, and F. H. H. Abdellatif. 2020. "Iota-carrageenan based magnetic aerogels as an efficient adsorbent for heavy metals from aqueous solutions." *Journal of Porous Materials* 27 (1):277–284.

Ahmed, N. S. E. 2005. "The use of sodium EDTA in the dyeing of cotton with reactive dyes." *Dyes and Pigments* 65:221–225.

Bafana, A., S. S. Devi, and T. Chakrabarti. 2011. "Azo dyes: past, present and the future." *Environmental Reviews* 19:350–370.

Benkhaya, S., S. El Harfi, and A. El Harfi. 2017. "Classifications, properties and applications of textile dyes: a review." *Applied Journal of Environmental Engineering Science* 3 (3):311–320.

Bomgardner, M. 2018. "These new textile dyeing methods could make fashion more sustainable." *Chemical and Engineering News* 96 (29), 1–3.

Cai, Z., Y. Sun, W. Liu, F. Pan, P. Sun, and J. Fu. 2017. "An overview of nanomaterials applied for removing dyes from wastewater." *Environmental Science and Pollution Research* 24 (19):15882–15904.

Cao, H., C. Scudder, C. Howard, K. Piro, H. Tattersall, and J. Frett. 2014. "Locally produced textiles: product development and evaluation of consumers' acceptance." *International Journal of Fashion Design, Technology and Education* 7 (3):189–197.

Carneiro, P. A., R. F. P. Nogueira, and M. V. B. Zanoni. 2007. "Homogeneous photodegradation of C.I. Reactive Blue 4 using a photo-Fenton process under artificial and solar irradiation." *Dyes and Pigments* 74:127–132.

Chang, W., G. Skandan, S. C. Danforth, B. H. Kear, and H. Hahn. 1994. "Chemical vapor processing and applications for nanostructured ceramic powders and whiskers." *Nanostructured Materials* 4 (5):507–520.

Chen, F, P. Fang, Y. Gao, Z. Liu, Y Liu, and Y Dai. 2012. "Effective removal of high-chroma crystal violet over TiO_2-based nanosheet by adsorption-photocatalytic degradation." *Chemical Engineering Journal* 204:107–113

Chequer, F., G. Oliveira, E. Ferraz, J. Cardoso, M. Zanoni, and D. Oliveira. 2013. "Textile dyes: dyeing process and environmental impact." In: *Eco-Friendly Textile Dyeing and Finishing*, edited by M. Gunay, 151–176. Rijeka: InTech.

Choi, J. -H. and M.-J. Kang. 2006. "Preparation of nano disperse dyes from nanoemulsions and their dyeing properties on ultramicrofiber polyester." *Fibers and Polymers* 7 (2):169–173.

Clark, M. 2011. "1- Fundamental principles of dyeing." In *Handbook of Textile and Industrial Dyeing*, 3–27. Woodhead Publishing.

Cooper, P. 1993. "Removing colour from dyehouse waste waters - a critical review of technology available." *Journal of the Society of Dyers and Colourists* 109 (3):97–100. doi: 10.1111/j.1478-4408.1993.tb01536.x.

Dang, T. M. D., T. T. T. Le, E. Fribourg-Blanc, and M. C. Dang. 2011. "Synthesis and optical properties of copper nanoparticles prepared by a chemical reduction method." *Advances in Natural Sciences: Nanoscience and Nanotechnology* 2 (1):015009.

Dastgerdi, Z. H., S. S. Meshkat, and M. D. Esrafili. 2019. "Enhanced adsorptive removal of Indigo carmine dye performance by functionalized carbon nanotubes based adsorbents from aqueous solution: equilibrium, kinetic, and DFT study."*Journal of Nanostructure in Chemistry* 9 (4):323–334.

Ding, X., M. Yu, Z. Wang, B. Zhang, L. Li, and J. Li. 2019. "A promising clean way to textile colouration: cotton fabric covalently-bonded with carbon black, cobalt blue, cobalt green, and iron oxide red nanoparticles." *Green Chemistry* 21 (24):6611–6621.

Elsahida, K., A. M. Fauzi, I. Sailah, and I. Z. Siregar. 2019. "Sustainability of the use of natural dyes in the textile industry." *IOP Conference Series: Earth and Environmental Science* 399:012065.

Emam, H. E., F. H. H. Abdellatif, and R. M. Abdelhameed. 2018. "Cationization of celluloisc fibers in respect of liquid fuel purification." *Journal of Cleaner Production* 178:457–467. doi: 10.1016/j.jclepro.2018.01.048.

Emam, H. E., N. H. Saleh, K. S. Nagy, and M. K. Zahran. 2016. "Instantly AgNPs deposition through facile solventless technique for poly-functional cotton fabrics." *International Journal of Biological Macromolecules* 84:308–318.

Foroutan, R., S. J. Peighambardoust, Z. Esvandi, H. Khatooni, and B. Ramavandi. 2020. "Evaluation of two cationic dyes removal from aqueous environments using CNT/MgO/CuFe$_2$O$_4$ magnetic composite powder: a comparative study." *Journal of Environmental Chemical Engineering* 9: 104752.

Ganachari, S. V., N. R. Banapurmath, B. Salimath, J. S. Yaradoddi, A. S. Shettar, A. M. Hunashyal, A. Venkataraman, P. Patil, H. Shoba, and G. B. Hiremath. 2017. "Synthesis Techniques for Preparation of Nanomaterials." In *Handbook of Ecomaterials*, edited by L. M. T. Martĭnez, O. V. Kharissova, and B. I. Kharisov, 1–21. Cham: Springer International Publishing.

Gao, W., M. Rigout, and H. Owens. 2017. "The structural coloration of textile materials using self-assembled silica nanoparticles." *Journal of Nanoparticle Research* 19 (9):303.

Gnanasekaran, L., R. Hemamalini, R. Saravanan, K. Ravichandran, F. Gracia, S. Agarwald, and V. K. Gupta. 2017. "Synthesis and characterization of metal oxides (CeO$_2$, CuO, NiO, Mn$_3$O$_4$, SnO$_2$ and ZnO) nanoparticles as photo catalysts for degradation of textile dyes." *Journal of Photochemistry & Photobiology, B: Biology* 173:43–49.

Grimes, C. A., D. Qian, E. C. Dickey, J. L. Allen, and P. C. Eklund. 2000. "Laser pyrolysis fabrication of ferromagnetic gamma'-Fe$_4$N and FeC nanoparticles." *Journal of applied physics* 87:5642–5644

Hasan, K. M. F., Md. N. Pervez, Md. E. Talukder, Mst. Z. Sultana, S. Mahmud, Md. M. Meraz, V. Bansal, and C. Genyang. 2019. "A novel coloration of polyester fabric through green silver nanoparticles (G-AgNPs@PET)." *Nanomaterials* 9 (4):569.

Hassan, M. M., and C. M. Carr. 2018. "A critical review on recent advancements of the removal of reactive dyes from dyehouse effluent by ion-exchange adsorbents." *Chemosphere* 209 (1):201–219.

Hu, C., C.Y. Jimmy, Z. Hao, and P. K. Wong. 2003. "Photocatalytic degradation of triazine-containing azo dyes in aqueous TiO$_2$ suspensions." *Applied Catalysis B: Environmental* 42 (1):47–55.

Islam, Md. A., I. Ali, S. M. A. Karim, Md. S. H. Firoz, Al-N. Chowdhury, D. W. Morton, and M. J. Angove. 2019. "Removal of dye from polluted water using novel nano manganese oxidebased materials." *Journal of Water Process Engineering* 32:100911.

Li, X., H. Lu, Y. Zhang, and Fu He. 2017. "Efficient removal of organic pollutants from aqueous media using newly synthesized polypyrrole/CNTs-CoFe$_2$O$_4$ magnetic nanocomposites." *Chemical Engineering Journal* 316:893–902.

Li, D., and G. Sun. 2007. "Coloration of textiles with self-dispersible carbon black nanoparticles." *Dyes and Pigments* 72 (2):144–149.

Mahmoud, H. R., S. M. Ibrahim, and S. A. El-Molla. 2016. "Textile dye removal from aqueous solutions using cheap MgO nanomaterials: adsorption kinetics, isotherm studies and thermodynamics." *Advanced Powder Technology* 27:223–231.

Maleki, A., U. Hamesadeghi, H. Daraei, B. Hayati, F. Najafi, G. McKay, and R. Rezaee. 2017. "Amine functionalized multi-walled carbon nanotubes: single and binary systems for high capacity dye removal." *Chemical Engineering Journal* 313:826–835.

Mohamed, F. M., Z. Li, and A. M. Zayed. 2020. "Carbon nanotube impregnated anthracite (An/CNT) as a superior sorbent for azo dye removal." *RSC Advances* 10 (43): 25586–25601.

Morris, K. F., D. M. Lewis, and P. J. Broadbent. 2008. "Design and application of a multifunctional reactive dye capable of high fixation efficiency on cellulose." *Coloration Technology* 124 (3):186–194. doi: 10.1111/j.1478-4408.2008.00140.x.

Narayanan, K. B. and N. Sakthivel. 2010. "Biological synthesis of metal nanoparticles by microbes." *Advances in Colloid and Interface Science* 156 (1):1–13.

Nguyen, C. H. and R.-S. Juang. 2019. "Efficient removal of cationic dyes from water by a combined adsorption-photocatalysis process using platinum-doped titanate nanomaterials." *Journal of the Taiwan Institute of Chemical Engineers* 99:166–179.

Pan, X., Y. Zhao, S. Liu, C. L. Korzeniewski, S. Wang, and Z. Fan. 2012. "Comparing graphene-TiO_2 nanowire and graphene-TiO_2 nanoparticle composite photocatalysts." *ACS Applied Materials & Interfaces* 4:3944–50.

Parashar, M., V. K. Shukla, and R. Singh. 2020. "Metal oxides nanoparticles via sol-gel method: a review on synthesis, characterization and applications." *Journal of Materials Science: Materials in Electronics* 31 (5):3729–3749.

Saharan, P., A. K. Sharma, V. Kumar, and I. Kaushal. 2019. "Multifunctional CNT supported metal doped MnO_2 composite for adsorptive removal of anionic dye and thiourea sensing." *Materials Chemistry and Physics* 221:239–249.

Shang, S. M.. 2013. *Process Control in Dyeing of Textiles*. Vol. 13. The Hong Kong Polytechnic University, China: Woodhead Publishing Limited.

Slepicka, P., N. S. Kasalkova, J. Siegel, Z. Kolska, and V. Svorcik. 2019. "Methods of gold and silver nanoparticles preparation." *Materials* 13 (1):1.

Tan, K. B., M. Vakili, B. A. Horri, P. E. Poh, A. Z. Abdullah, and B. Salamatinia. 2015. "Adsorption of dyes by nanomaterials: recent developments and adsorption mechanisms." *Separation and Purification Technology* 150:229–242.

Tang, B., M. Zhang, X. Hou, J. Li, L. Sun, and X. Wang. 2012. "Coloration of cotton fibers with anisotropic silver nanoparticles." *Industrial & Engineering Chemistry Research* 51 (39):12807–12813.

Tavakoli, A., M. Sohrabi, and A. Kargari. 2007. "A review of methods for synthesis of nanostructured metals with emphasis on iron compounds." *Chemical Papers* 61 (3):151–170.

Unser, S., I. Bruzas, J. He, and L. Sagle. 2015. "Localized surface plasmon resonance biosensing: current challenges and approaches." *Sensors* 15 (7):15684–15716.

Wang, C., X. Zhang, F. Lv, and L. Peng. 2012. "Using carbon black nanoparticles to dye the cationic-modified cotton fabrics." *Journal of Applied Polymer Science* 124 (6):5194–5199. doi: 10.1002/app.34067.

Wang, J., Q. Zhang, X. Shao, J. Ma, and G. Tian. 2018. "Properties of magnetic carbon nanomaterials and application in removal organic dyes." *Chemosphere* 207:377–384.

Wardman, R. H.. 2017. *An introduction to textile coloration: Principles and practice, An introduction to textile coloration: Principles and practice*. Hoboken: John Wiley & Sons.

Wu, M., B. Ma, T. Pan, S. Chen, and J. Sun. 2016. "Silver-nanoparticle-colored cotton fabrics with tunable colors and durable antibacterial and self-healing superhydrophobic properties." *Advanced Functional Materials* 26 (4):569–576. doi: 10.1002/adfm.201504197.

Yeung, K. W. and S. M. Shang. 1999. "The infl uence of metal ions on the aggregation and hydrophobicity of dyes in solutions." *Journal of the Society of Dyers and Colourists* 115:228–232.

Yin, X., W.-F. Xu, Y. Pu, J. Zhai, D. Wang, and J.-X. Wang. 2020. "Preparation of aqueous nanodispersions of disperse dye by high-gravity technology and spray drying." *Chemical Engineering & Technology* 43 (10):2118–2125. doi: 10.1002/ceat.201900570.

Yusuf, M., M. Shabbir, and F. Mohammad. 2017. "Natural colorants: historical, processing and sustainable prospects." *Natural Products and Bioprospecting* 7 (1):123–145. doi: 10.1007/s13659-017-0119-9.

Zhang, Y.-R, S.-Q. Wang, S.-L. Shen, and B.-X. Zhao. 2013. "A novel water treatment magnetic nanomaterial for removal of anionic and cationic dyes under severe condition." *Chemical Engineering Journal* 233:258–264.

Zhao, Y., Z. Xie, H. Gu, C. Zhu, and Z. Gu. 2012. "Bio-inspired variable structural color materials." *Chemical Society Reviews* 41 (8):3297–3317.

Zollinger, H.. 1987. Synthesis, Properties of Organic Dyes and Pigments. In: *Color Chemistry*, 92–100. New York, NY: Wiley-VCH, Weinheim.

8 Nanotechnology-Enabled Approaches for Textile Applications
Solutions to Eliminate the Negative Impacts of Dyeing Materials and Processes

Azadeh Bashari
Amirkabir University of Technology
Industrialization Center for Applied Nanotechnology (ICAN)

Mina Shakeri
Tarbiat Modares University
Industrialization Center for Applied Nanotechnology (ICAN)

CONTENTS

8.1 Introduction 133
8.2 Nanotechnology in Textile Dyeing 134
    8.2.1 Nanopigments 134
    8.2.2 Microbial Nanopigments 137
    8.2.3 Nanodisperse Dye 137
8.3 Nanobubble Dyeing 144
    8.3.1 Principle of Nanobubble Technology 145
8.4 Nanofiltration in Textile Dyeing 148
8.5 Conclusion and Future Scope 152
Acknowledgment 152
References 152

8.1 INTRODUCTION

The textile dyeing is a wet process that uses dyes, chemicals, and large amounts of water. This process continues to expand every year using new technologies. The need for cleaner, more cost-effective, and value-added textiles is the driving force for these developments. The art of using color to enhance the appearance of textiles had been known since many years ago. The use of natural dyes extracted from vegetables, fruits, flowers, certain insects, and fish dates back to 3500 BC. Although color is the most attractive property of fabrics, the fabrics were earlier being dyed with a limited and dull range of colors. Besides, they showed low color fastness against washing and sunlight. As a result, they needed a mordant to form a bridge between the fiber and dye, thus making the dyers' work complicated. The mordants (such as chromium) may be very toxic and may have a high impact on the wastewater quality. On the other hands, natural dyestuffs require large quantities of water for dyeing. About 80% of the dyestuffs stay on the textile, while the rest go to the drain. Consequently, natural dyes prepared from wild plants and lichens can have a very high impact on the environment (Kant 2012). The discovery of synthetic dyes by W. H. Perkins in 1856 has provided a wide range of dyes that are color fastness and come in a wider color range and brighter shades (Whittaker and Wilcock 1949). The workers were involved in the manufacture of synthetic dyes and were exposed to the dyes and the chemical intermediates used in the production process. Years later, it became clear that workers were involved in the manufacture of certain colors such as fuchsine and auramine, dyes based on benzidine and 2-naphthylamine, developed a high incidence of bladder cancer (Christie 2007).

The production and use of synthetic dyes for fabric dyeing have become a massive industry today. The use of synthetic colors has an adverse effect on the environment. The presence of sulfur, naphthol, vat dyes, nitrates, acetic acid, soaps, enzymes, chromium compounds, and heavy metals such as Cu, As, Pb, Cd, Hg, Ni, and Co and some auxiliary chemicals make the textile effluent highly toxic. Other harmful chemicals in the water may be chlorinated stain removers, formaldehyde-based dye fixing agents, hydrocarbon-based softeners, and nondegradable dyeing chemicals. These organic materials form by products that are often carcinogenic and therefore undesirable (Kant 2012). The development and utilization of nanotechnology in the textile industry can reduce the use of harmful and toxic chemicals that are harmful for the environment and are very useful in the production of textiles in an environmentally friendly way (Conway 2016).

8.2 NANOTECHNOLOGY IN TEXTILE DYEING

Nanotechnology is the science and engineering of materials with dimensions of approximately 1 to 100 nm (1 billion nm = 1 m) in length. In the last decades, there have been numerous advances in nanotechnology, and its many applications in the textile industry. The traditional textile dyeing techniques in which water serves as the medium for carrying dyes to polymeric fibers produces a large amount of colored and/or toxic wastewater and spent a lot of money for wastewater treatment. Different dyeing mechanisms via new technologies such as nanotechnology are certainly needed. Since there are various applications for nanotechnology in the textile industry, only the applications related to textile dyeing are mentioned here.

8.2.1 NANOPIGMENTS

The use of pigments in the textile dyeing has many advantages such as less water and energy consumption, less effluent load, and more efficiency, so pigments are suitable

Nanotechnology-Enabled Approaches

options for environmentally friendly coloring textiles. Pigments are a group of textile colorants, can be categorized as organic or inorganic, and insoluble in water and not interacting with polymeric textiles.

Pigments, especially inorganic pigments, have been widely used in applications including coatings, printings, and paintings. When pigments are used for the coloration of textiles, various processes are employed, including pigment padding, batch exhaustion, and pigment printing. These processes sometimes include binders, solvents, and other optional additives. Using these techniques, the coloration occurs at the surface of the fabrics between the textile polymers and the pigment molecules, and the binders serve as a bridge. For industrial applications, a pigment coloration system in which pigment particles are finely and stably separated is considered a good dispersion because particle size and stability are key factors in determining their physicochemical properties (Lin 2003). One possible application of nanotechnology is to directly use pigment nanoparticles in textile coloration. If the pigment particles can be reduced to a small size, the particles can be dispersed well to avoid aggregation in dye baths (Ahmed and El-Shishtawy 2010). Generally, since nanopigment suspensions are submicron particles, they are classified into the category of colloidal systems (Kobayashi 1996). There are three stages, including wetting, mechanical disintegration, and stabilization for fabricating a nanopigment suspension (Figure 8.1).

- In the wetting stage, the clusters of pigment particles are soaked by the liquid and replaced by the air absorbed on the surface of the hydrophobic pigment.
- The particle size of wet pigments is reduced into smaller clusters by mechanical forces from a ball grinder, high pressure homogenizer, or ultrasonic disruptor.
- Single pigment particles interact with an appropriate surfactant to stabilize the suspension system to avoid the reagglomeration of the smaller pigment particles (Agbo et al. 2018).

Exhaust dyeing of cationised cotton with nanoscale-pigment dispersion prepared using a microfluidizer has recently been studied (Fang et al. 2005).The nanoscale pigment dispersion was prepared with a conventional pigment, CI Pigment Red 22 (Figure 8.2) was stirred with an aqueous solution of an anionic polymeric dispersant XG-1 (3% on weight of pigment) on high-speed mixer for 30 minutes at 10000 rpm. The resulting conventional pigment dispersion had an average particle diameter of around 1500 nm. This dispersion was converted to a nanoscale dispersion via a

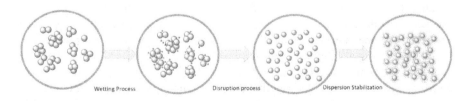

FIGURE 8.1 Fabrication process of the nanopigment suspension.

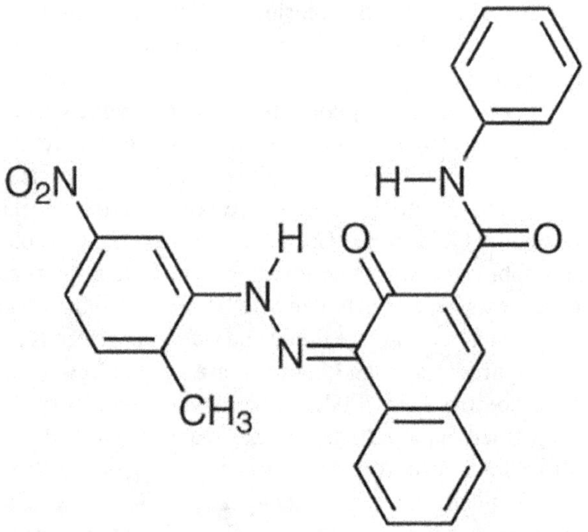

FIGURE 8.2 Chemical structure of CI Pigment Red 22.

microfluidizer, processing at a pressure of 22,000 Pa for 2.5 Hours, which resulted in an average particle diameter of 128 nm and the final pigment dispersion had 40% solid content (Fu et al. 2005).

The smaller the particle size, the higher the surface energy, accounting for the greater adsorption of nanoscale pigment particles on the surface of the cotton, and the results indicated that the dyeings obtained have better soft handle and more brilliant shade with reduced pigment requirement than those obtained with a conventional pigment dispersion (Fang et al. 2005).

The use of ultrasound energy in fabric dyeing is well known in recent years, the use of ultrasonic energy can increase dyeing yield, reduce dyeing temperature, time and reduce the use of dyeing auxiliaries (Kan and Yuen 2005).

In order to investigate the effect of ultrasound on dyeing cationized cotton with nanopigments, Hao et al., prepared a stable nanoscale pigment dispersion by ultrasound. For the insoluble pigments, the increase in pigment adsorption can also be achieved with the ultrasound via the effect of acoustic cavitation on the dispersion of nanopigments that make pigment particle deposition on the cotton surface or diffusion only into the large pores between the yarns. They found the ultrasound is an effective technique to improve the equilibrium adsorption of the nanopigments. The higher equilibrium adsorption due to the smaller size of the pigment particles was in result of higher cavitation effect and the higher ultrasonic power on the other hand, and they found that ultrasound can boost the adsorption rate of nanopigments by accelerating the movement of nanoparticles through the fiber–liquid interface and thus reduce the time needed to reach adsorption equilibrium. SEM images (Figure 8.3) showed that the nanoscale pigments can be more evenly deposited and distributed on the cotton surface when using ultrasound (Hao, Liu, and Liu 2012).

Nanotechnology-Enabled Approaches

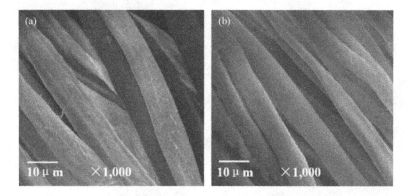

FIGURE 8.3 SEM photographs of cotton fibers (a) with conventional and (b) with ultrasonic dyeing method (Hao, Liu, and Liu 2012).

8.2.2 Microbial Nanopigments

Microbial pigments as the most potential biomass pigments have recently attracted more attention in textile dyeing due to their sustainability, natural functionality, biocompatibility, and cleaner production. Microbial pigments, unlike natural pigments obtained from plants and insects, have a very good prospect for industrial applications because the resources are not limited by seasonal, climatic, and geographical factors (Tuli et al. 2015). Prodigiosin as a pyrrole structure microbial pigment has recently attracted the attention of researchers due to its bright colors and antibacterial effect (Liu et al. 2013). Gong et al. found that the pyrrole structure of red pigments produced by microorganisms synthesized in the cells is in the form of nanoscale particles in extracellular culture. In this study, a mimic phospholipid bilayer membrane was constructed as an in vitro platform to research the trans-membrane behavior. For the first time, the biomimetic membranes were used to assess the transmembrane transport of microbial nanopigments for preparation of dye suspension. In Figure 8.4, schematic pictures of the biomimetic membrane platform for transmembrane transport studies of microbial pigments are shown (Gong et al. 2019).

This research has found the efficient preparation of microbial dye liquor based on fermentation broth and has promoted the use and development of microbial pigments (Gong et al. 2019).

8.2.3 Nanodisperse Dye

Disperse dye as a water insoluble dye with an affinity to hydrophobic fibers is usually utilized in the textiles to impart a color to synthetic fibers, for example, polyester, acrylic, and acetate. Polyester is the most hydrophobic of all common fibers because of a highly compact and crystalline structure. Therefore, its aqueous dyeing is carried out at high temperature and high pressure using disperse dyes. Before using, disperse dye is converted to a low particle size, dispersed in water, and added to treatment bath during the textile coloration process. Mashaly et al. studied on absorption behavior of ball-milled nanodisperse dye onto polyester fabrics (Mashaly

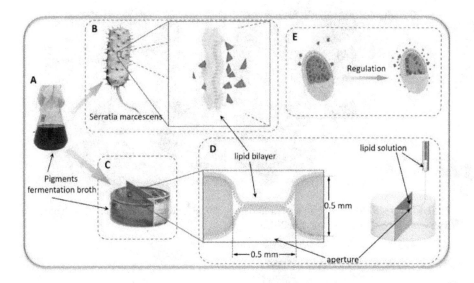

FIGURE 8.4 Schematic picture of the biomimetic membrane platform for transmembrane transport studies of microbial pigments. (A) Pyrrole structure nanopigment fermentation broth. (B) Schematic of Serratia marcescens (left) and lipid bilayers (right). (C) Schematic of pigments transmembrane transport experiment. (D) Preparation process of the biomimetic membrane. (E) Schematic illustration of intracellular pigment distribution (Gong et al. 2019).

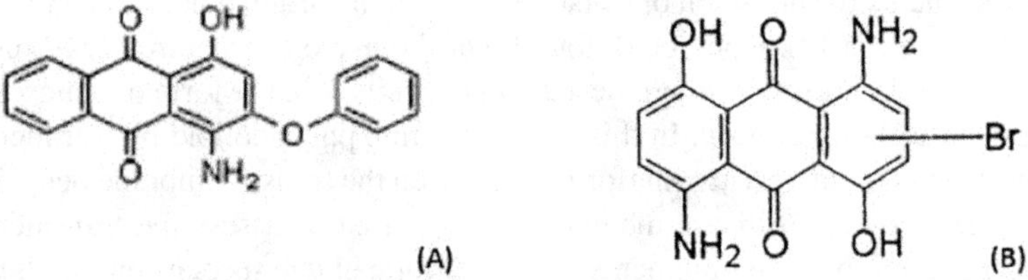

FIGURE 8.5 A figure showing the chemical structure of (A) Disperse red 60 and (B) Disperse blue 56.

et al. 2014). In this work, nano-Disperse red 60 and Disperse blue 56 (Figure 8.5) produced via ultrasonic homogenizer technique were used. Dyeing of fabrics was carried out with (15–90 minutes) and without sonication at the temperature range of 70°C–100°C. On the other hand, salicylic acid, HC, and ZnO nanopowder were chosen as carriers and the agent for improving the light fastness of the dyed polyester fabrics, respectively.

The obtained results confirmed that simultaneous use of disperse nanosize dye suspension with ultrasonic has a synergistic effect. The most obvious effect of ultrasonic is the dispersing of nanodye particles in liquids and preventing from particle agglomerates. This process leads to smaller particles with much more size uniformity. Ultrasonic cavitations improve the transferring of dye particles into fiber, too. As shown in Figure 8.6, increasing time of sonication from 15 to 90 minutes leads to

Nanotechnology-Enabled Approaches 139

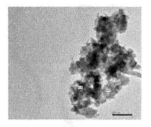

TEM of Disperse blue 56 before sonication   TEM of Disperse blue 56 after sonication

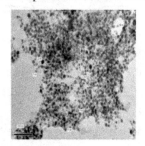

TEM of Disperse red 60 before sonication   TEM of Disperse red 60 after sonication

FIGURE 8.6 Transition electron microscope (TEM) obtained for the used dyes (Mashaly et al. 2014).

increase in presence of dye particles in dyed polyester compared with the nonsonicated dyes (at 0 time) (Mashaly et al. 2014).

During ultrasonic process, the same-charged particles strongly collide with each other, and therefore, agglomeration reduces. On the other hand, the high energy and temperatures during the collision of particles cause the decomposition of particles and reduction of the particle size, and eventually leads to stability of the dye suspension (El-Sayed et al. 2001). The nano-ZnO powder used in post treatment in comparison with pretreatment showed better results in light fastness which increased the UV protection and decreased the rate of fading (Mashaly et al. 2014).

Another method to solve the agglomeration problem in dyeing bath is the encapsulation of the dyes within the polymer to produce nanocolorants. Polymer-encapsulated dyes could find applications in a wide range. Polymerization methods have many types such as suspension polymerization (Ma et al. 2005), dispersion polymerization (Horák and Benedyk 2004), emulsion polymerization (Yanase et al. 1993), soap-free emulsion polymerization (Shibuya et al. 2014), miniemulsion polymerization (Schork et al. 2005), and microemulsion polymerization (Pavel 2004). In recent years, a large number of nanohybrid particles were prepared by mini emulsion polymerization because of its attractive advantages.

One of the advantages of the miniemulsion polymerization method is encapsulation of nanoparticles. The basic requirements of dyes to be suitable for the introduction into minidroplets are high hydrophobicity, high solubility in monomer, and large absorption coefficient. As shown in Figure 8.7, El-Sayed et al prepared

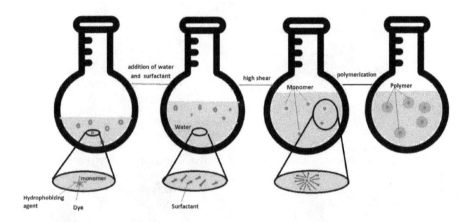

FIGURE 8.7 Principle of encapsulation by modified miniemulsion polymerization.

TABLE 8.1
The Recipe Used to Prepare Nano C.I. Disperse Red 60 Particles

Ingredient	Role	Concentration
Styrene	Monomer	80 wt%[a]
Methyl Methacrylate	Monomer	12 wt%[a]
Divinyl Benzene and Glycol Dimethacrylate (1:1)	Crosslinkers	8 wt%[a]
C.I. Disperse Red 60		40 wt%[a]
Sodium Dodecyl Sulfate, Ultravon PL, Cetyl trimethyl Ammonium Bromide	Surfactants	20–80 mmoL[b]
NaHCO$_3$		10 mmoL[b]
Potassium Persulfate, 2,2-Azobis (2-Methyl Propionitrile)	Initiator	10 mmoL[b]
Deionized Water		78–79 wt%[c]
Polystyrene	Hydrophobic agent	0.2 wt%[c]
HV 3050		0.15 wt%[c]

Source: El-Sayed et al. 2012.
[a] Based on the monomer 12.5 g.
[b] Based on aqueous phase.
[c] Based on the total recipe.

nanocolorant, nano-Disperse Red 60, via a modified miniemulsion polymerization process (El-Sayed et al. 2012).

The recipe used for the synthesis of nano-C.I. Disperse Red 60 particles via modified miniemulsion polymerization is summarized in Table 8.1 (El-Sayed et al. 2012).

A modified miniemulsion polymerization process was successfully used to prepare nano-Disperse Red 60 particles with particle size ~ 8–40 nm. In addition, characterization of nanodye was investigated by UV–vis spectra which showed a shift to higher frequencies with decreasing particle size. FTIR spectra results proved the formation of composite nanoparticles of nanodye. The conductance profiles were found to be a useful tool in describing the various intervals in miniemulsion polymerization. Kinetic analysis of the polymerization process was studied by a first-order-kinetic model. The surface tension and surface charge measurements proved that nanodye has a good stability at all pH ranges while Disperse Red 60 has an excellent stability only in neutral and alkaline media and an incipient instability in acidic medium (El-Sayed et al. 2001). In conventional dyeing processes, the disperse dyes are applied from fine aqueous dispersion at 130°C on polyester and at 100°C on nylon fabrics (Choi and Kang 2006). In textile dyeing where the substrate solution reactions are mainly governed by the adsorption phenomenon, excellent chromatic properties are achieved due to the nanoscale effects of homogeneous nanocolorants (Hu et al. 2008). In this regard, Kale et al., prepared oil in water nanoemulsion containing three disperse dyes with different energy levels using high-speed homogenizer in order to dyeing nylon fabric. Table 8.2 shows the properties of disperse dyes used in this study (Kale et al. 2017).

TABLE 8.2
The Properties of Disperse Dyes

Dye Name	Dye Structure	M_W (g/mol)	Energy Level	Solid Content (%)	λ_{max} (nm)
C. I. Disperse Blue 79.2		639.4	High	53.7	587
C. I. Disperse Red 73		348.35	Medium	24.4	528
C. I. Disperse Orange 25		323.35	Low	36.6	481

Source: Kale et al. 2017.

Oil in water nanoemulsion was prepared by using high-speed homogenizer. Five parts of Tween 80 were dissolved in 80 parts of distilled water in which 15 parts of paraffin oil were added. This mixture was subjected to homogenization at 12,000 rpm for 20 minutes. In order for preparation of the nanodisperse dye solution, the required quantity of dried crude dyes was taken for 3% shade according to the solid content of the dyes obtained and were dissolved in 10 mL DMF. This solution was then added dropwise to 90 mL nanoemulsion and homogenized at 12,000 rpm for 15 minutes.

The dyeings of nylon fabric with nanodisperse dyes were carried out using nanoemulsions without adding any auxiliaries. The dyed samples were then subjected to reduction clearing treatment using 2 g/L of caustic soda, sodium hydrosulfite, and Auxipon NP at 70°C for 20 minutes. All fabric samples were thoroughly washed with cold water and then air dried. Actual dyed samples and the result of the washing, rubbing, sublimation, and light fastness of the dyed samples are summarized in Table 8.3 (Kale et al. 2017).

Nanoemulsion provides very fine dispersion by reducing dye particle size, which also acts as a transporting medium for the dye into the fiber core without any addition of the auxiliary. The dyeing results showed comparable color strength and fastness to

TABLE 8.3
Color and Fastness Properties of Nanodisperse vs Commercial Dyed Samples

Samples		Color	K/S	Washing	Rubbing Dry	Rubbing Wet	Sublimation at 210°C	Light
C.I. Disperse Red 73	Commercial dye		27.14	4	4–5	4	4	8
C.I. Disperse Red 73	Nanoemulsion dye		30.18	4–5	4–5	4	3	8
C.I. Disperse Blue 79.2	Commercial dye		18.97	4–5	5	3–4	4	8
C.I. Disperse Blue 79.2	Nanoemulsion dye		20.23	5	5	3–4	3–4	8
C.I. Disperse Orange 25	Commercial dye		11.64	4–5	4–5	3–4	3–4	8
C.I. Disperse Orange 25	Nanoemulsion dye		12.59	4–5	4–5	3–4	3	8

Source: Kale et al. 2017.

commercial dyeing samples. Thus nanodisperse dyes provide an alternative for dyeing nylon fabric with disperse dyes, and save a considerable amount of energy, time, and cost (Kale et al. 2017).

Cellulose fibers are annually dyed with 120,000 tons (about of the 11% dyestuff market) of vat dyes (Roessler et al. 2002). Vat dyes are insoluble in water. In the dyeing process, vat dyes are reduced in an alkaline medium and converted to the soluble leuco form, which has affinity to cellulosic fibers and penetrates into the fibers. Then insoluble structure is restored after oxidation, and the dye is trapped within the fibers (Aspland 1997).

Today in most industrial processes, vat dyes (such as indigo for dyeing denim fabric) are reduced by sodium hydrosulfite as a reducing agent to attain a water-soluble form of the dye with more affinity to the cellulosic fiber. After diffusion into the fiber, dye will be reoxidized to the water-insoluble form. The residual of reducing agents in dyed wastewater will finally be oxidized into species such as dithionite, sulfite, sulfate, thiosulfate, and toxic sulfide that can cause excess heavy contamination of waste water from dyeing. In this regard, much research has been made to replace the environmentally unfriendly reduction and/or oxidation agents by ecological alternatives using chemical, electrochemical, electrocatalytic hydrogenation, and biological methods (Božič and Kokol 2008). In one study, Hakeim et al. prepared aqueous dispersions of three nanoscale vat dyes (Figure 8.8) through ball milling and ultrasonication of dyes in the presence of a dispersing agent. The nanoscale vat dyes have been applied in dyeing and printing of cotton to evaluate the effect of nanoscale dispersion on the reducing agent amount and the difference in coloration performance of a nanoscale and conventionally dispersed vat dyes (Hakeim et al. 2013).

For vat dispersion preparation, vat dye was mixed with sodium dodecyl sulfate (SDS) and water. This mixture was vigorously stirred with a magnetic stirrer. The resulting mixture was then ball milled. Nanoscale vat dyes with particle sizes ~20 nm were prepared by using ball milling and ultrasonication in the presence of SDS. By this method, crystal form of vat dyes was obtained. Ultrasonication played a significant role in decreasing vat particle size. The nanosize vat dyes in dyeing and printing can act as a leuco form, without needing a reducing agent and could penetrate and fix in the cotton fibers and gave satisfactory color yield. Using the nanosize vat dye is ecologically and economically acceptable and reduces the use of environmentally unfavorable reducing agents by improving dyeing and printing with vat dyes (Hakeim et al. 2013).

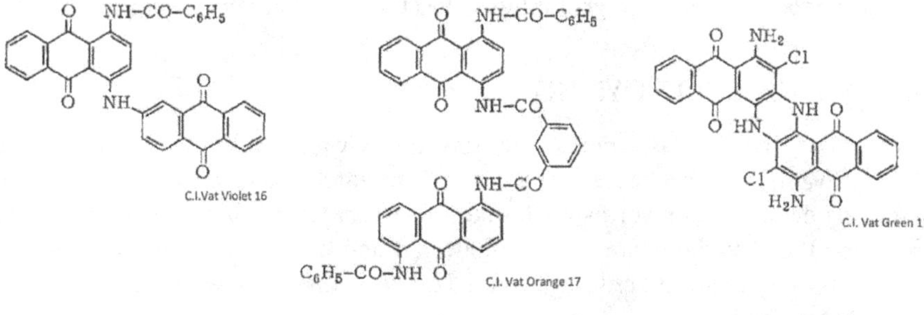

FIGURE 8.8 Chemical structures of vat dyes (Hakeim et al. 2013).

In other research, hybridization of nanosized chitosan with nanoparticles of a disperse dye under the sonication was studied. Three techniques were used in the nanoparticle preparations of chitosan, disperse dye, and chitosan/disperse dye nanoparticles. The obtained nanocolorant had different hydrophobic/hydrophilic characteristics as well as color intensity due to the ratio of the disperse dye in the mixture. The colored hybrid was applied on textile (flax, polyester, and polyester/flax blend) printing using the silk screen technique. Nanodisperse dye solution and chitosan nanoparticles were prepared as follows:

Disperse dye and nonionic dispersing agent (Twin) were suspended in water, and then the solution was subjected to vigorous stirring with ultrasonic stirrer at about 80°C. Chitosan was dissolved in acetic acid. The pH was raised to 4.6–4.8 with NaOH, and sodium tripolyphosphate solution was dropped slowly to chitosan solution with stirring. Chitosan nanoparticles as a suspension were collected and stored in deionized water and air dried (Hebeish et al. 2019).

For preparation of nanodisperse dye/chitosan nanoparticle hybrids, the as-prepared nanodisperse dye was added to the prepared nanochitosan solution at different ratios 25:15, 25:25, and 15:25 and stirred for 30 minutes to form a printing paste. Then the printing paste was used in printing flax and polyester and their blend fabrics. To this end, the printed fabrics were fixed by thermofixation then washed. During sonication, chitosan nanoparticles and dispersed dyes are subjected to very energetic conditions that cause significant changes in both types of nanoparticles. Thus, the chemical bonds are broken, and new bonds are formed. At the same time, the interaction and intercalation between the hybrid nanoparticles occur while maintaining the molecular structure of the compound. Increasing the ratio of the dye to the chitosan is accompanied by increment in K/S values of the printed fabrics, regardless of the method of color fixation (microwave or steaming fixation) used. The K/S values of prints brought about using the nanohybrid dye follow the order: flax > flax/polyester > polyester, reflecting the effect of loading the hydrophobic nanodye on the hydrophilic chitosan nanoparticles. The combination of the two opposite entities in one would help establish certain hydrophilicity to the synthesized colorant, and so creates affinity for the hydrophilic textile such as flax. On the other hands, because of the presence of chitosan, color hybrid containing chitosan nanoparticles exhibits higher antibacterial activity against gram-positive bacteria (*Bacillus subtilis*, *Staphylococcus aureus*) than against gram-negative bacteria (*Escherichia coli*, *Pseudomonas aeruginosa*) (Hebeish et al. 2019). The structure of the bacteria displays higher negative charge on the cell surface of gram-positive than on the cell surface of gram-negative bacteria, leading to a greater adsorption and higher inhibitory activity of chitosan (Kong et al. 2010).

8.3 NANOBUBBLE DYEING

The textile industry is among the top three water-wasting industries (Shu, Pannirselvam, and Jegatheesan 2019). Recent research has developed a zero-liquid discharge method that involves wastewater treatment (Bahadur and Bhargava 2019). However, these methods are very complicated and expensive. Therefore, the best solution is to move toward reducing colored effluents instead of wastewater treatment (Gulzar et al. 2019).

Advanced technologies or methods are trending in dyeing in recent times owing to their improved results over the conventional dyeing (Yusuf, Shabbir, and Mohammad 2017). One way to eliminate or reduce the dyed waste discharge is to use the modern low liquor ratio machines in the textile dyeing industry (Raja et al. 2019). In the textile industry, the exhaust dyeing is an important technique that uses significant water, energy, and chemicals. Liquor to goods ratio (L:R) reflects how much water and chemicals are required for the specific weight of the fabric, and it is basically the ratio of a dye-bath volume to the dry material weight. Typically, the L: R ratio for the exhaust dyeing ranges between 1:5 and 1:40 (Roy Choudhury 2013). The higher liquor ratio exhibits more even dyeing vs. more water, energy, and chemical consumption (Shang 2013).

Nanobubble technology has emerged as a water efficient exhaust dyeing technique, which offers the dyeing with extremely low L: R ratio of 1:1. This technology can cause to the reduction of the water consumption up to 86%, chemical reduction up to 50%, a reduction of 44% in energy consumption, and elimination of 97% of wastewater (Mohsin et al. 2020).

In the textile processing, the nanobubbles act as chemical carriers on the fabric surface (Garcia 2015).

8.3.1 Principle of Nanobubble Technology

A new technology based on nanobubbles, developed and patented by Jeanologia, is known as e-flow. The e-flow "breaks up" the surface of the garment, achieves a soft feeling, and controls shrinkage. A small amount of water is required, and there is zero discharge from the process. In this technique, air from the atmosphere is introduced into an e-flow reactor and subjected to an electromechanical shock, creating nanobubbles and a flow of wet air (Garcia 2015).

Nanobubbles (NBs) are tiny bubbles with a respective diameter of <200 nm and have been explored for various applications. Nanobubbles (NBs) are very tiny bubbles with a diameter less than 200 nm and various potential applications. The high Laplace pressure inside NBs would likely cause them to dissolve into solution quickly, so the NBs are known as the unstable entities (Ljunggren and Eriksson 1997). As shown in Figure 8.9, micro bubbles (MBs) gradually shrink in size and subsequently collapse due to prolonged stagnation and dissolution of internal gases into the surrounding waters, while NBs remains as such for months and do not burst out at once (Tasaki et al. 2009). The interface of NBs consists of hard hydrogen bonds similar to those found in ice and gas hydrates. This in turn reduces the penetration of NBs, which helps maintain a sufficient kinetic balance of NBs against high internal pressure (Takahashi, Chiba, and Li 2007).

Cavitation method is often used for formation of MBs and NBs in solution. Based on the mode of generation, cavitation is broadly classified into acoustic, hydrodynamic, optic, and particle cavitation. Millions of hotspots in the reactor can be generated through hydrodynamic (cavitation due to the pressure variations in the flowing liquid) and acoustic (cavitation induced by the passage of ultrasonic waves) cavitations due to very high localized

energy density, which in turn results in extremely high pressure and temperatures in the range of 10–500 MPa and 1000–10,000 K, respectively (Suslick 1990).

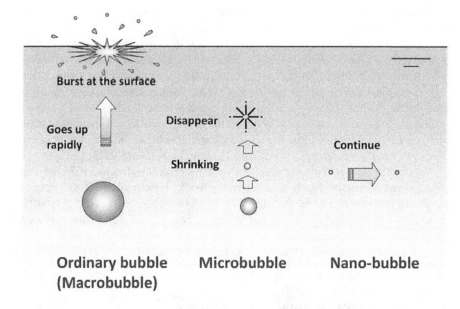

FIGURE 8.9 Schematic diagram showing macro, micro and nanobubbles (Takahashi, Chiba, and Li 2007).

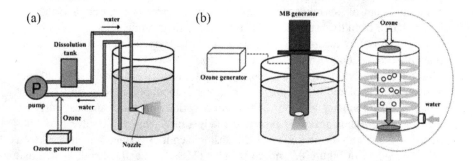

FIGURE 8.10 Schematic diagrams used for MBs/NBs generation. (A) The decompression-type generator and (B) the gas–water circulating-type generator (Ikeura, Kobayashi, and Tamaki 2011).

The two widely used methods for the preparation of NBs are based on decompression and gas-water circulation. As shown in Figure 8.10, for the decompression generator, a supersaturated condition for gas dissolution is created at high pressure of 304–405 kPa. At such high pressure, supersaturated gas is highly unstable and eventually escapes out from the water. As a result, a large number of MBs would be generated instantly. However, for gas-water circulation type generator, the gas is introduced into the water vortex, and gas bubbles are subsequently broken down into MBs by breaking up the vortex (Takahashi 2009).

Mohsin et al., attempted to use nanobubble technology for exhaust dyeing of cotton fabric with 11 different reactive dyes at L: R. Table 8.4 shows the shade depth of

TABLE 8.4
Color and K/S Value of the Fabric with Different Liquor Ratios and Dyes

Liquor Ratios	1:1 K/S	1:5 K/S	1:10 K/S
Dye Name			
Drimaren Blue Cl-Br	13	14.1	14.3
Drimaren Blue K-Br	24	25	26.2
Drimaren HF-br	18.1	19.7	22
Drimaren Ultimate Navy gr	22.5	23.7	23.9
Drimaren Ultimate Yellow gr	17.5	18.6	20
Drimaren Ultimate Yellow	5	6	6.9
Drimaren Ultimate Red Fuchsia gr	33	36	39
BEZAKTIV Yellow GO	24	27	29
BEZAKTIV Red GO	21.5	24.9	28
BEZAKTIV Navy GO	18	20.1	23
BEZAKTIV Golden GO	13	15	19

Source: Kale et al. 2017.

the cotton fabric samples with three liquor ratios for 11 dyes. The uniformity of the all 11 dyes for the nanobubble dyeing was excellent (Mohsin et al. 2020).

The uniformity of dyeing is the biggest challenge in the ultra-low liquor dyeing, and it was successfully achieved in this research by using the optimized dye solution pressure, flow rate, and time. In addition, fastness properties like dry rubbing, wet rubbing, and washing fastness for all 11 dyes were in the acceptable range, and the highest difference of fastness rating among different liquor ratio was only half a rating. Air permeability of the fabric dyed with nanobubble mechanism was superior for all 11 dyes. In addition, no water was discharged during the nanobubble dyeing during the main dyeing cycle (Mohsin et al. 2020).

8.4 NANOFILTRATION IN TEXTILE DYEING

The filtration process is one of the most common treatments used in the textile industry. Although filtration techniques require an initial setup cost, they are offset by significant cost savings through of salts and water reuse. Ultrafiltration (UF), reverse osmosis (RO), and nanofiltration (NF) have been widely used for treatment and reuse of chemicals and water in industry (Tang and Chen 2002). NF has been known for having the properties between UF and RO and therefore offers significant advantages, e.g., lower osmotic pressure difference, higher diffusion flux, higher retention of multivalent salts and molecular weight compounds (> 300), relatively low investment, and low operation and maintenance costs (Hilal et al. 2004). Many researchers have evaluated the performance of NF membranes in terms of color retention, salt rejection, diffusion flux, and COD retention. The results have proven that NF membranes are the suitable separation process to be employed for the treatment of textile wastewater and generally showed an acceptable rejection (Akbari, Remigy, and Aptel 2002). A list of the top 25 materials for fabrication and modification of nanofiltration membranes, ranked by number of papers, is provided in Figure 8.11 (Oatley-Radcliffe et al. 2017).

The textile effluents typically contain many types of pollutants. Table 8.5 shows the typical characteristics of wastewater from the effluents of the dyeing and finishing processes that contain a variety of components of varying concentrations (Marcucci et al. 2003; Barredo-Damas et al. 2006).

Compared to other classes of dyes, the rate of fixation of reactive dyes on fibers is still very low, where about 5%–50% of dyes remain in textile wastewater due to incomplete exhaustion and hydrolysis of the dye in the alkaline dye bath during the dyeing processes. The loss of dyes in the effluent depends on the degree of fixation of the combination of different dyes and fibers. Therefore, in order to reduce water pollution, it is important to decolorize the effluent before discharging it into the environment. On the other hand, higher salt concentrations in the wastewater may be a major problem in some environment due to salinizing the soil. Therefore, the waste stream treatment system is a process to solve with the environmental problem and a way to recover rinsed water and minimize the volume of waste discharged. Table 8.6 reviews the application of the available polymeric NF membranes used for textile wastewater (Lau and Ismail 2009).

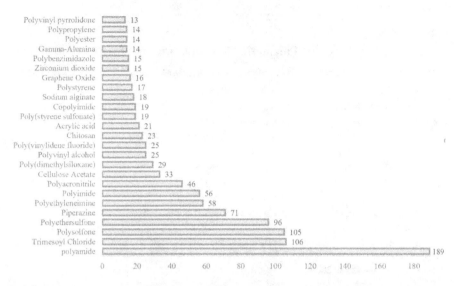

FIGURE 8.11 The top 25 chemicals identified to fabricate or modify nanofiltration membranes found in this review (Oatley-Radcliffe et al. 2017).

TABLE 8.5
Typical Characteristics of Wastewater from a Textile Dyeing Process

Aspect/Component	Value
pH	2–10
Temperature, °C	30–80
Chemical oxygen demand (COD), mg/L	50–5000
Biological oxygen demand (BOD), mg/L	200–300
Total suspended solids (TSS), mg/L	50–500
Organic nitrogen, mg/L	18–39
Total phosphorus, mg/L	0.3–15
Total chromium, mg/L	0.2–0.5
Color, mg/L	>300

Source: Lau and Ismail 2009.

Among the commercialized NF membranes, thin-film composite nanofiltration (TFC-NF) membranes are the most widely used filter. Its significant permeability and selectivity to asymmetric NF membranes offer competitive improvement of this type of membrane. The available TFC–NF membranes are mainly prepared by forming a very thin polyamide (PA) active layer on the polysulfone (PSf) or polyethersulfone (PES) porous support layer. The substrate membrane is prepared via a dry-wet phase inversion method, and the active layer is formed through the interfacial polymerization (IP) technique. Through this process and the formation of ultra-thin layer (300–400 nm)

TABLE 8.6
Commercially Available Polymeric NF Membranes for Textile Applications

Membrane (Manufacturer)	Configuration (Polymer Material)	MWCO (Da)
MPS 31 (Weizmann)	Spiral wound (NA[a])	NA[a]
NF45 (Dow/Film Tec)	Spiral wound (PA)[b]	200
DK 1073 (Osmonics)	Spiral wound (PA)[b]	300
ATF 50 (Adv. Membr. Tech.)	Spiral wound (TFC[c] of PIP[d] On PSf[e])	340
TFC–SR2 (Fluid System)	Flat sheet (TFC[c] of PSf[e])	200–400
DK 2540 (Osmonics)	Spiral wound (NA)[a]	NA[a]
NTR 7450 (Nitto–Denko)	Flat sheet (sPES)[f]	600–800
UTC 20 (Toray Ind.)	Flat sheet (PA)[b]	180
NF 70 (Dow/Film Tec)	Flat sheet (PA)[b]	250
DL 4040F (Osmonics)	Spiral wound (NA)[a]	150–300
Desal 5 DK (Osmonics)	Flat sheet (TFC[c] –PAb)	150–300

Source: Lau and Ismail 2009.

P_{ave} = average permeability; R = rejection.
[a] Not available.
[b] Polyamide.
[c] Thin-film composite.
[d] Piperazineamide.
[e] Polysulfone.
[f] Sulfonated polyethersulfone.

onto the porous membrane, TFC–NF membranes have more ability to salt rejection over asymmetric NF membrane (Song, Liu, and Sun 2005; Du and Zhao 2004).

In a study of color and COD retention by NF at pilot-plant scale, three different types of spiral wound membranes, DK 1073, NF 45 and MPS 31 were used simultaneously in the same unit (Figure 8.12). Characteristics of the nanofiltration membranes used in this work are shown in Table 8.7.

According to the obtained results, NF membranes such as NF 45 and DK 1073 exhibited good performance in terms of dye retention. The maximum dye rejection against a dye concentration of 450–500 mg/L, was up to 99.2% and 99.8%, respectively. As well as, the performance of MPS 31 was also investigated and the dye retention was varied from 90.1% to 97.3%. However, the average of color rejection of MPS 31 was slightly higher than NF 45 and DK 1073. This may be due to its smaller molecular weight cut off (Lopes, Petrus, and Riella 2005).

On the other hand, Sungpet et al. found that secondary layers formed by dye and the Donnan effect may be responsible for dye removal instead of membrane MWCO. Fouling layer occurred resulting from the absorption of the dye onto the membrane, resulting in an increase of dye rejection (Sungpet, Jiraratananon, and Luangsowan 2004). The results prove that wastewater characteristics, molecular weights of the dyes used, hydraulic conditions, volume reduction factor, temperature, pH, pressure, etc. affect the efficiency of color filtration (Ismail, Irani, and Ahmad 2013).

Nanotechnology-Enabled Approaches

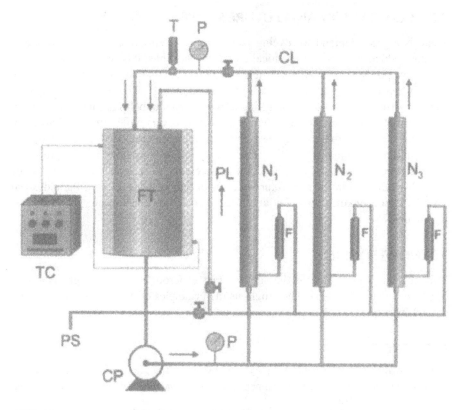

FIGURE 8.12 Nanofiltration pilot plant schematic diagram: FT: feed tank -550 L capacity; CP: centrifugal pump; P: pressure gauge; N_1, N_2, N_3: nanofiltration modules; F: flowmeters; CL: concentrate line; PL: permeate line; T: thermometer; TC: temperature control; PS: permeate (Lopes, Petrus, and Riella 2005).

TABLE 8.7
Characteristics of the Nanofiltration Membranes

Membrane Type	Manufacturer	Material	Membrane Properties	Filtering Area	MWCO (Da)[a]	Max. Operating Pressure (bar)
MPS 31	Weizmann	NA[b]	Hydrophobic	1.60 m²	NA[b]	45
NF 45	Dow/FilmTec	Polyamide	Hydrophilic	2.40 m²	200	41
DK 1073	Osmonics	Polyamide	Hydrophilic	1.77 m²	300	40

Source: Lopes, Petrus, and Riella 2005.
[a] Molecular weight cut-off.
[b] Not available.

8.5 CONCLUSION AND FUTURE SCOPE

There is a great demand for dyeing fabrics using environmentally friendly methods and reducing water and chemical consumption. Thus, it is worthwhile to create innovative production techniques. In this regard, various benefits have been shown with further exploration and research in nanotechnology. The use of nanopigment and nanodisperse dye systems in dyeing as well as nanobubbles has had a significant effect on reducing water consumption and creating toxic textile effluents. Further, it is anticipated that in the coming years, the environmental problems caused by dyeing plants in the textile industry will be resolved by use of physical treatment processes such as nanofiltration systems; hence, it is necessary to introduce these on a huge scale. It seems that in the near future, the use of nanotechnology onto the industrial scale in dyeing textiles will play an effective role in preserving the world's water resources.

ACKNOWLEDGMENT

This chapter is dedicated to the sublime and generous soul of our great master, Mr. Firoozmehr Mazaheri, who taught us the principles of dyeing and printing and passed away on 2020-11-30.

REFERENCES

Agbo, Christiana, Wizi Jakpa, Bismark Sarkodie, Andrews Boakye, and Shaohai Fu. 2018. "A review on the mechanism of pigment dispersion." *Journal of Dispersion Science and Technology* 39 (6):874–889.

Ahmed, N. S. E., and R. M. El-Shishtawy. 2010. "The use of new technologies in coloration of textile fibers." *Journal of Materials Science* 45 (5):1143–1153.

Akbari, A., J. C. Remigy, and P. Aptel. 2002. "Treatment of textile dye effluent using a polyamide-based nanofiltration membrane." *Chemical Engineering and Processing: Process Intensification* 41 (7):601–609.

Aspland, J. R. 1997. *Textile dyeing and coloration*. Research Triangle Park: American Association of Textile Chemists and Colorists, NC.

Bahadur, Nupur, and Nipun Bhargava. 2019. "Novel pilot scale photocatalytic treatment of textile & dyeing industry wastewater to achieve process water quality and enabling zero liquid discharge." *Journal of Water Process Engineering* 32:100934.

Barredo-Damas, Sergio, María Isabel Alcaina-Miranda, María Isabel Iborra-Clar, A Bes-Pia, José Antonio Mendoza-Roca, and Alicia Iborra-Clar. 2006. "Study of the UF process as pretreatment of NF membranes for textile wastewater reuse." *Desalination* 200 (1–3):745–747.

Božič, Mojca and Vanja Kokol. 2008. "Ecological alternatives to the reduction and oxidation processes in dyeing with vat and sulphur dyes." *Dyes and Pigments* 76 (2):299–309.

Choi, Jae-hong, and Min-Ju Kang. 2006. "Preparation of nano disperse dyes from nanoemulsions and their dyeing properties on ultramicrofiber polyester." *Fibers and Polymers* 7 (2):169–173.

Christie, Robert M. 2007. *Environmental aspects of textile dyeing*. Elsevier.

Conway, Roy. 2016. "Technical textile finishing." In *Handbook of Technical Textiles (2nd Ed.) Vol.1: Technical Textile Processes*, Horrocks, A. R., Anand, S. C. (eds.), 189–210. Amsterdam: Elsevier.

Du, Runhong, and Jiasen Zhao. 2004. "Properties of poly (N, N-dimethylaminoethyl methacrylate)/polysulfone positively charged composite nanofiltration membrane." *Journal of membrane science* 239 (2):183–188.

El-Sayed, G. M, M. M. Kamel, N. S. Morsy, and F. A. Taher. 2012. "Encapsulation of nano Disperse Red 60 via modified miniemulsion polymerization. I. Preparation and characterization." *Journal of applied polymer science* 125 (2):1318–1329.

El-Sayed, Mohamed, Olfat Y Mansour, Ibrahim Z Selim, and Maha M Ibrahim. 2001. "Identification and utilization of banana plant juice and its pulping liquor as anticorrosive materials." *Journal of Scientific and Industrial Research* 60, 738–747.

Fang, Kuanjun, Chaoxia Wang, Xia Zhang, and Yi Xu. 2005. "Dyeing of cationised cotton using nanoscale pigment dispersions." *Coloration Technology* 121 (6):325–328.

Fu, S.-H., X. Zhang, H. M. Zhu, and K. J. Fang. 2005. "Preparation of ultra-fine pigment and its application." *Dyestuffs Coloration* 42:32–34.

Garcia, Begoña. 2015. "Reduced water washing of denim garments." In *Denim*, Paul, R. (ed.), 405–423. Amsterdam: Woodhead Publishing.

Gong, Jixian, Jiayin Liu, Xueqiang Tan, Zheng Li, Qiujin Li, and Jianfei Zhang. 2019. "Biopreparation and regulation of pyrrole structure nano-pigment based on biomimetic membrane." *Nanomaterials* 9 (1):114.

Gulzar, Tahsin, Tahir Farooq, Shumaila Kiran, Ikram Ahmad, and Arruje Hameed. 2019. "Green chemistry in the wet processing of textiles." In *The Impact and Prospects of Green Chemistry for Textile Technology*, Shahid-ul-Islam and Butola, B.S. (eds.), 1–20. Amsterdam: Elsevier.

Hakeim, O. A., S. H. Nassar, A. A. Raghab, and L. A. W. Abdou. 2013. "An approach to the impact of nanoscale vat coloration of cotton on reducing agent account." *Carbohydrate polymers* 92 (2):1677–1684.

Hao, Longyun, Jingquan Liu, and Rongzhan Liu. 2012. "Ultrasound-assisted adsorption of anionic nanoscale pigment on cationised cotton fabrics." *Carbohydrate polymers* 90 (4):1420–1427.

Hebeish, A., A. A. Ragheb, M. Rekaby, H. M. El-Hennawi, and A. A. Shahin. 2019. "Chitosan/disperse dye nanoparticles for concomitant printing and antibacterial finishing." *Nanotechnologies in Russia* 14 (9):462–470.

Hilal, Nidal, Habis Al-Zoubi, N. A. Darwish, A. W. Mohamma, and M. Abu Arabi. 2004. "A comprehensive review of nanofiltration membranes: treatment, pretreatment, modelling, and atomic force microscopy." *Desalination* 170 (3):281–308.

Horák, Daniel, and Nataliya Benedyk. 2004. "Magnetic poly (glycidyl methacrylate) microspheres prepared by dispersion polymerization in the presence of electrostatically stabilized ferrofluids." *Journal of Polymer Science Part A: Polymer Chemistry* 42 (22):5827–5837.

Hu, Zhenkun, Minzhao Xue, Qing Zhang, Qiaorong Sheng, and Yangang Liu. 2008. "Nanocolorants: a novel class of colorants, the preparation and performance characterization." *Dyes and Pigments* 76 (1):173–178.

Ikeura, H., F. Kobayashi, and M. Tamaki. 2011. "Removal of residual pesticide, fenitrothion, in vegetables by using ozone microbubbles generated by different methods." *Journal of Food Engineering* 103 (3):345–349.

Ismail, Hanafi, Maryam Irani, and Zulkifli Ahmad. 2013. "Starch-based hydrogels: present status and applications." *International Journal of Polymeric Materials and Polymeric Biomaterials* 62 (7):411–420. doi: 10.1080/00914037.2012.719141.

Kale, Ravindra D, Amit Pratap, Prerana B Kane, and Vikrant G Gorade. 2017. "Dyeing of nylon fabric using nanoemulsion." *Indian Journal of Scientific Research* 14 (2):257–261.

Kan, C. W., and C. W. M. Yuen. 2005. "Use of ultrasound in textile wet processing." *Textile Asia* 36 (5):47–52.

Kant, Rita. 2012. "Textile dyeing industry an environmental hazard." *Natural Science* 4 (1):22–26. doi: 10.4236/ns.2012.41004.

Kobayashi, Toshikatsu. 1996. "Pigment dispersion in water-reducible paints." *Progress in organic coatings* 28 (2):79–87.

Kong, M., X. G. Chen, K. Xing, and H. J. Park. 2010. "Antimicrobial properties of chitosan and mode of action: a state of the art review." *International Journal of Food Microbiology* 144 (1):51–63. doi: 10.1016/j.ijfoodmicro.2010.09.012.

Lau, W.-J, and A. F. Ismail. 2009. "Polymeric nanofiltration membranes for textile dye wastewater treatment: preparation, performance evaluation, transport modelling, and fouling control—a review." *Desalination* 245 (1–3):321–348.

Lin, L. 2003. "Mechanisms of pigment dispersion." *Pigment & Resin Technology* 32(2):78–88. https://doi.org/10.1108/03699420310464784.

Liu, Xiaoxia, Yujie Wang, Shiqing Sun, Changjun Zhu, Wei Xu, Yongdoo Park, and Haimeng Zhou. 2013. "Mutant breeding of Serratia marcescens strain for enhancing prodigiosin production and application to textiles." *Preparative Biochemistry and Biotechnology* 43 (3):271–284.

Ljunggren, Stig, and Jan Christer Eriksson. 1997. "The lifetime of a colloid-sized gas bubble in water and the cause of the hydrophobic attraction." *Colloids and Surfaces A: Physicochemical and Engineering Aspects* 129: 151–155.

Lopes, Cristiane N, José Carlos C Petrus, and Humberto G Riella. 2005. "Color and COD retention by nanofiltration membranes." *Desalination* 172 (1):77–83.

Ma, Z. Y., Y. P. Guan, X. Q. Liu, and H. Z. Liu. 2005. "Synthesis of magnetic chelator for high-capacity immobilized metal affinity adsorption of protein by cerium initiated graft polymerization." *Langmuir* 21 (15):6987–6994.

Marcucci, Manuele, Ingrid Ciabatti, Alessandro Matteucci, and Guido Vernaglione. 2003. "Membrane technologies applied to textile wastewater treatment." *Annals of the New York Academy of Sciences* 984 (1):53–64.

Mashaly, H. M., R. A. Abdelghaffar, M. M. Kamel, and B. M. Youssef. 2014. "Dyeing of polyester fabric using nano disperse dyes and improving their light fastness using ZnO nano powder." *Indian Journal of Science and Technology* 7 (7):960–967.

Mohsin, Muhammad, Shaheen Sardar, Muhammad Hassan, Nusrullah Akhtar, Ahmad Hassan, and Muhammad Sufyan. 2020. "Novel, sustainable and water efficient nano bubble dyeing of cotton fabric." *Cellulose* 27 (10):6055–6064.

Oatley-Radcliffe, Darren L, Matthew Walters, Thomas J Ainscough, Paul M Williams, Abdul Wahab Mohammad, and Nidal Hilal. 2017. "Nanofiltration membranes and processes: a review of research trends over the past decade." *Journal of Water Process Engineering* 19:164–171.

Pavel, Florentina M. 2004. "Microemulsion polymerization." *Journal of dispersion science and technology* 25 (1):1–16.

Raja, A. S. M, A. Arputharaj, S. Saxena, and P. G. Patil. 2019. "Water requirement and sustainability of textile processing industries." In *Water in Textiles and Fashion*, Muthu, S. S. (ed.), 155–173. Amsterdam: Elsevier.

Roessler, Albert, David Crettenand, Otmar Dossenbach, Walter Marte, and Paul Rys. 2002. "Direct electrochemical reduction of indigo." *Electrochimica Acta* 47 (12):1989–1995.

Roy Choudhury, Asim Kumar. 2013. "Green chemistry and the textile industry." *Textile Progress* 45 (1):3–143.

Schork, F Joseph, Yingwu Luo, Wilfred Smulders, James P Russum, Alessandro Butté, and Kevin Fontenot. 2005. "Miniemulsion polymerization." In *Polymer particles*, Okubo, M. (ed.), 129–255. Berlin, Heidelberg, New York: Springer.

Shang, S. M. 2013. "Process control in dyeing of textiles." In *Process control in textile manufacturing*, Majumdar, A., Das, A., Alagirusamy, R. and Kothari, V. K. (eds.), 300–338. Oxford: Elsevier.

Shibuya, Kazuhiro, Daisuke Nagao, Haruyuki Ishii, and Mikio Konno. 2014. "Advanced soap-free emulsion polymerization for highly pure, micron-sized, monodisperse polymer particles." *Polymer* 55 (2):535–539.

Shu, Li, Muthu Pannirselvam, and Veeriah Jegatheesan. 2019. "Nanofiltration of dye bath towards zero liquid discharge: a technical and economic evaluation." In *Water Scarcity and Ways to Reduce the Impact*, Pannirselvam, M., Shu, L., Griffin, G., Philip, L., Natarajan, A., and Hussain, S. (eds.), 47–61. Cham: Springer.

Song, Yujun, Fuan Liu, and Benhui Sun. 2005. "Preparation, characterization, and application of thin film composite nanofiltration membranes." *Journal of applied polymer science* 95 (5):1251–1261.

Sungpet, Anawat, Ratana Jiraratananon, and Piyanoot Luangsowan. 2004. "Treatment of effluents from textile-rinsing operations by thermally stable nanofiltration membranes." *Desalination* 160 (1):75–81.

Suslick, Kenneth S. 1990. "Sonochemistry." *Science* 247 (4949):1439–1445.

Takahashi, Masayoshi. 2009. "Base and technological application of micro-bubble and nano-bubble." *Materials Integration* 22 (5):2–19.

Takahashi, Masayoshi, Kaneo Chiba, and Pan Li. 2007. "Free-radical generation from collapsing microbubbles in the absence of a dynamic stimulus." *The Journal of Physical Chemistry B* 111 (6):1343–1347.

Tang, C., and V. Chen. 2002. "Nanofiltration of textile wastewater for water reuse." *Desalination* 143 (1):11–20. doi: 10C/S0011-9164(02)00216-3.

Tasaki, Tsutomu, Tsubasa Wada, Kanji Fujimoto, Shinji Kai, Kaoru Ohe, Tatsuya Oshima, Yoshinari Baba, and Masato Kukizaki. 2009. "Degradation of methyl orange using short-wavelength UV irradiation with oxygen microbubbles." *Journal of Hazardous Materials* 162 (2–3):1103–1110.

Tuli, Hardeep S, Prachi Chaudhary, Vikas Beniwal, and Anil K Sharma. 2015. "Microbial pigments as natural color sources: current trends and future perspectives." *Journal of Food Science and Technology* 52 (8):4669–4678.

Whittaker, C. M., and C. C. Wilcock. 1949. "Dyeing with coal-tar dyestuffs." *Tindall and Cox Baillière* 5:1–7.

Yanase, Noriko, Hiromichi Noguchi, Hideki Asakura, and Tatsuo Suzuta. 1993. "Preparation of magnetic latex particles by emulsion polymerization of styrene in the presence of a ferrofluid." *Journal of Applied Polymer Science* 50 (5):765–776.

Yusuf, Mohd, Mohd Shabbir, and Faqeer Mohammad. 2017. "Natural colorants: historical, processing and sustainable prospects." *Natural Products and Bioprospecting* 7 (1):123–145. doi: 10.1007/s13659-017-0119-9.

9 Gamma Ray Irradiation Technology for Textile Surface Treatment/Modification
An Overview

*Mohd Gulfishan, Rayees Afzal Mir,
Syed Aasif Hussain, Sajjad Khan,
Fatima Shabir, and Eyram Khan*
Glocal University Saharanpur

CONTENTS

9.1 Introduction	157
9.2 Basics of Gamma Ray Irradiation Technology	159
9.2.1 Computerized Control System	160
9.3 Applications of GRIT for Textile Surface Treatment	161
9.3.1 Improved Physical and Mechanical Properties	161
9.3.1.1 Physical Properties	161
9.3.1.2 Mechanical Properties	161
9.3.2 Enhanced Dye Uptake and Dyeability	164
9.3.2.1 Effect of Gamma Radiation on the Fabric Dyed with Natural Dyes	164
9.3.3 Water Repellency	165
9.3.3.1 Water Repellent Finish	166
9.3.3.2 Mechanism of Water Repellency	166
9.3.4 Flame Retardant Property	166
9.3.5 Others	167
9.4 Conclusion and Future Dimension	168
References	168

9.1 INTRODUCTION

The textile industry is a major branch of the world's economy, some of its major producers and exporters being China, India, and Italy. Textiles have been found to originate as early as the Neolithic period from evidence found in excavations, with some

DOI: 10.1201/9781003140467-9

indication that weaving was known even in the Paleolithic. Over time, the material and techniques evolved and became more sophisticated such as the creation of synthetic textiles. Along with manufacturing industries, the textile industry is among the most complex industries. Wastewater treatment is one of the major problems faced by textile manufacturers. From textile plant, the arrangement of the processes and products makes the wastewater which contains many types of pollutants such as sizing agents, dyes, pigments, softening agents, complexing agents, stiffening agents, surfactants, oils, fluorocarbon, wax, and many other additives which are used throughout the processes. These pollutants supply to high suspended solid (SS), chemical oxygen demand (COD), acidity, heat, color, and basicity. This wastewater can exert great environmental problems due to their high color, high chemical oxygen demand, and large amount of suspended solids. Textile industries are the major sources of water pollution in terms of releasing highly colored waste streams in surface water bodies. Wastewater generated in textile processing plants contaminates by releasing toxic synthetic colorants and various perilous chemicals. In a study, Bhuiyan et al. pointed to degrade the organic pollutants and dye molecules of textile wastewater by using gamma irradiation followed by the examination of physicochemical parameters of the irradiated water as well as looking into the scope for using treated wastewater for dyeing and irrigation purposes. The wastewater samples were submitted to Cobalt-60 gamma radiation source (BogdanLungu and IoanaStanculescu 2018).

To add high value on deferent major textile materials, the advancement of new technology in the textile world has led scientists and researchers to invent narrative finishes. It was a new pane for researchers to advent and explores new research fields which include geotextiles, flame retardant textiles, insect hideous textiles, medical textiles, aroma textiles, smart textiles, antibacterial textiles, nanotextiles, etc. On the surface of fibers and fabrics, recent developments in the textile industry are mainly focused on physical and chemical modifications. To improve or impart permanent useful functional properties, different chemical and biological methods have been used electively on the surface of major high demanded textile materials. However, some of these chemicals are toxic and sometimes expensive (Shahid-ul-Islam and Faqeer Mohammad 2015).

Gamma radiation which is produced during the radioactive decay of the atomic nucleus is an electromagnetic penetrative ionizing radiation. The excitation of the electron is one of the main results of gamma-irradiation which results in the process of ionization from which may cause bond breaking, the effects being the short life radicals, creation of excited states, and finally formation of new bonds. Two of the main zones where it is used are research and industry. In research field by using gamma irradiation for textile operations, the optimization of textile dyeing, cultural heritage artifacts' biological decontamination, and reducing the ecological impact. In industry, it can be used for DNA alteration in plants, sterilization of structures and property material modification (Ahmad Bhatti et al. 2012; Bashar et al. 2018).

Nowadays, the superior value of the functional clothes in textile industries is because of their enhanced properties and performance. For example, materials such as zinc oxide (ZnO) nanoparticles can be a used electromagnetic interference-shielding material. Moreover, to protect human beings from high microwave absorptions, ZnO-coated cotton is used as a wearable cloth which could be a potential

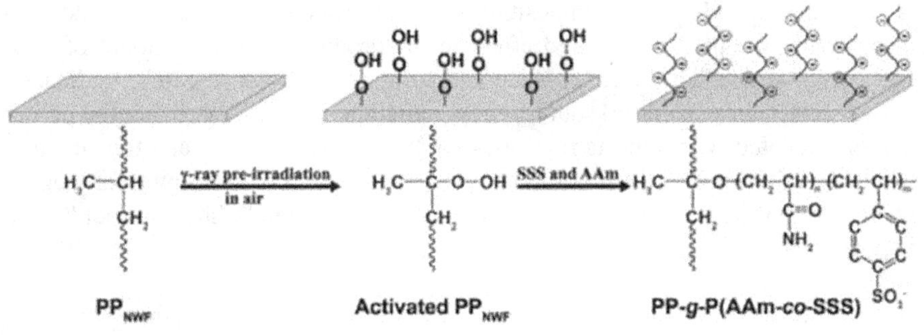

FIGURE 9.1 Gamma-ray preirradiation-induced graft polymerization of sodium styrenesulfonate (Shahidi and Weiner 2016).

application. Previously, as a flame retardant agent, ZnO-coated cotton fabrics have been engaged (Wang et al. 2018). However, the interactions between ZnO and cotton have not yet been studied in detail. Gamma ray irradiation imparts radical creation on materials, which is well known to cause cross-linking or chain scission. To understand whether gamma ray irradiation on both cotton and ZnO nanostructures can improve the linkage or dominate the scission process is thus noteworthy.

Primarily, related to ZnO films, textile substrates have been used on novel applications (Kołodziejczak-Radzimska and Jesionowski 2014). The cellulose substance amounts are counted to be around 88%–96% among the other fiber which usually contains many constituents. The applications of nanomaterials using flexible textile are to get various functional properties, (Riaz et al. 2018), while Wang et al. (2014) gave a suggestion on the development in photocatalyst applications by means of modified textiles. Verbič et al. (2019) presented the practice of ZnO on a textile substrate for applications, such as hydrophobicity, flame retardancy, thermal insulation, and so forth. In recent times, Gupta et al. (2020) have established that composites of reduced graphene oxide/zinc oxide over cotton fabric can be engaged for the application of high microwave absorption. Although several applications on this material have been carried out, this kind of material still has a greater attraction in considering the gamma ray exposure property on cotton fibers. Presently, polymeric membranes are extensively employed in the process of biomedical materials contacting with blood. Sulfonated polypropylene nonwoven fabric (PPNWF) has been successfully formed through gamma-ray preirradiation-induced graft polymerization of sodium styrene sulfonate (SSS) and acrylamide (AAm) by Shahidi and Weiner (2016) (Figure 9.1).

9.2 BASICS OF GAMMA RAY IRRADIATION TECHNOLOGY

In a process of radiation, a product or material is intentionally irradiated to conserve, modify, or improve its characteristics. By placing the product in the vicinity of a radiation source, this process is carried out (such as cobalt-60) for a fixed time interval whereby the product is exposed to radiation emanating from the source. A portion of the radiation energy that goes up to the product is absorbed by the product;

the amount depends on its composition mass and time of exposure. For each type of product, to realize the desired effect in the product a certain amount of radiation energy is needed; the exact value is determined through research. Radioactive material, such as a cobalt-60 source, emits radiation. On the other hand, the product that is irradiated with gamma rays does not become radioactive, and thus it can be handled on the whole. This is similar to X-ray examination in a hospital for diagnostic purposes; the patient is exposed to radiation (X rays) but he/she does not become radioactive.

9.2.1 Computerized Control System

On-line organization computer software with visual exhibit is now a standard tool on many irradiators. It provides several aspects of the operation and safety of the irradiator by providing instant and continuous information, for example, all the key parameter values that can affect dose to the product; this information is necessary for processes like status of the source position, control, and all interlocks; this information is necessary for safe operation of the facility, location, and status of the product containers in the facility. Also this information is necessary for product control.

Gamma-irradiated ZnO-based functional materials over cotton fabrics were considered to reveal the change in the crystallinity, chemical and electronic properties, and the enhancement of linkage between the ZnO nanoparticle and cotton fabric after gamma ray irradiation. XRD studies revealed the decrease in crystallinity, reducing to almost 59.6% after exposure to 9 kGy, whereas in the surface morphology, SEM does not reveal any significant change. In lattice parameters of ZnO particles, XRD results also revealed shrinkage after irradiation, and in comparison with the pristine one, they become more compact. The gradual decrease in the XRD integrated intensity and lattice parameters could be due to the oxygen defects or deficiency created in the sample after gamma ray exposure.

An increase in the dose level results in the increased optical band gap energy, which may be endorsed to the decrease in lattice parameters, due to oxygen deficiency after irradiation, in the ZnO nanostructure. Moreover, these results are reliable with NEXAFS results obtained from the Zn L3 edge and O–K edge, in the samples after irradiation there are suggestions of creating the possibility of oxygen deficiency. However, the results from NEXAFS and XPS are contradictory with each other due to the fact that the former is bulk sensitive and the latter is surface sensitive. The B.E. in the case of XPS spectra, shift by 2 eV at 1.5 kGy to 4 eV at 9 kGy in comparison with the pristine sample. In this kind of ZnO-based system, this observed phenomenon of B.E. shift by more than 2 eV is totally different from other reported works. Such a difference might be due to the defects created in the surface of ZnO particles after gamma ray exposure. It could be due to oxygen ions that transfer from the surface of Zn nanoparticles to the radical formation on the surface of cotton fabrics after gamma ray irradiation. Overall, this work suggests that it could be subjugated for enhancing the binding between the coated functional nanoparticles with cotton fabrics, which can be served as an optional approach for the functional cloth in the textile industry.

9.3 APPLICATIONS OF GRIT FOR TEXTILE SURFACE TREATMENT

9.3.1 IMPROVED PHYSICAL AND MECHANICAL PROPERTIES

9.3.1.1 Physical Properties

Gamma rays are high-energy electromagnetic radiations with wavelengths less than 10 picometers and energies greater than 100 keV. One of the promising methods is gamma ray surface alteration of textiles. The effects of gamma radiation on the physical properties of natural lignocelluloses (jute) polymer were studied. At a specific dose rate, samples were irradiated to achieve the necessary total doses. Changes in parameters such as tensile power, elongation at split, and work done at rupture for lignocelluloses samples were measured after irradiation with γ-rays from a cobalt-60 source. According to the XRD results, the substance under investigation is an isotactic polypropylene (iPP) with a monoclinic structure. The presence of a methyl group (which acts as a fingerprint for PP) peak in its spectra also helps to identify iPP. γ-irradiation of polypropylene resulted in significant changes in the material's physical properties due to chain scission, oxygen effects, and cross linking activities. These enhancements, according to Raman and UV-visible spectroscopic studies, result in the formation of degradation products such as water vapor and carbon monoxide from hydroxyl and carbonyl groups. The absorption limit of the second peaks undergoes a hypsochromic change as a result of the formation of the carbonyl group.

Because of their excellent mechanical properties, heat and oxidation resistance, and environmental stability, carbon composite fabrics are suitable reinforcing materials in advanced composite fields such as solar panels for space stations and electric vehicle bodies. All of these excellent properties are heavily dependent on crosslink density, which ultimately influences the overall properties of the resulting composites. Radiation-induced grafting has recently become popular as a low-cost method of developing new usable materials. It was confirmed that γ-ray irradiation grafting can be used to evenly functionalize fabric surfaces. This novel technique is simple, environmentally friendly, and adaptable. Compared to the traditional electrochemical process, the functionalization was much more uniform. The composites' interfacial strength increased dramatically (Javed et al., 2009).

9.3.1.2 Mechanical Properties

Martins et al. (1999) discovered that increasing the adsorbed dose when gamma irradiating polyethylene in air at dose rates ranging from 10 to 2000 kGy and a dose rate of 2.5 kGh/h improved mechanical resistance parameters. There was an increase in ultimate tensile stress and a decrease in ultimate elongation in the tensile studies. There was a 50% offset yield stress and an improvement in yield stress in compression simulations. The literature, in general, indicates a strong correlation between absorbed dose and polymer mechanical resistance parameters. The tendency of this relationship, on the other hand, is difficult to predict. The fact that mechanical resistance is a macroscopic result of several microscopic phenomena can be linked to the various assumptions about the effect of radiation on polymers found in the papers reviewed. To put it another way, cross-linking and oxidative degradation seem to happen at the same time and in a nonhomogeneous manner across the sample, while mechanical resistance parameters display the effects of integrating these phenomena across the sample. McKellop (1996) provided an example of this fact, stating that the predominance of oxidation is related to the availability of oxygen to react with free radicals. The oxygen renewal rate through the sample during irradiation in the presence of air is proportional to its size. Furthermore, only Wündrich (1985) addressed the relationship between mechanical strength and dose rate; most papers on this subject focused solely on the difference in absorbed dose. In air irradiation, the effect of dose rate on mechanical parameters revealed a greater predominance of oxidation, which is accentuated by a decrease in dose rate. The effects of irradiation in the presence of air, on the other hand, are dependent on other factors, such as sample measurements, and assumptions about these effects cannot be adequately discussed without taking these factors into account.

Polypropylene (PP) is a common thermoplastic with low manufacturing costs, good mechanical properties, and a broad range of industrial and consumer applications, including textile fibers, medical devices (operation gowns and covers, and drug packaging), hygiene (diapers, disposals), food packaging, filters, absorbents, wipes, battery separators, etc. The use of ionizing radiation on polymeric materials will continue to be of great interest because it is the only technique that allows you to introduce energy into materials and produce beneficial improvements, as long as the appropriate doses are used. When polymeric materials are exposed to ionizing radiations (e.g., γ-rays), they ionize and excite. According to, this irradiation produces highly reactive intermediates, free radicals, ions, and excited states. γ-irradiation of polypropylene resulted in significant changes in the material's physical properties due to chain scission, oxygen effects, and cross linking activities. These enhancements, according to Raman and UV-visible spectroscopic studies, result in the formation of degradation products such as water vapor and carbon monoxide from hydroxyl and carbonyl groups. The absorption limit of the second peaks undergoes a hypsochromic change as a result of the formation of the carbonyl group. It was also discovered that as the dose of radiation increases, the melting temperature and crystallinity decrease. The mechanical properties of the polymer deteriorated as the dosage increased, as evidenced by a decrease in the materials' elongation to fracture and tensile strength.

The effects of radiation on polymer structure and, as a result, physical properties are well understood in the plastics industry. Depending on the irradiation conditions, the literature on the mechanisms underlying these effects reveals two contradictory trends: cross-linking of polymer molecules, which increases mechanical power, and oxidative degradation, which usually causes material weakening. The amount of oxygen available on the material, as well as its ability to substitute it as it is consumed by chemical reactions with radicals formed during irradiation, appear to be related to which tendency will prevail. Many articles discuss the effect of radiation on mechanical strength as a result of the interplay of cross-linking and oxidative degradation. Wündrich (1985) summarizes the literature on mechanical resistance parameter values for a variety of plastics and elastomers exposed to radiation under various conditions. Instead of directly presenting parameter values, the radiation effects were compared using the half-dose value principle which is defined as the absorbed dose required to reduce a parameter value to 50% of its initial value According to the author, increases in parameter values were observed, mostly in cases involving high doses, but these effects were not realistic. According to the abovementioned results, when irradiation is performed in the absence of oxygen (vacuum or inert atmosphere), the effect of radiation on mechanical resistance parameters is independent of dose intensity. However, in the presence of air, the relationship between impact and dose rate is clear. For high-density polyethylene, for example, as the dose rate is reduced, the half-dose value for ultimate tensile stress decreases. This dosage is higher in the absence of air than in the presence of air for irradiation. According to Wündrich, this action is due to the role of oxygen in polymer degradation (oxidative degradation). Because it is a time-dependent process caused by two mechanisms: oxygen diffusion in the polymer and the disintegration reaction of the peroxides produced, the effect increases as the dose rate is reduced. Singh (1999) describes the processes that support these claims. According to the author, cross-linking and oxidative degradation are two phenomena that occur as a result of the HDPE gamma irradiation process.

In general, equation below depicts the reaction between HDPE free radicals (PE•) and cross-linking between polymer molecules, which is usually caused by the loss of a hydrogen atom in HDPE (represented as PE).

This recombination reaction is predominant in cases of irradiation in vacuum or inert atmospheres.

$$PEi + PEi \rightarrow PE - PE$$

in the presence of air during irradiation, peroxide radical formation and the reaction of the PE free radical and the peroxide radical.

$$PEi + O2 \rightarrow PEO2\ i$$

$$PEi + PEO2\ i \rightarrow PEOOPE$$

The cross-linking formation (equation 1) at very high dose rates is the predominant reaction (103 Gy/s, induced in electron accelerators), according to Singh (1999). This occurs due to oxygen is rapidly absorbed, restricting the formation of peroxide radicals (equation 2) to the rate of oxygen diffusion. The problem of the consequences of HDPE irradiation is also discussed (Premnath et al., 1999). According to the scientists, the polyethylene irradiation precise effects of will vary depending on the factors including polymer molecular weight, temperature, additives and storage under atmospheric conditions before, during, and after irradiation, and sample size.

I. The 5% offset yield stress and yield stress for density tests increased continuously as the absorbed dose increased, and this effect was more marked at lower dose rates, the changes in the mechanical properties of HDPE, according to an analysis of subjected to gamma irradiation.
II. Trends were not well defined; the modulus of elasticity values for compression tests mostly were higher than those for nonirradiated samples.

III. The yield stress for the tensile tests decreased as the dose was increased at the lowest dose rate, but as the dose was increased at higher dose rates, this parameter began to increase; and for the tensile measures,
IV. The modulus of elasticity decreased across all the doses and dose rate intervals, with the effect becoming more distinct at lower dose rates. It's worth noting that the experimental setup used in this study was created to allow for a comparison of the mechanical strength properties of nonirradiated and irradiated samples of a particular HDPE material. As a result, the assumptions and content used restrict the validity of the data and conclusions reached.

Extensive knowledge of the variety of mechanical properties (tensile strength, tearing strength, stiffness) explained by various authors.

Weave Structures	Rating	Comments	Fabric Quality
1/1 Plain	4	Slight pilling	Very good
2/1 Twill	2–3	Significant moderate pilling	Good to medium
2/2 Twill	2–3	Significant moderate pilling	Good to medium

Pilling and abrasion resistance: When the other factors (EPI, PPI, Warp Count, Weft Count, and Fabric Width) remain constant, resistance) of woven fabric because weave structure variation was discovered. Finally, it can be concluded that plain weave has higher tensile strength, abrasion resistance, stiffness, and pilling resistance than twill weave because these mechanical properties improved with increasing the amount of interlaced warp and weft yarns and decreasing the number of floating in the weave. However, tearing strength is higher in twill weave since this property increases as the number of floating threads in the weave decreases, i.e. as the number of interlacements decreases. A clear understanding of a woven fabric's mechanical properties can lead to further development of a woven fabric structure, which will be beneficial to many ends by use applications, especially protective clothing that performs complex as well as apparels. The effects of mechanical properties of varying weave structures are examined in the current study. It's worth noting that while the fabric parameters remain the same, more complex fabric mechanical behaviors can be applied and researched.

9.3.2 Enhanced Dye Uptake and Dyeability

9.3.2.1 Effect of Gamma Radiation on the Fabric Dyed with Natural Dyes

When unlike extracts of irradiated and un-irradiated turmeric powder were introduced to dye the irradiated and unirradiated cloth, the result shows a significant difference in color of the power (Adeel et al. 2012). The color strength of the methanol solubilized extract was higher than that of the aqueous (heat) and alkali solubilized extracts. The alkaline deprivation of curcumin into products like vanilic acid, vaniline, feruloylmethane, ferulic acid, and other fission products, which sorb on the fabric immediately with the colorant and impart dull redder shades, results in low color potency when using alkali solubilized extract (Tonnesen and Karlsen, 1985a).

The colorant can undergo hydrolytic degradation when using (heat) aqueous solubilized extract because it is insoluble in water. As a result, the actual colorant concentration on the fabric becomes low, resulting in low color intensity (Tonnesen and Karlsen, 1985b). The actual colorant has a better possibility to sorb onto fabric and impart yellow color with some dark shreds by using methanol solubilized extract. Fabric irradiation is another aspect that influences the color strength of the fabric. Previous research has shown that UV irradiation improves coloration and increases the dye absorption capacity of cotton fabrics by oxidizing cellulose surface fibers (Millington 2000; Javed et al. 2008). The colorants from the methanol solubilized extract penetrate the fiber vicinities, and when the color intensity was experienced with a spectraflash SF 650, a dark yellow shade was observed. In the absence of irradiation, insoluble impurities and colorant have a greater risk of sorbing on the matrix, resulting in dull redder shades. With varying quantities of chitosan and dyed silk yarns were treated with mono and bifunctional reactive dyes for this purpose. It was discovered that pretreatment of silk yarns with aqueous chitosan solutions improved dye absorption as well as antibacterial activity when compared to untreated silk yarn. Based on the depth of shade values, it was revealed that raising the chitosan absorption up to 3% (w/v) did not result in a noticeable color difference between the samples for both reactive dyes. Increased chitosan concentration did not all the times result in increased depth of shade (K/S), but it did result in dye precipitation and uneven dyeing. Furthermore, the colorfastness to light and washing properties of the chitosan-treated silk samples were improved.

9.3.3 Water Repellency

One of the most important functional properties for protective clothing that does not compromise comfort. The ability of a textile material to resist wetting is known as water repellency. The contact angle between the fabric surface and water droplet determines the water droplet's tendency to spread out over the fabric surface. Water repellent textiles have a wide range of applications, including automotive, consumer, and clothing. This repellency can be achieved by coating textile fibers with a thin layer of the water repellent chemicals. Water repellency can be achieved by altering the surface energy of textiles with little impact on other functional properties such as weight, durability, breathability, softness, and so once (Bongioivanni 2011; Ferrero 2012).

Because of its comfort and breathability, water-repellent fabric is becoming increasingly popular. The basic requirement for garments in a wet environment is that they maintain the wearer dry by being waterproof and/or water repellent. water-repellent fabrics and waterproof fabrics have a wide variety of uses, and are commonly used in garment manufacturing in traditional garments for men, women, and children in uniforms and work wear, and clothing for sport and leisure. When water repellent materials come into contact with water, they form drops that are easily removed from the fabric surface. A fabric that is water-repellent is resistant of being wet by water droplets and to water dispersal over its surface. The water repellency of a fabric prevents water from entering the fabric's macro-structure, which has a positive effect on garment weight and breathability. Textile engineers are continually worried about improving fabric quality and gaining professional qualifications. Water repellent is an effective textile finish or property that gives the fabric an extra effect. Water repellents are chemical finishes that prevent textiles from getting wet while also preserving other comfort features.

9.3.3.1 Water Repellent Finish

The properties of repellent finishes are achieved by decreasing the free energy at fiber surfaces. The drop will spread if the adhesive interactions of fiber and a drop of liquid put on it are greater than the liquid's internal cohesive interactions. Low energy surfaces are those that have a low number of interactions with liquids. Their vital surface energy or surface tension, γ-C, must be less than the surface tension of the repelled liquid, L (internal coherent interaction).

9.3.3.2 Mechanism of Water Repellency

Low energy surfaces can be applied to textiles in different ways Such as-

1. Water-repellent materials are mechanically incorporated into or on the fiber and fabric surface, in the fiber pores, and in the space between the fibers and yarns. Paraffin emulsions are an example of this.
2. The repellent material's chemical reaction with the fiber surface. Fatty acid resins are an example of this.
3. A repellent film forms on the fiber surface. Silicone and fluorocarbon goods are examples of these. Finally, unique fabric constructions such as stretched polytetrafluoroethylene films (Goretex), hydrophilic polyester films (Sympatex), and micro porous coatings may be used.

9.3.4 FLAME RETARDANT PROPERTY

Flame retardants are an effective source of material safety enhancement. The use of flame retardant (FR) and flame retardant textiles is influenced by the safety and preservation of human lives and valuables.

On the bases of their effectiveness in presentation after several washing cycles, flame-retardant textile finishes can be classified as nondurable, semi durable, or durable. Inorganic salts, borax and boric acid mixture, diammonium phosphate, and urea are some of the most popular nondurable flame-retardant chemicals on the market (Schindler 2004; Horrocks 2011). According to the literature, phosphorus-based flame retardants combined with nitrogenous compounds offer the most effective management for all forms of textiles due to their synergistic impact. In the mid-nineteenth century, halogen compounds based on chlorine and bromine were introduced to the market, either alone or in conjunction with an antimony compound to provide synergism. Flame-retardant formulations dependent on phosphorus, nitrogen, and halogen, such as tetrakis phosphonium salt and N-alkyl phosphopropionamide derivatives, have been commonly used in industrial applications over the past six decades (Horrocks 2011; Katovic et al. 2009). On cotton textiles when such formulations are used, however, the fabric tear and tensile strength are substantially reduced, as well as the fabric becoming stiffer. This was largely due to the need for an acidic atmosphere for the application of the aforementioned formulations, as well as the use of high-temperature drying method or curing methods. Although the process is successful for a variety of fibers, it is harmful, unsafe, and expensive (Banerjee 1985). To counter these disadvantages, a number of environmentally safe and cost-effective formulations using butane tetra carboxylic acid (BTCA) as the binding agent have been shaped in recent decades to minimize the amount of formaldehyde released from the treated fabric (Horrocks 2011; Kandola et al. 1996). There have also been attempts to develop halogen-free phosphorus-nitrogen-based flame retardants to facilitate further char formation during cellulosic substrate combustion (Horrocks 2011). (ZnO) Nanozinc oxide particles, TiO_2, different clay compounds, and polycarboxylic acid have recently been renowned on cotton and other textiles in an environmentally pleasant manner, thanks to the quick growth of nanoscience and technology (Hady 2013; Kan 2012). Plasma treatment has also been formulated as a pretreatment for enhancing the absorption of fire-retardant chemicals, graft polymerization of acrylate phosphate and phosphonate derivatives, and as a post treatment for improved reaction. The demand for cellulosic textiles manufactured and finished with natural products, such as natural dyes, enzymes, and fashionable textiles with functional values, is growing due to an ever-increasing knowledge of human health, hygiene, and fashionable textiles with functional values (Sarvanan et al. 2013). Antimicrobial, mosquito-repellent, and well-being textiles, as well as aromatic and medicinal plant extracts, are gaining prominence in academic research and industrial product production for both conventional and exciting upcoming markets (Joshi 2007; Alongi et al. 2013). To increase their thermal stability, cellulosic textiles have been treated with bio-macromolecules like DNA from herring sperm and salmon fish (Alongi et al. 2013). Similar uses have been studied for whey proteins, casein, and hydrophobias (Bosco et al. 2012; Carosio et al. 2013). Efforts to use agro-residues and the other plant molecules for flame-retardant finishing of cellulosic and lignocelluloses textiles have also been formed in recent years. In this regard, the use of banana pseudo stem sap (BPS) and the spinach extracts for flame-retardant ultimate of cotton and jute fabrics is becoming more common due to their environmental benefits, cost effectiveness, and ease of application (Basak et al. 2014).

9.3.5 OTHERS

Gamma decay from naturally occurring radioisotopes like potassium40 and secondary radiation from multiple atmospheric encounters with cosmic-ray particles, are natural sources of gamma rays on Earth. Gamma rays provide data on some of the universe's most energetic events, but they are mostly absorbed by earth's atmosphere. Our only vision of the universe in gamma rays is provided by instruments onboard high-altitude balloons and satellite flights, like as the Fermi Gamma-ray Space Telescope.

The Gamma-induced molecular changes can also be experienced to modify the properties of semi-precious stones, such as turning white topaz blue. For the amount of levels, density, and thicknesses, noncontact industrial sensors typically use gamma radiation sources in the manufacturing, mining, chemicals, food, soaps and detergents, and pulp and paper industries. Gamma-ray sensors are also used in the water and oil industries to measure fluid levels. The radiation source is usually Co-60 or Cs-137 isotopes. Gamma ray detectors are being used as part of the Container Security Initiative in the United States (CSI). These machines are advertised as having a scanning speed of 30 containers per hour. In a process known as irradiation, gamma radiation is frequently used to kill living organisms. This has applications such as sterilizing medical equipment (as an alternative to autoclaves or chemical mean), the removal of decay-causing bacteria from many foods and the avoidance of fruit and vegetable sprouting in order to preserve freshness and flavor. Although their cancer-causing properties, gamma rays are used to treat some types of cancer because they kill cancer cells. In order to kill the cancerous cells multiple concentrated beams of gamma rays are directed to the growth in the procedure known as gamma-knife surgery.

9.4 CONCLUSION AND FUTURE DIMENSION

In the textile industry, fiber surface modification plays a pivotal role. The surface properties such as adhesion and water repellency are of much concern. There are different surface modification techniques used by different scientists around the world to improve fiber's surface and subsequently texture of the materials produced. Gamma ray irradiation technology has been recognized as one of the most promising techniques among other surface modification techniques. During the past decades, GIT has been frequently used to treat the surfaces of materials. Though the surface of the material treated with GIT results in changes of chemical and physical properties of the material, it causes changes in other properties of the material slightly. In the 21st century, it may be assumed that GRIT will come with more efficiency and accuracy to change the surfaces of the materials. The irradiation treatments have been in front of recognition as a surface modification technique. Much attention has been paid to treating the surfaces of materials with irradiation methods in the past few decades. In a broad sense, treating the surface of a material with irradiation methods can cause physical and chemical changes to occur at the surface but affect the bulk properties of the material little.

REFERENCES

Ahmad Bhatti, I., Adeel, S., Nadeem, R., and Asghar, T. Improvement of colour strength and colourfastness properties of gamma irradiated cotton using reactive black-5. *Radiation Physics and Chemistry* (2012) 81(3): 264–266, https://doi.org/10.1016/j.carbpol.2014.11.023.

Alongi, J., Carletto, R.A., Balsio, A.D. et al. Intrinsic intumescent like flame retardant properties of DNA treated cotton fabrics. *Carbohydrate Polymers* (2013) 96(1): 296–304.

Banerjee, S.K., Day, A., and Ray, P.K. Fire proofing jute. *Textile Research Journal* (1985) 56: 338–343.

Basak, S., Samanta, K.K., Saxena, S. et al. Flame resistant cellulosic substrate using banana pseudostem sap. *Polish Journal of Chemical Technology* (2014) 17(1): 123–133.

Bashar, M., Siddiquee, M., and Khan, M. Preparation of cotton knitted fabric by gamma radiation: a new approach. *Carbohydrate Polymers* (2018) 120: 92–101. https://doi.org/10.1177/0040517512449045.

Bongiovanni, R., Zeno, E., Pollicino, A., Serafini, P.M., and Tonelli, C. UV-light induced grafting of fluorinated monomer onto cellulose sheets. *Cellulose* (2011) 18: 117–126.

Bosco, F., Carletto, R.A., Alongi, J. et al. Thermal stability of flame resistance of cotton fabrics treated with whey proteins. *Carbohydrate Polymers* (2012) 94(1): 372–377.

Carosio, F., Blasio, A.D., Cuttica, F. et al. Polyester and polyester cotton blend fabrics have been treated with caseins. *Industrial and Engineering Chemistry Research* (2013) 53(10): 3917–3923.

Ferrero, E., Periolatto, M., and Udrescu, C., Water and oil-repellant coatings of perfluoropolyacrylate resins on cotton fibres: UV curing in comparison with thermal polymerization. *Fibres Polymers* (2012) 13: 191–198.

Gupta, S., Chang, C., Anbalagan, A.K., Lee, C.-H., and Tai, N.-H. Reduced graphene oxide/zinc oxide coated wearable electrically conductive cotton textile for high microwave absorption. *Composites Science and Technology* (2020) 188: 107994, DOI:10.1016/j.compscitech.2020.107994.

Hady, A.A.E., Farouk, A., and Sharaf, S. Flame retardancy and UV protection of cotton based fabrics using nanoZnO and polycarboxylic acid. *Carbohydrate Polymers* (2013) 92(1): 400–406.

Horrocks, A.R. Flame retardant challenges for textiles and fibres. *Polymer Degradation and Stability* (2011) 96(3): 377–392.

Javed, I., Bhatti, I.A., and Adeel, S. Effect of UV radiation on dyeing of cotton fabric with extracts of henna leaves. *Indian Journal of Fiber and Textile Research* (2008) 33: 157–162.

Joshi, M., Ali, S.W., and Rajendran, S. Antibacterial finishing of polyester cotton blend fabric using neem: a natural bioactive agent. *Journal of Applied Polymer Science* (2007) 106: 793.

Kan, C.W., Lam, Y.L., and Yuen, C.W. Fabric handle of plasma-treated cotton fabrics with flame-retardant finishing catalyzed by titanium dioxide. *Green Processing and Synthesis* (2012) 1(2): 195–204.

Kandola, B.K., Horrocks, A.R., Price, D. et al. Flame retardant treatments of cellulose and their influence on the mechanism of cellulose pyrolysis. *Journal of Macromolecular Science* (1996) 36: 794–796.

Katovic, D., Vukusic, S.B., Gragac, S.F. et al. Flame retardancy of paper obtained with environmentally friendly agents. *Fibres& Textiles in Eastern Europe* (2009) 17(3): 90–94.

Kołodziejczak-Radzimska, A. and Jesionowski, T. Zinc oxide-from synthesis to application: a review. *Materials* (2014) 7: 2833–2881, DOI: 10.3390/ma7042833.

Lungu, Bogdan and Ioana Stanculescu. Application of gamma irradiation for the functionalization of textile materials. (2018), DOI: 10.24264/icams–2018.I.6.

Martins, A.G., Suarez, J.C.M., and Mano, E.B. Produtos poliolefínicos reciclados com desempenho superior aos materiais virgens correspondentes. *Polímeros* (1999) 9: 27–32.

McKellop, H.A. Does gamma radiation speed or slow wear? *Bulletin, The American Academy of Orthopedic Surgeons* (1996) 44–50.

Millington, K.R. Comaprison of effect of gamma and ultraviolet radiation on wool keratin. *Coloration Technology* (2000) 116: 266–272.

Premnath, V., Bellare, A., Merrill, E.W., Jasty, M., and Harris, W.H., Molecular rearrangements in ultra high molecular weight polyethylene after irradiation and long-term storage in air. *Polymer* (1999) 40: 2215–2229.

Riaz, S., Ashraf, M., Hussain, T., Hussain, M. T., Rehman, A., Javid, A., Iqbal, K., Basit, A., and Aziz, H. Functional finishing and coloration of textiles with nanomaterials. *Coloration Technology* (2018) 134: 327–346, DOI: 10.1111/cote.12344.

Sarvanan, D., Lakshmi, S.N.S., Raja, K.S. et al. Biopolishing of cotton fabric with fungal cellulose and its effect on the morphology of cotton fibres. *International Journal of Fibre and Textile Research* 38 (2013): 156–159.

Schindler, W.D. and Hauser, P.J. (ed.). Flame retardant finishes. In *Chemical Finishing of Textiles*. Woodhead Publishing Limited, Boca Raton, FL (2004).

Shahid-ul-Islam and Faqeer Mohammad, High-energy radiation induced sustainable coloration and functional finishing of textile materials. *Industrial & Engineering Chemistry Research* (2015), 54, 3727–3745.

Shahidi, Sheila and Jakub Wiener. Radiation Effects in Textile Materials, Radiation Effects in Materials, Waldemar A. Monteiro, IntechOpen (July 20th 2016). DOI: 10.5772/63731. Available from: https://www.intechopen.com/books/radiation-effects-in-materials/radiation-effects-in-textile-materials.

Singh, A., Irradiation of polyethylene: some aspects of cross-linking and oxidative degradation. *Radiation Physics and Chemistry* (1999) 56: 375–380.

Tonnesen, H. H. and Karlsen, J. Studies on curcumin and curcuminoids. V. Alkaline degradation of curcumin. *Zeitschrift für Lebensmittel-Untersuchung und -Forschung* (1985a) 180: 132–134.

Tonnesen, H. H. and Karlsen, J. Studies on curcumin and curcuminoids.VI. Kinetics of curcumin degradation in aqueous solution. *Zeitschrift für Lebensmittel-Untersuchung und -Forschung* (1985b) 180: 402–404.

Verbič, A., Gorjanc, M., Simončič, B. Zinc oxide for functional textile coatings: recent advances. *Coatings* (2019) 9: 550. DOI: 10.3390/coatings9090550.

Wang, J., Zhao, J., Sun, L., and Wang, X. A review on the application of photocatalytic materials on textiles. *Textile Research Journal* (2014) 85: 1104–1118. DOI: 10.1177/0040517514559583.

Wang, Y.-W., Shen, R., Wang, Q., Vasquez, Y. ZnO microstructures as flame-retardant coatings on cotton fabrics. *ACS Omega* (2018) 3: 6330–6338. DOI: 10.1021/acsomega.8b00371.

Wündrich, K. A review of radiation resistance for plastic and elastomeric materials. *Radiation Physics and Chemistry* (1985) 24: 503–510.

Zhao, F. and Huang, Y. Uniform modification of carbon fibers in high density fabric by X-ray irradiation grafting. *Materials Letters* (2011) 65: 3351–3353.

10 Overview on Significant Approaches to Waterless Dyeing Technology

K. Murugesh Babu
Bapuji Institute of Engineering and Technology

CONTENTS

10.1	Introduction	171
10.2	Conventional Dyeing and Sustainability Issues	172
10.3	Waterless Dyeing Techniques	174
10.4	Airflow Dyeing (Air Dyeing)	175
	10.4.1 Principle of Air-Dyeing	175
	10.4.1.1 Some Specific Benefits of Airflow Dyeing	176
10.5	Electrochemical Dyeing	176
	10.5.1 Direct Electrochemical Dyeing	178
	10.5.2 Indirect Electrochemical Dyeing	178
	10.5.2.1 Advantages of Electrochemical Dyeing	178
10.6	Plasma Dyeing	179
10.7	Supercritical CO_2 Dyeing	180
	10.7.1 Advantages of SC-CO_2 Dyeing	182
10.8	Ultrasonic Dyeing	182
	10.8.1 Equipment for Ultrasonics	183
	10.8.2 Dyeing Process	183
	10.8.2.1 Advantages	183
	10.8.2.2 Disadvantages	184
10.9	Microwave-Assisted Dyeing	184
10.10	Comparison between Conventional and Waterless Dyeing Techniques	185
	10.10.1 Conventional Dyeing	185
	10.10.2 Waterless Dyeing	185
10.11	Conclusion	185
References		186

10.1 INTRODUCTION

The worldwide textile wet processing sector produces large amounts of wastewater and has detrimental effects on potable water and aquatic life if not treated properly and discharged into the environment. Pollution control and environmental sustainability lead to green technology resulting in the reduction or elimination of the use of harmful carcinogenic dyes, which balances prevention of pollution and water conservation (Zaidy et al., 2019). There is a huge consumption of water in the dyeing industry, especially using conventional exhaust dyeing methods which is a serious concern for the environment. As a result, wastewater treatment costs will pose severe threats in the future. The highest concern is about the presence of residual dyes, electrolytes, and metal ions in dye wastewater, which leads to various intolerable health risks and environmental pollution. Textile dyeing is a water-consuming process that weighs stress on the freshwater resources and requires approximately 100–200 L of water to dye 1 kg of textile material. Hence, the conservation of water and chemicals required for dyeing would be a real breakthrough for the dyeing industry. Dyes are considered as important coloring substances in any textile industry, but they are discharged as harmful effluents, posing direct threat to aquatic life and food chain (Hussain and Wahab, 2018). Suspended effluents in textile wastewater can cause hemorrhage and skin diseases. The chemicals also block the sunlight and increase the biological oxygen demand, increasing environmental toxicity and allergic reactions and health complexities (Ghaly et al., 2014).

New techniques and processing sequences are being developed worldwide, for the dyeing of textile fibers and fabrics with the main aim of improving energy requirements and maintaining eco-sustainability. In this regard, emerging techniques for dyeing processes are being introduced to dye different textile materials, which demonstrates their own merits and demerits. Recently, waterless dyeing techniques have been developed, facilitating the processing of pretreated fabric with less amount of dye solution. These techniques require few washing steps resulting in saving at least 90% water and energy. In addition, they eliminate the wastewater treatment and discharge, making them environment friendly and sustainable (Zaidy et al., 2019).

In recent years, there has been successful utilization of various emerging dyeing technologies, and some of them such as ultrasonic dyeing, electrochemical dyeing, microwave, plasma dyeing, and supercritical carbon dioxide dyeing have been tested and are gradually finding their way toward commercialization. In this chapter, an attempt has been made to review different approaches and methods for water conservation and reduction of wastewater in the textile industry, emphasizing the advantages and limitations of waterless dyeing techniques like air dyeing, electrochemical dyeing, plasma dyeing, ultrasonic dyeing, supercritical CO_2 dyeing, etc.

10.2 CONVENTIONAL DYEING AND SUSTAINABILITY ISSUES

Wet processing which consists of pre-treatment of textile materials, dyeing, printing, and finishing is an important stage in any textile industry, and coloring and finishing processes add aesthetic value to a product. Different types of fibrous materials like cotton, wool, silk, polyester, nylon, and acrylic classified among three groups, i.e., cellulosic, protein, and synthetics require different methods and recipes for their dyeing (Arputharaj et al., 2016). The input and output structure of the dyeing process is depicted in Figure 10.1. Inputs are raw material, water, dye, chemicals, and energy consumption. Output is a dyed fabric, generating wastewater with effluents. The dye

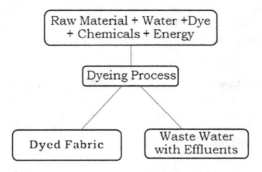

FIGURE 10.1 Input-output structure in the dyeing process.

waste water contains residual dyes and chemicals which are harmful to the environment. It becomes necessary to treat this wastewater to control environmental pollution.

Major sustainability issues in textile wet processing are (i) usage of water, (ii) energy consumption, and (iii) wastewater generation.

- **Usage of water:** Water is a major input in the dyeing process, and the process consumes nearly 100–200 L of water to dye 1 kg of fibers. Most of the water used in the process is discharged as waste water containing toxic dyes and chemicals. The water consumed and released also depends upon the type of fabrics used. Average water consumption for different types of materials is presented in Table 10.1.
- **Energy consumption:** Energy is used in every dyeing process; many of the exhaust dyeing processes are carried out using high temperature consuming lots of energy per day, leading to increased carbon footprints in the environment.

TABLE 10.1
Water Consumption for Various Materials

Type of Material Processed	Maximum Water Consumption (m^3/ton Fiber Material)
Wool	285
Woven	114
Knit	84
Carpet	47
Stock/Yarn	100
Nonwoven	40
Felted Fabric Finishing	213

Source: Adapted from Karthik and Gopalakrishnan, 2014.

TABLE 10.2
Ingredients of Effluents of Dyes

Dye Class	Type of Fiber Dyed	Major Ingredients
Direct, Reactive	Cotton, Viscose and other cellulosic fibers	Common salt/Glauber's salt, surfactants, urea, and alkali
Vat, Sulfur	Cotton, Viscose and other cellulosic fibers	Sodium hydrosulfite, sodium sulfide, alkali, salts
Disperse	Polyester, Nylon, Acrylic and other synthetic fibers	Carrier, acetic acid, sodium hydroxide, sodium hydrosulfite, dispersing agents
Acid	Silk, Wool and other protein fibers	Glauber's salt, acetic acid/sulfuric acid, and surfactants

Source: Adapted from Arputharajet al., 2016.

- **Waste water generation:** Dyeing is an important coloring process for textile materials and plays a significant role in the whole wet processing section. Water used in textile dye houses can generate large quantities of dye wastewater daily. The effluents in wastewater generated by dye houses depends on the type of dyes, auxiliaries, machines, and dyeing techniques used (Arputharaj et al., 2016). The unnecessary and excess quantity of water used and its discharge and effluent treatment costs add to the cost of finished textile products substantially. Since most of the textile materials are dyed with synthetic dyes, which are derived from petroleum products, waste water released after dyeing will contain residues of harmful dyes and chemicals such as heavy metals, acids, and alkalis. Various ingredients present in the effluents of different classes of dyes are listed in Table 10.2.

Large amounts of wastewater contain organic and inorganic compounds, and not all the dyes get fixed on to the materials; some remaining portions of the dyes are unfixed and get washed out. These unfixed dyes are called astextile effluents (Ghaly et al., 2014). To overcome the sustainability issues in textile processing, it is required to adopt approaches such as use of biodegradable chemicals, reducing water consumption, reducing energy consumption, recycling wastewater, and creating awareness to GO GREENCALL (Kant, 2016).

10.3 WATERLESS DYEING TECHNIQUES

Traditional dyeing is a multistep process that uses chemicals and auxiliaries, consequently producing large quantities of waste water that cannot be reused. Waterless dyeing techniques use radically less water than traditional methods; these techniques vary: some use no water at all, while others reduce water usage by about 95% (Davies, 2016). In the following sections, some of the important approaches to waterless dyeing have been discussed.

10.4 AIRFLOW DYEING (AIR DYEING)

Air dye technology manages the application of color to textiles without the use of water. It was developed and patented by Colorep, a California-based sustainable technology company. Depending on the fabric, and type of dyeing, air dyeing uses up to 95% less water, and up to 86% less energy, contributing 84% less to global warming. The damage caused to fabrics is only about 1% as compared to damage caused by conventional dyeing methods (10%). In addition, this method doesn't require any post-treatment or finishing treatments (Babu and Anekonda, 2019).

10.4.1 Principle of Air-Dyeing

In this dyeing machine, an atomized dye liquor which is combined with a high-pressure airflow is finally sprayed on fabric to be dyed (Figure 10.2). This results in requirement of a small quantity of water as it just serves as a solvent while the other dyeing chemicals strikes the fabric directly. Hence, airflow dyeing machine demonstrates energy-saving, high efficiency and environmental protection compared to other overflow dyeing machines. Neither a conventional dyebath nor an aqueous medium is required for fabric movement in dyeing as the fabric is carried by an air mixture which emerges from a blower. Hence fabric movement can be carried out without the liquor. As a result, the fabric is in constant movement from the beginning to the end of the dyeing process, as well as during drainage and filling processes (Babu and Anekonda, 2019).

Air is an important element of this technology because the dye liquor is replaced with air as a transporting medium, which is a major step toward reducing water

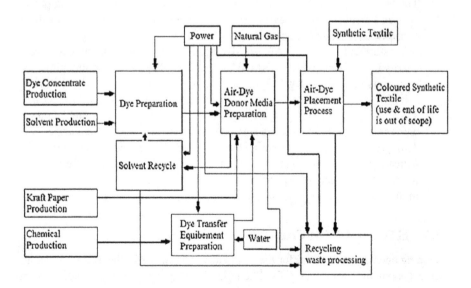

FIGURE 10.2 Air-dyeing process flow.

and chemical consumption. Even dyeing conditions are maintained as moisture-saturated air-stream ensures uniform distribution of temperature on the fabric and in the machine. In addition, higher fabric speeds can be achieved due to less amount of dye liquor in the system resulting in a lightweight fabric.

The main advantages of airflow dyeing can be summarized as follows:

- Depending on the nature of the fiber, it is possible to achieve lowest liquor ratio of approximately 1:2 to 1:4.
- Since other conventional systems use water as the transporting medium, this machine use only air, energy savings up to 40% can be achieved.
- The overall process time can be reduced by approximately 25%.
- The process results in lowest water and wastewater levels.
- Similar dyeing effects, as with conventional dyeing, such as single color, different color on each side of the fabric, a pattern on one side, and a solid on the other side or a pattern on both sides of the fabric can be easily obtained.

10.4.1.1 Some Specific Benefits of Airflow Dyeing

- Production of pollutant water and hazardous waste is prevented during the application of color as air is used as transport medium. This means less quantity of water is required for dyeing.
- The process results in substantial reduction of energy required, thereby lowering overall dyeing costs and contributing to ecological sustainability.
- Since the process eliminates the use of boilers, screen printing machines, drying ovens, or unnecessary auxiliary chemicals, the major sources of pollution are prevented.
- The process eliminates hazardous wastewater as a byproduct of dyeing process and results in huge savings in water. In addition, it is possible for the consumers to achieve 'print to order' sequences.
- It is estimated that using Airflow dyeing technology, a company can save energy to the extent of 1,132,500 MJ, about 157,500 ga of water, greenhouse emissions of 57,500 (Kg CO_2 equiv. emissions) for every 25,000 T-shirts sold.
- This dyeing technology paves a way for the consumers at the point of purchase to choose their style and sustainability practice at a realistic price.
- This process is especially suitable for woven and knitted goods made of delicate cellulose and man-made-fibers with a high percentage of elastane fibers. Elastane, such as spandex incorporated fabrics find wide application in the automotive, lingerie, sports-leisure and swimwear segments.

10.5 ELECTROCHEMICAL DYEING

Reducing agents are essential for the reduction of dyestuffs in the dyeing of vat and sulfur dyes on cellulosic materials. The reduction step using strong reducing agents to attain the reduced form of dyestuff is a common step. The conventional reducing agents such as sodium hydrosulfite, sodium sulfide used in dyeing of cotton with vat, and sulfur dyes produce undesirable oxidized byproducts that remain in the dyebath. It is difficult to use these dyebaths again as the reducing power of the reducing agents is lost and cannot be recovered. This necessitates the use of excess quantities of NaOH and hydrosulfites (nearly three times more than the theoretical requirement) in the dyebath to maintain the reduced conditions. In addition, only a small quantity of these reducing agents is required for dye reduction and the excess chemicals are released to the wastewaters. The wastewaters from these dyebaths contain harmful decomposition residues such as sulfite and sulfates from hydrosulfites, sulfides from vat and sulfur dyes. These non-ecofriendly chemicals which have a high COD cause severe threat to the environment. The suphur compounds (e.g., Na_2S, NaHS, etc.), present in the decomposed products of hydrosulfites pollute the atmosphere through the formation of H_2S. In addition, the salts of sulfur in the form of sulfates and sulfites such as Na_2SO_3, $NaHSO_4$, Na_2SO_4, $Na_2S_2O_3$, etc. result in contamination of sewage and corrosion of concrete pipes which transport the sewage waters. Hence, enough precaution and care have to be exercised in selecting the reducing agents for the above classes of dyes (Kulandainathan et al., 2007).

With a view to improve eco-efficiency, electrochemical dyeing method has been introduced to reduce the vat and sulfur dyes which avoids the addition of harmful reducing agents such as sodium hydrosulfite (Sala and Gutiérrez-Bouzán, 2012). The concept of electrochemical dyeing which substitutes the non-regenerable agents is a new strategy to improve the eco-friendliness of the dyeing processes of cellulosic fibers using vat, sulfur, and indigo dyes. Laboratory-scale experiments and full-scale prototype tests are being carried out to commercialize this technique during the last few years.

In electrochemical dyeing, the dyes such as vat and sulfur dyes are reduced by the process of electrochemical reduction. Electrons from the electric current replace the chemical reducing agents, and effluent-contaminating substances are completely eliminated (Senthil Kumar et al., 2009). The process is more efficient due to the continuous regeneration of the reducing agent at the electrode. In addition, this process results in complete recycling of the dye bath water (Nieminen et al., 2007). The first attempt in this direction was to reduce the quantities of sodium hydrosulfite required for reducing the vat dyes by the application of a direct voltage. The result of this reduction was the conversion of sodium hydrosulfite at the cathode into a form that exhibits increased reducing power. By varying suitable conditions such as cathode potential, sodium dithionite concentration, pH, and temperature, it is possible to generate powerful reducing species from sodium hydrosulfite with redox potential higher than sodium hydrosulfite itself (Sala and Gutiérrez-Bouzán, 2012). This behavior of sodium dithionite is due to the decomposition of hydrosulfite to produce a free radical ion. Regeneration of the reducing agents at an applied voltage, however, is not possible after dyeing with conventional methods, preventing recycling of the bath liquors. In the electrochemical dyeing technique, the same concept is adopted one step ahead making the liquor recycling possible (Bechtold and Turcanu, 2009; Bechtold, 1993).

Electrochemical dyeing can be carried out by two different methods and they are direct electrochemical dyeing and indirect electrochemical dyeing (Anon: International Dyer, 1999, Li et al., 2020).

In earlier methods, the required quantities of sodium hydrosulfite required for the reduction step was obtained by the application of a direct voltage. Thus, sodium dithionite at the cathode is converted into a form demonstrating an increased reducing power. By adapting appropriate reduction at the cathode, under suitable conditions such as cathode potential, sodium dithionite concentration, pH, and temperature, powerful reducing species can be generated from sodium hydrosulfite with higher redox potential (Sala and Gutiérrez-Bouzán, 2012). This behavior of sodium dithionite is due to the decomposition of hydrosulfite to produce a free radical ion. After conventional method of dyeing, however, these products cannot be regenerated at an applied voltage, making recycling of the bath liquor difficult. In the electrochemical dyeing technique, the same concept is adopted one step ahead making the liquor recycling possible (Li et al., 2020).

10.5.1 Direct Electrochemical Dyeing

In this method, the dye is reduced at the surface of cathode directly without the addition of reducing agent or mediator. Contrary to incomplete chemical reduction of the dye with reducing agents in conventional methods, in this method, complete chemical reduction of the dye is possible with an increased stability of leuco dye. A small quantity of soluble reducing agent is required at the beginning to generate some quantity of leuco dye which will then act as an electron-shuttle between the electrode and the surface of the dye pigment, and the subsequent process is self-sustaining (Roessler and Jin, 2003). However, dye has to come in contact with the surface of electrode in order to get reduced. Sulfur dyes can be successfully reduced by this technique (Sahoo and Gupta, 2007; Chavan and Chakraborty, 2001; Senthil Kumar et al., 2009).

10.5.2 Indirect Electrochemical Dyeing

In this method, the dye is not directly reduced at the electrode surface, like in direct electrochemical reduction, rather, a mediator (iron complexes with triethanolamine or gluconic acid as ligand) that reduces the dye is added. The mediator which undergoes reduction and oxidation cycles is oxidized after dye reduction and subsequently reduced at the cathode surface so that it can be used again for further reduction of dyes. This cycle repeats continuously during the dyeing operation (Bechtold et al., 1997; Merk et al., 2003; Merk et al., 2004).

10.5.2.1 Advantages of Electrochemical Dyeing

1. Savings in the chemical costs as chemical wastes are reduced by 80% and water savings around 85%.
2. Reduction of waste water recycling costs.
3. Several health benefits can be gained due to the fact that the dyeing is an eco-friendly process with less toxic nature of effluents as there is no presence of sulfates and sulfites due to which there is adverse effect on aquatic life.

4. Maximum process reliability can be achieved through control of reducing potential as required ranging from 0 to 1200 mV just by varying the current which also results in complete control over various dyeing parameters. Dye reduction rate is very good (10 mg dye/minutes), and dye pick up may go up to 85%–90%.
5. The process results in better quality dyed products with overall good fastness properties compared to conventional dyeing process.
6. Electrochemical dyeing can be very useful for the application of vat, indigo, and sulfur dyes due to the elimination of reducing agents in the dye bath (Sala and Gutiérrez-Bouzán, 2012).

10.6 PLASMA DYEING

Plasma treatment of textiles is considered as an eco-friendly process which has a huge potential of water and energy savings and has attracted attention of many researchers today. Plasma is the fourth form of matter comprising partially ionized gas (i.e., electron density is balanced by that of positive ions) first discovered by Irving Langmuir in 1929. Plasma is electrically conductive due to the presence of free electric charges and is internally interactive, and shows strong response to electromagnetic fields (Alexandar Friedman, 2008). Plasma, being highly reactive, is able to react with other substances leading to various chemical fusions and fissions. Depending upon the current, temperature, pressure, gaseous matter, and the type of textile fiber used, plasma can be used to modify various fiber and fabric surfaces to achieve different functionality (Kumar and Gunasundari, 2018).

Plasma treatment on textile fiber and fabric surfaces can be done to induce significant surface modifications. This can result in the removal of natural or synthetic greasy substances and waxy material present in these materials. As a result, there is a great improvement in the diffusion of dye molecules leading to improvement in rate of dyeing (Gorjanc et al., 2020). Hence, properties such as intensity of color and fastness to washing can be enhanced for several fabrics such as cotton, polyamide, polyester, polypropylene, silk, and wool dyed with different classes of dyes. In addition, plasma treatment results in higher dye exhaustion levels leading to uniform dyeing properties. The amount of dye and water required for a desired shade are also reduced with a significant reduction in effluent load, cost of dyeing and environment impact. Some of the main reasons for the improvement of dyeability are improvement in capillary action, enhancement of surface area, reduction of crystallinity, development of reactive functional sites on the fibers, creation of roughness (etching), etc. (Haji and Naebe, 2020).

Four different types of plasma such as corona discharge, dielectric barrier discharge (DBD), atmospheric-pressure glow discharge (APGD), and atmospheric-pressure plasma jet (APPJ) can be used in the industry. For textile applications, low-temperature plasma has been found most suitable depending upon the fiber nature, processing type, speed, etc. (Zille et al., 2014). Therefore, low-temperature plasma can be used instead of the chemical method for the treatment of fibers such as polyester fibers. By using argon and air plasma, an increase in color depth can be obtained for PET fabrics due to the increase of surface roughness and surface area.

In addition, the introduction of hydrophilic groups, induced by both reactive and chemically inert plasmas, may increase the water swelling capability and the affinity of PET fibers for dyes containing polar groups which also results in an increased K/S value of dyed specimens.

It has been demonstrated that, the wicking rate, absorbency and dyeability of cotton fiber fabrics can be substantially improved by treatment with O_2 plasma. This also reduces the time of dyeing (Sun and Stylios, 2004). In another study it has been found that the hydrophilicity and dyeability of bleached and mercerized cotton fabrics treated with low pressure plasma can also be greatly improved. In case of wool fibers, it can be observed that, its surface morphology determines the diffusion of dye molecules in the fibers. Due to the presence of a high number of disulfide cystine crosslinks (–S–S–) in the A-layer of the exocuticle, and of fatty acids on the fiber's surface, the fiber is hydrophobic in nature. It is observed that low-temperature plasma treatments help to improve the dyeing behavior of these fibers using different classes of dyes. Plasma induces oxidation of cystine linkages reducing the number of crosslinks and facilitates transcellular and intercellular dye diffusion.

10.7 SUPERCRITICAL CO_2 DYEING

Supercritical carbon dioxide ($ScCO_2$) is an alternative dyeing method that eliminates the use of water while achieving results comparable to current dyeing processes. It is an advanced technique that offers a water-free solution. The use of supercritical fluids as a solvent in the dyeing process has attracted considerable attention in the recent years. This dyeing method is referred to as waterless dyeing due to the use of supercritical carbon dioxide as a dyeing medium in place of water. Main benefits of this method are no use of water, lower consumption of energy, no drying after dyeing, and shorter dyeing times. The advantages of the fluid are both economic and ecological. It is used for dyeing synthetic fibers which replaces water for dyeing and also reduces the air effluents (Agrawal, 2020). A supercritical dyeing fluid can easily dissolve solid dyestuffs and penetrate even the smallest pores of the fibers. Supercritical carbon dioxide is a suitable solvent for organic dyes and the dyes become miscible with supercritical state of carbon dioxide and are transported to the fibers because carbon dioxide is available easily, and it is harmless to environment, non-toxic, and non-explosive (Miah et al., 2013).

Carbon dioxide is a nontoxic, chemically inert, nonflammable, and recyclable material (Abou Elmaaty and El-Aziz, 2017). Above its critical point (31.1°C and 73 atmospheres) (Figure 10.3), it becomes a remarkable solvent, displaying solvent properties similar to those of liquid hydrocarbons and can be used for many natural and synthetic dyes, which are utilized for the dyeing of cotton and various synthetic fibers. The dye solution dispersed in the supercritical CO_2 can be easily carried to the fiber to be dyed due to the gas-like diffusion of supercritical CO_2 which disperses the dye evenly into the minute interstices of the fiber (Bach et al., 2002). Auxiliary chemicals such as exhausting or leveling agents are not required during dyeing and washing-off step can be completely eliminated. In addition, left-over dyes can also be reused after dyeing. Low viscosity of supercritical carbon dioxide and its high

Waterless Dyeing Technology

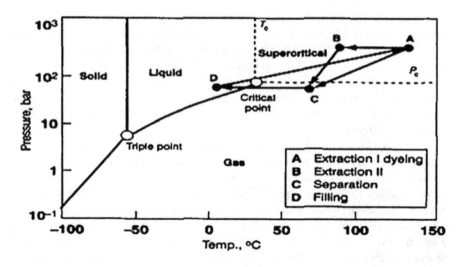

FIGURE 10.3 Pressure-temperature phase diagram for carbondioxide. (Adapted from Bach et al., 2002.)

diffusion rates with dissolved dye molecules are highly promising for the dyeing techniques (Agrawal, 2020).

In a machine (Figure 10.4) used for dyeing with SCCO$_2$ has a temperature controller, a stainless-steel dyeing vessel, a manometer (measuring the pressure of fluid), a pump for CO$_2$ and contains a cooler for cooling the head of carbon dioxide pump. The main advantage of using ScCO$_2$ is that, the dye after the process can easily be removed by removing the air pressure. In this way, CO$_2$ recycled up to 95%. Hence, waste is not generated in this process, and water is not consumed. The machine

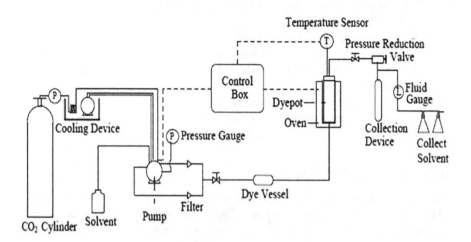

FIGURE 10.4 Schematic diagram of the SCCO$_2$ dyeing process.

is designed to minimize the volume of vessel, relative to the amount of fabric, so beam-dyeing principle was chosen. In this beam-dyeing process, instead of water, CO_2 is circulated with help of pump, dyes are filled in the separate vessel containing with pressure pump. Through a heat exchanger and finally through the dyeing vessel, CO_2 is circulated through the beam dyeing roll. During the circulation, the dye gets dissolved and diffuses into fibers due to high temperature (120°C) and pressure up to (250 bar). The penetration of CO_2 molecules into the polymer reduces its glass-transition temperature (T_g).

In this machine, two costly parts are dyeing vessel and centrifugal circulation pump. Other parts are a pressure vessel to pressurization, a regulating valve to lower the pressure inconstantly, a chiller to CO_2 gas cooling, air-driven piston-cylinder pump to transfer liquid CO_2 to dyeing vessel, a dye vessel to kept in require dye powder with contain a filter to prevent stain formation, a shell and tube heat exchanger for heating $ScCO_2$. Another separate vessel is provided with a steam jacket to enable vaporization of CO_2 to place and to prevent entrainment of dye particles into this storage vessel (Nalan, 2015). An air-driven gas booster is provided to lower the pressure of dyeing vessel up to four bars at the end of the dyeing process.

Supercritical dyeing requires higher pressure, and it becomes necessary to use autoclaves with large holding capacities above temperature 120°C. Normally, the excess dye removed after dyeing of CO_2 is converted to gaseous form so that the dye particles precipitate, and by pumping, clean CO_2 can be recirculated back to the dyeing vessel. Later, the fabric can be rinsed with acetone to remove residual dye particles (Agrawal, 2020).

10.7.1 Advantages of SC-CO$_2$ Dyeing

1. Organic dyes which are slightly polar or nonpolar in nature can be easily dissolved in super critical CO_2 and can be used as a medium of dyeing in place of water for effective transfer and penetration of dye particles in to the core of fibers.
2. This dyeing technique is comparatively simple because exhaustion and diffusion of dye, removal of excess dye, recycling of the dye and carbon dioxide can all be carried out only by changing the conditions such as temperature and pressure.
3. 90% of the carbon dioxide can be recovered and recycled after each dyeing process no water is used in this process and all the carbon dioxide is recovered in its gaseous state at the end of the dyeing process.

10.8 ULTRASONIC DYEING

Use of the ultrasound technique for dyeing of textile materials was started way back in 1941. Ultrasonic frequencies lie between 20 kHz and 500 MHz, and ultrasound is an oscillating sound pressure wave with a frequency greater than the upper limit of the human hearing range. In other words, ultrasound energy refers to sound waves with frequencies above 20,000 oscillations per second. The acoustic waves are generated when this high-intensity ultrasound is subjected to the liquid medium. When these high-frequency ultrasonic waves are absorbed in liquid system, they cause the formation of microscopic bubbles, or cavitations resulting in enormous heating of the liquid when used in liquids. The formation of cavities depends upon several factors such as the frequency and intensity of waves, temperature, and vapor pressure of liquids. Tiny, powerful shock waves are created when these bubbles oscillate and collapse generating higher pressure and higher temperature of about >1000 atm and 5000°C.

Different configurations of whistles, magnetostrictive and piezo- electric transducers, hooters, and sirens have been used to generate the ultrasonic waves. While the working of sirens and whistles permits efficient transfer of the ultrasonic sound to the ambient air, magnetostrictive and piezo-electric transducers will only produce low oscillation amplitudes, which are difficult to transfer to gases.

10.8.1 Equipment for Ultrasonics

The main parts of an ultrasound equipment are a generator and a converter (cleaning bath). The main function of the generator is to convert 50–60 Hz alternate current to high-frequency electrical energy. This electrical energy is fed to the transducer where it is transformed to mechanical vibration. The transducer changes this electrical high-frequency signal from the generator into ultrasonic sound waves. It vibrates longitudinally transmitting waves into liquid medium, and as these waves propagate, cavities are formed. Prototype dyeing machines which have been designed for continuous dyeing of yarns and fabrics generally consist of units such as ultrasonic tank, transport system, and a microprocessor to monitor the process parameters. The ultrasonic tank can have a dimension of 92 × 60 cm with the help of a thermostat, and the temperature can be varied up to 100°C.

10.8.2 Dyeing Process

Various factors like radiation pressure, streaming, cavitations, heat, agitation, interface instability, friction, diffusion, and mechanical rupture affect the efficiency of ultrasonic sound energy. Important steps in ultrasonic-assisted dye adsorption of an adsorbate onto the adsorbent are intra-particle transport, film diffusion, and pore diffusion. Ultrasonic waves cause increased swelling, reduction in Tg of fibers, and an increase in the fiber/dye partition coefficient. They enhance the transfer of dye particles to the fiber surface and accelerate the rate of diffusion of dye inside the fiber by reducing the boundary layer. The dyeability is improved due to break up of micelles and high-molecular weight aggregates into uniform dispersions in the dye bath.

10.8.2.1 Advantages

The main advantages of ultrasonic dyeing are savings in energy, enhancement in dyeing process, reduction in use of auxiliary chemicals, and processing times. The dyeing process improves the product quality substantially and is highly beneficial in the dyeing of hydrophobic dyes such as insoluble disperse dyes on synthetic fibers. It is a low-temperature dyeing process and with a reduction in effluent load.

10.8.2.2 Disadvantages

The difficulty of this process is production of high intensity and uniform ultrasound in a large vessel.

10.9 MICROWAVE-ASSISTED DYEING

Microwave radiations are electromagnetic waves having frequencies between 103 and 106 MHz. Microwave heating is an alternate method to conventional heating due to its rapid, efficient, and homogeneous heating characteristics. Since microwaves can easily penetrate the material particles, heating occurs instantaneously and consistently eliminating the problems in conventional heating systems. The dye exhaustion, dye affinity, and dyeing rate can be improved due to induced microwave irradiation which influences the kinetic energy of the water and dye molecules in most of the dyeing processes. In addition, microwave irradiation causes the dye molecules to collide with the fibers at a fast rate resulting in improvement in color strength of dyed goods in a shorter time period compared to the conventional pad-batch dyeing method (Khattab et al., 2017, Kim et al., 2003, Öner et al., 2013, Xu and Yang, 2002).

Several studies indicate that microwave dyeing gives better results. It was reported that there was a substantial reduction in dyeing time, about 75% reduction in salt consumption, and 20% reduction in alkali consumption when cotton fabrics were dyed using reactive dyes by batch-wise method using microwave irradiation compared to conventional methods (Khatri et al., 2015; Haggag et al., 2014). Study on dyeing of flax fabrics with reactive dyes by the pad-dry-bake method using microwave radiation reveals that when the fabric is soaked in urea and then baked in microwave gives better results in terms of major improvement in its dyeability. This was due to the significant role played by microwave irradiation resulting in increased color depth (L^*) in comparison to the microwave heating time. Microwave heating causes urea to melt and embed within the flax fibers. Reactive dyes were then adsorbed by the molten urea in the padding step promoting better fixation within the substrate. As observed in SEM analysis, the microwave heating step modifies the surface morphology of the flax fibers, creating a rough surface resulting in better adsorption and diffusion of dye molecules by the flax fibers during the padding step. Hence, greater dye exhaustion and color depth (L^*) were achieved on the dyed substrate. Interestingly, there was no change in crystallinity and tensile strength in urea/microwave-treated fabric (Sun et al., 2005). In another study on microwave-assisted dyeing of wool fibers, it was observed that the dye uptake and diffusion coefficient of treated wool were improved after microwave pretreatment, but the adsorption behavior was found to be constant. The study also reported that the reduction of the concentration of S–S bonds in keratin and the surface damage were the main causes for improvement in the absorption of dye molecules by the wool fibers (Xue, 2016).

Microwave-assisted dyeing of synthetic fibers has been reported. Polyester fabrics that were impregnated in aqueous solutions of urea and sodium chloride for 10 minutes and then dyed for 7 minutes by microwave apparatus (2450 MHz, 700 W) under optimum conditions showed better dye exhaustion (Kale and Bhat, 2011). It was found that solvents such as n-hexane, acetone, dimethylformamide, and aqueous solutions of urea and sodium chloride as padding solutions were more effective than 100% water for microwave dyeing. Similarly, microwave-assisted dyeing of PBT fibers (Öner et al., 2013) results in better color depth and K/S as compared to the conventional dyeing methods. There was a drastic reduction in total dyeing time resulting in good color fastness and tensile properties for dyed samples.

10.10 COMPARISON BETWEEN CONVENTIONAL AND WATERLESS DYEING TECHNIQUES

10.10.1 Conventional Dyeing

- Conventional dyeing processes consume large quantities of water and produce highly polluted dye wastewaters that contain residual dyes and chemicals, contributing to 15%–20% of the total wastewater flow that must be subjected to expensive effluent treatment processes prior to their discharge into rivers.
- Conventional textile dyeing technology is highly water and energy intensive in terms of pretreatment, dyeing, and post-treatment (drying) operations.
- Conventional dyeing processes involve prolonged process duration of up to 3–4 hours per batch.

10.10.2 Waterless Dyeing

- In waterless dyeing techniques, no water or very less water is required for most of the dyeing processes.
- Reduction in waste water resulting in elimination of wastewater discharges and wastewater treatment processes.
- Reduction in energy consumption with only about 20% energy requirements.
- Significant reduction in dyeing times: The total time required for entire dyeing and washing varies between 15 and 60 minutes.
- Most processes eliminate separate drying of fabrics after dyeing.
- The utilization of dyes is very high, and the formation of residual is very less. The unused dye can be easily recovered and reused.
- Dyeing occurs with high degree of evenness.

10.11 CONCLUSION

The textile industry is considered one of the most water-dependent industries and is also the backbone of many developing economies. The introduction of new technologies such as waterless dyeing techniques, bio-processing, digital printing, etc. can lead to a great saving in water consumption and help mills remain competitive while reducing their dependence on water and contributing to improve the environment. Depending on the type of dye used, the dyeing section consumes water to the extent of 30–50 L/kg of cloth and about 60 L/kg of yarn contributing to about 15%–20% of the total wastewater produced. There lies a great responsibility on governments,

NGOs, and consumers to apply a proactive approach to reducing the environmental footprint of this industry to protect the environment. The textile around the world industries will face severe regulations with a new set of legislation imposed on them by the governments to reduce water pollution. There is an urgent need to adapt sustainable dyeing practices to reduce the environmental impact. In this direction, the abovementioned waterless dyeing technologies can lead to the saving of large quantities of freshwater, at the same time reducing the water pollution due to prevention of wastewater disposal into freshwater resources.

REFERENCES

Abou Elmaaty, T. and Abd El-Aziz, E. (2017). "Supercritical carbon dioxide as a green media in textile dyeing: A review", *Textile Research Journal*, 88(10): 1184–1212.

Agrawal, B.J. (2020). "Supercritical carbon-dioxide assisted dyeing of textiles: an environmental benign waterless dyeing process", *International Journal of Innovative Research and Creative Technology*, 1(2): 201–206.

Alexander, A.F. (2008). *Plasma Chemistry*, Cambridge University Press, New York.

Anon (1999). International Dyer, 184, November.

Arputharaj, A., Raja, A.S.M., and Saxena, S. (2016). Developments in sustainable chemical processing of textiles. In: Muthu, S. and Gardetti, M. (eds), *Green Fashion. Environmental Footprints and Eco-design of Products and Processes*, 217–252, Springer, Singapore.

Babu, K.M. and Anekonda, S.M. (2019). https://www.textileschool.com/5234/air-flow-dyeing-an-eco-friendly-water-preserving-fabric-dyeing-technology/2/ May 10.

Bach, E., Cleve, E., and Schollmeyer, E. (2002). "Past, present and future of supercritical fluid dyeing technology an overview", *Review of Progress in Coloration and Related Topics*, 32(1): 88–102.

Bechtold, T., Burtscher, E., Turcanu, A., and Bobleter, O. (1997). "Dyeing behavior of indigo reduced by indirect electrolysis", *Textile Research Journal*, 67(9): 635–642.

Bechtold, T. and Turcanu, A. (2009). "Electrochemical reduction in vat dyeing: greener chemistry replaces traditional processes", *Journal of Cleaner Production*, 17(18): 1669–1679.

Bechtold, T., and von Farbstoffen, V.R. (1993). European patent WO 90/15182, US patent 5 244 549, September 14.

Chavan, R.B. and Chakraborty, J.N. (2001). "Dyeing of cotton with indigo using iron (II) salt complexes", *Coloration Technology*, 117(2): 88–94.

Davies, N. (2016). "The sustainability of waterless dyeing", *AATCC Review*, 16(1): 36–41.

Ghaly, A.E., Ananthashankar, R., Alhattab, M., and Ramakrishnan, V.V. (2014). "Production, characterization and treatment of textile effluents: a critical review", *Journal of Chemical Engineering & Process Technology*, 5(1): 1–18.

Gorjanc, M., Bukosek, V., and Gornsek, M. (2020). "Influence of water vapour plasma treatment on specific properties of bleached and mercerized cotton fabric", *Textile Research Journal*, 80(6): 557–567.

Haggag, K., El-Molla, M.M., and Mahmoued, Z.M. (2014). "Dyeing of cotton fabrics using reactive dyes by microwave irradiation technique", *Indian Journal of Fibre& Textile Research*, 39(4): 406–410.

Haji, A. and Naebe, M. (2020). "Cleaner dyeing of textiles using plasma treatment and natural dyes: a review", *Journal of Cleaner Production*, 265: 121866.

Hussain, T. and Wahab, A. (2018). "A critical review of the current water conservation practices in textile wet processing", *Journal of Cleaner Production*, 198: 806–819.

Kale, M.J. and Bhat, N.V. (2011). "Effect of microwave pretreatment on the dyeing behaviour of polyester fabric", *Coloration Technology*, 127(6): 365–371.

Kant, R. (2016). "Textile dyeing industry an environmental hazard", *Natural Science*, 4(1): 22–26.

Karthik, T. and Gopalakrishnan, D. (2014). Environmental analysis of textile value chain: an overview. In: Muthu, S.S (ed), *Roadmap to Sustainable Textiles and Clothing, Textile Science and Clothing Technology*, 153–188, Springer Science and Business Media, Singapore.

Khatri, A., Peerzada, M.H., Mohsin, M., and White, M. (2015). "A review on developments in dyeing cotton fabrics with reactive dyes for reducing effluent pollution", *Journal of Cleaner Production*, 87: 50–57.

Khattab, T.A., Elnagdi, M.H., Haggaga, K.M., Abdelrahmana, A.A., and AbdelmoezAly, S. (2017). "Green synthesis, printing performance, and antibacterialactivity of disperse dyes incorporating arylazopyrazolopyrimidines", *AATCC Journal of Research*, 4(4): 1–8.

Kim, S.S., Leem, S.G., Do Ghim, H., Kim, J.H., and Lyoo, W.S. (2003). "Microwaveheat dyeing of polyester fabric", *Fibers and Polymers*, 4(4): 204–209.

Kulandainathan, M., Patil, K., Muthukumaran, A., and Chavan, R. (2007). "Review of the process development aspects of electrochemical dyeing: its impact and commercial applications", *Coloration Technology*, 123(3): 143–151.

Kumar, P.S. and Gunasundari, E. (2018). Sustainable wet processing - An alternative source for detoxifying supply chain in textiles. In: Muthu, S. (ed), *Detox Fashion, Textile Science and ClothingTechnology*, 37–60, Springer Nature, Singapore.

Li, X., Wang, K., Wang, M., Zhang, W., Yao, J., and Komarneni, S. (2020). "Sustainable electrochemical dyeing of indigo with Fe (II)-based complexes", *Journal of Cleaner Production*, 276(3): 123251.

Merk et al. (2004). US Patent 20040069653, April.

Merk et al. (2003). US Patent Application 20030098246, May.

Miah, L., Ferdous, N., and Azad, M.M. (2013). "Textiles material dyeing with supercritical carbon dioxide (CO_2) without using water", *Chemistry and Materials Research*, 3(5): 38–40.

Nalan, D. (2015)."Waterless Dyeing (Review)", Annals of The University of Oradea Fascicle of Textiles, Leatherwork, 23.

Nieminen, E., Linke, M., Tobler, M., Beke, and Vander, B. (2007). "EU COST action 628: life cycleassessment (LCA) of textile products, eco-efficiency and definition of best availabletechnology (BAT) of textile processing", *Journal of Cleaner Production*, 15: 1259–1270.

Öner, E., Büyükakinci, Y., and Sökmen, N. (2013). "Microwave-assisted dyeing ofpoly (butylene terephthalate) fabrics with disperse dyes", *Coloration Technology*, 129(2): 125–130.

Roessler, A. and Jin, X. (2003). "State of the art technologies and new electrochemical methods for the reduction of vat dyes", *Dyes and Pigments*, 59(3): 223–235.

Sahoo, A. and Gupta, K.K. (2007). Asian Dyer, April.

Sala, M. and Gutiérrez-Bouzán, M. (2012). "Electrochemical techniques in textile processes and wastewater treatment", *International Journal of Photoenergy*, 2012: 1–12.

Senthil Kumar, R., FirozBabu, K., Noel, M., and AnbuKulandainathan, M. (2009). "Redox mediated electrochemical method for vat dyeing in ferric-oxalate-gluconate system: process optimization studies", *Journal of Applied Electrochemistry*, 39(12): 2569.

Sun, D. and Stylios, G.K. (2004). "Effect of low temperature plasma treatment on the scouring and dyeing of natural fabrics", *Textile Research Journal*, 74(9): 751–756.

Sun, H., Lin, L., Jiang, X., and Bai, X. (2005). "The improvement of dyeability of flax fibre by microwave treatment", *Pigment & Resin Technology*, 34(4): 190–196.

Xu, W. and Yang, C. (2002). "Hydrolysis and dyeing of polyester fabric usingmicrowave irradiation", *Coloration Technology*, 118(5): 211–214.

Xue, Z. (2016). "Study of dyeing properties of wool fabrics treated with microwave", *The Journal of The Textile Institute*, 107(2): 258–263.

Zaidy, S., Vacchi, F., Umbuzeiro, G., and Freeman, H. (2019). "Approach to waterless dyeing of textile substrates - use of atmospheric plasma", *Industrial & Engineering Chemistry Research*, 58(40): 18478–18487.

Zille, A., Oliveira, F., and Souto, A. (2014)."Plasma treatment in textile industry", *Plasma Processes and Polymers*, 12(2): 98–131.

11 Nonaqueous Dyeing of Textile Materials

Sushant S. Pawar, Mohammad Shahid, and Ravindra V. Adivarekar
Institute of Chemical Technology

CONTENTS

11.1 Introduction 189
11.2 Nonaqueous Dyeing of Textiles 191
    11.2.1 Supercritical Carbon Dioxide (Sc-CO_2) Dyeing 191
    11.2.2 Nonaqueous Solvent Dyeing 194
11.3 Conclusion and Future Outlook 198
Acknowledgment 198
References 198

11.1 INTRODUCTION

The textile industry, particularly textile wet-processing, is a massive consumer of water and one of the major contributors of industrial effluents [1]. Water used in wet processing procedures is contaminated by a range of chemicals as a result of the use of various chemicals in diverse industrial processes such as pretreatment, dyeing, printing, and finishing, and this effluent, if discharged untreated, pollutes the environment. Water is an essential component in textile wet processing. As the world's population grows, so does the demand for textile materials. As a result, the production of textile materials rises. Conventional processes and machines use a lot of water, especially when it comes to pretreatment and coloration of textiles. Among various wet processing stages in the textile production chain (Figure 11.1), dyeing is one of the most water-intensive, polluting, and energy consuming steps in the textile manufacturing process [2,3]. The water requirement for the processing of fibers depends upon the type of fibers. Operations such as spinning, weaving, and garmenting use very little water. However, wet textile processing, such as preparatory operations, dyeing, and finishing, requires a large volume of water. A considerable amount of water is also needed for steam drying, boilers, ion exchange, and cleaning in addition to these operations [4]. With many types of fibers, dyes, chemicals, and machineries used in wet processing, estimating the amount of water required is difficult. However, dyeing textile fibers typically requires 20%–30% of total water utilized in textile wet processing, whereas preparatory operations often require 38%–50% [5]. Traditional aqueous dyeing of textiles is estimated to use approximately 100–150L of water to dye 1 kg of fiber [6]. Approximately 280,000

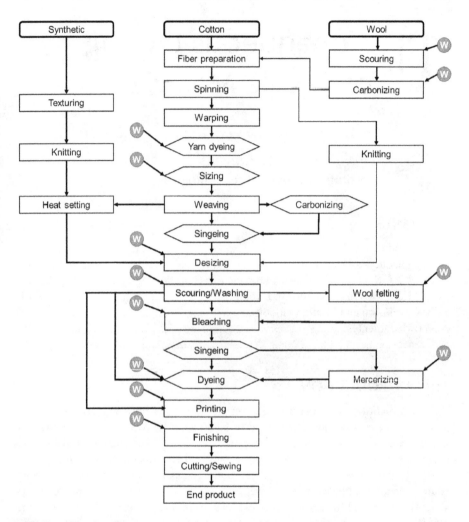

FIGURE 11.1 Flowchart for the general steps in textile fabrication [2,3].

tons of textile dyes are discharged in wastewater industrial effluents each year, posing a direct threat to aquatic life and human health [7].

Consumer environmental awareness, stricter environmental legislation, and global water scarcity have compelled the textile industry to assess, restructure, and mitigate its water use and associated effluent risks [7]. The textile industry is currently the world's second-largest polluter of clean drinking water, and it should embrace new technologies that use little or no water [8]. In recent years, the textile industry has taken steps to reduce pollution and waste by implementing lower liquor-ratio dyeing equipment, low/no-water textile processes, and the replacement or abandonment of hazardous chemicals [9]. Researchers are actively working to improve the conventional water-based dyeing system by reducing water and chemical consumption while increasing dyeing and fixation efficiency and efficacy and developing alternative dyeing systems by switching the dyeing medium from an aqueous to a nonaqueous system to mitigate the ecological problems arising from conventional aqueous dyeing [10].

11.2 NONAQUEOUS DYEING OF TEXTILES

Although water is the most frequent dissolving medium for dyeing textiles, nonaqueous dyeing of textiles has been extensively researched with the goal of reducing water usage and wastewater minimization in textile manufacturing. In recent years, dyeing with nonaqueous media, such as air, supercritical CO_2 fluid, and liquid nonaqueous solvent, has received a lot of interest [11]. The primary focus of these nonaqueous approaches to dyeing has been the application of disperse dyes to polyester, although other classes of dye and other fiber types have also been considered.

11.2.1 Supercritical Carbon Dioxide (Sc-CO$_2$) Dyeing

Supercritical fluids, with properties intermediate between liquid and gas, are defined as substances heated and pressurized above their critical temperature and pressure, where liquid and vapour states become indistinguishable [12]. CO_2 has a low critical temperature and pressure of 31°C and 7.4 MPa, making it one of the most economically feasible and widely used supercritical fluids. CO_2 in a supercritical state has low diffusivity and low viscosity. These properties of s-CO_2 help to acquire high solvating power by varying temperature and system pressure. The solubility of disperse dye in Sc-CO_2 is moderate; however, it can be increased by adding cosolvents such as acetone and methanol [13]. Supercritical CO_2(Sc-CO_2) dyeing is one of the approaches to dyeing for textile materials to reduce the consumption of water in the textile sector. Water-free dyeing technology such as s-CO_2 dyeing can reduce a significant amount of water, chemicals, and energy. A large amount of freshwater, energy, chemical usage, and a huge burden of effluent discharge are the primary concerns for textile dyeing industries, which could be avoided by using s-CO_2. Sc-CO_2 is a viable alternative to organic solvents and traditional aqueous textile dyeing because it is a cost-effective, nontoxic, nonflammable, and green solvent with distinct physical properties. In many industrial applications, environmentally safe Sc-CO_2 has been offered as a replacement for water and other solvents [14–17]. The success of SC-CO_2 in manufacturing has stimulated research in using it in textile processing, particularly dyeing.

The swelling and plasticizing effect of hydrophobic fibers and the dissolution of disperse dye in Sc-CO_2 make it suitable for dyeing of hydrophobic fibers such as polyester and other synthetic fibers [18]. Since the very beginning of Sc-CO_2 dyeing of textiles, dyeing of polyester has been the focus of numerous investigations all over the world [19]. Generally, the solubility of disperse dye in Sc-CO_2 is moderate, and it varies between 10^{-4} and 10^{-7} mole fractions under dyeing parameters 120°C–140°C at 30 MPa [20]. The dissolution of disperse dyes in Sc-CO_2 is dependent on the particle size of the dye [21]. Except for the dye solubility, the dyeing process of Sc-CO_2 is fast and simple. Because of the low viscosity, high diffusivity, and faster homogeneous penetration of CO_2 into the fiber, dissolved dyes can achieve a uniform shade in a short period of time. At higher pressure, CO_2 helps to decrease the T_g of polyester fabric [22]. Dyeing auxiliaries such as carriers, dispersion agents, and surfactants are not usually required in Sc-CO_2 dyeing. After dyeing polyester fabric, there is no need to dry it because the residual dye and CO_2 may be recovered [23,24]. The use of Sc-CO_2 for disperse dyeing of polyester fabrics has been investigated by many researchers [17,19,20,25]. The first commercial Sc-CO_2 dyeing was carried out by DyeCoo Textiles Systems B.V. [26–28]. The use of Sc-CO_2 in textile processes would have a significant impact on freshwater consumption and wastewater discharge. The Sc-CO_2 dyeing process would be more beneficial to reduce the consumption of harmful dyeing auxiliaries. Figure 11.2 depicts a simplified schematic of a typical Sc-CO_2 textile dyeing process that demonstrates the basic operation processes [29]. Due to technological advancements in recent years, the Sc-CO_2 of polyester has reached industrial scale, with multiple plants running in different parts of the world [17,30]. There is now a large amount of experimental data available on the uptake and solubility of many dyes, especially disperse dyes, for Sc-CO_2 dyeing [31–36]. Although Sc-CO_2 dyeing of PET is now commercially viable due to tremendous efforts in understanding the dyeing mechanism and resolving technological issues associated with industrialization [17,19], research is still ongoing aimed at identifying suitable dyes for improved coloration [37–39], procedure optimization [40–42], uniform dye distribution [43], oligomer removal [44,45], establishment of efficient color matching system [46,47], and novel technologies [48].

The Sc-CO_2 dyeing has also been extended to other types of textiles such as nylon [49–52], polypropylene [53], polylactide [54], acrylic [49,55], aramid [29], cotton [54,56,57], wool [52,54,58], silk [52,54,59], etc. Here are a few examples of representative reports:

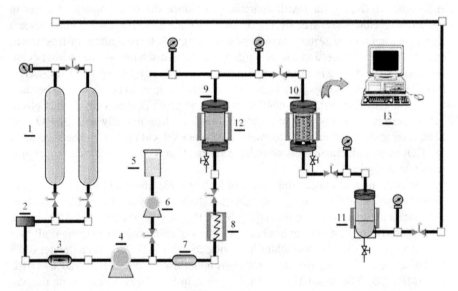

FIGURE 11.2 Schematic diagram of the supercritical carbon dioxide dyeing apparatus. (1) CO_2 cylinder, (2) purifier, (3) refrigerator, (4) high-pressure pump, (5) cosolvent tank, (6) cosolvent pump, (7) mixer, (8) heat exchanger, (9) dye vessel, (10) dyeing vessel, (11) separator, (12) heat compensating jacket, and (13) computer. (Reproduced with permission from Ref. [29].)

Zheng et al. [29] investigated the dyeing properties of meta-aramid textiles in Sc-CO_2 using dimethyl terephthalate, ethyl alcohol, and CINDYE DNK as carriers and presented the functional mechanism of carriers in supercritical carbon dioxide carrier. Improved dyeability of meta-aramid fabrics with Disperse Blue Black 79, Disperse Rubine H-2GL, and Disperse Yellow EC-3G may be achieved by adjusting the dyeing temperature, dyeing pressure, dyeing time, dye concentration, carrier concentration, and carbon dioxide flow. The carriers chosen in supercritical carbon dioxide could improve the dyeing properties of meta-aramid samples by intensifying the movement of macromolecular chains, lowering the glass transition temperature of the fiber, and increasing the diffusibility of dye molecules into the amorphous phase of the fiber. The meta-aramid samples dyed in supercritical carbon dioxide had good fastness toward washing, rubbing, and light. Furthermore, carriers had minimal effect on the flammability of meta-aramid. Penthala et al. [50] synthesized a series of azo and anthraquinone dye derivatives from low-cost starting materials and used these dyes in primary dyeing experiments on nylon 6,6 fabric to confirm the feasibility of scaling up Sc-CO_2 dyeing to the factory level. All of the dyes had good color strength values. The effect of dye concentration on color strength values was also explained. The results of the thermal analysis showed that the Sc-CO_2 had no effect on the nylon 6,6 fabric. The dyed fabrics demonstrated excellent fastness toward washing and sweat fastness, as well as good to moderate light fastness. Elmaaty et al. [60] developed a simple and effective method for dyeing polypropylene fiber with hydrazonopropane nitrile disperse dyes in a water and sc-CO_2 medium. The researchers looked at the effects of dyeing parameters such as dye concentration, temperature, time, and pressure in both water and Sc-CO_2 media. In both water and sc-CO_2, the dyes demonstrated a strong affinity for polypropylene fabrics. The color strength and fixation of dyed polypropylene fibers in Sc-CO_2 were significantly higher than in water. The fastness properties of dyed samples in Sc-CO_2 (washing, rubbing, light, and sublimation) were comparable to those of water. Wen and Dai [61] successfully dyed polylactide fibers with Disperse Blue 79 in Sc-CO_2. Temperature, time, and pressure in the dyeing process all had an impact on the fibers' dyeing qualities. The dye sorption increased as the dyeing temperature increased. The dye uptake of the PLA fiber increased before 17 MPa but decreased when the pressure exceeded 17 MPa. Sc-CO_2 dyeing exhibited a faster rate of dyeing, lesser fiber damage, and comparable fastness toward light and washing to those of water. Zheng et al. [62] investigated the feasibility of Sc-CO_2 dyeing acrylic fibers on an industrialized unit, as well as the effect of dyeing temperature, dyeing pressure, dyeing time, and dye concentration on the K/S values of acrylic fibers. In the supercritical CO_2 dyeing process, the dyed acrylic fibers demonstrated good color fastness to washing, rubbing, and light, as well as commercially acceptable levelness and reproducibility for acrylic fibers. Gao et al. [63] investigated the Sc-CO_2 solubilities of reactive disperse dyes synthesized with the reactive group of 1,3,5-trichloro-2,4,6-triazine dyeing cotton fabrics. The dyes' solubility increased with increasing pressure at the same temperature and decreased with increasing pressure at the same temperature. The polarity of the dye molecules also had an effect on the solubilities. Less polar molecules had higher solubility, while more polar molecules had lower solubility. They demonstrated the great effectiveness of reactive disperse dyes in the Sc-CO_2

dyeing of cotton fibers by comparing disperse dye with reactive disperse dye. Li et al. [64] proposed a dyeing mechanism for dyeing wool fibers in sc-CO_2 using Disperse Red 153 and Blue 148, and based on this mechanism, they designed and synthesized four novel dyes with similar structures: azo thiazole-OH, azo thiazole-OCH_3, azo thiazole-$N(CH3)_2$, and azo thiazole-NHS. Azo thiazole-$N(CH_3)_2$ and azo thiazole-NHS dyeing performed better than azo thiazole–OH and azo thiazole-OCH_3 dyeings with acceptable washing fastness. The fixation rate of azo thiazole-NHS neared 100%, demonstrating that adding a succinimidyl ester reactive group enhanced the fixing rate of disperse dyes. Yan et al. [59] synthesized a red-hued unique disperse reactive SCFX-AnB3L dye with three moieties: an anthraquinonoid matrix, a bridge group, and a dichlorotriazine reactive group, which they used in an ecological dyeing of silk in Sc-CO_2. Adsorption, uptake, and fixation behaviors were all excellent, with high color fastness ratings. Although dyeing of different types of fiber have been reported, Sc-CO_2 has found the most applicability in the dyeing of polyester with disperse dyes. Shorter dyeing times, lower water, energy, and dye effluent, greater productivity, dispersant-free dyeing, and CO_2 recovery are all considered as advantages of the technology over traditional aqueous dyeing [11].

11.2.2 Nonaqueous Solvent Dyeing

Textile wet processing, such as desizing, scouring, bleaching, mercerization, dyeing, and finishing, has traditionally employed solvents. As conventional water-based dyeing causes negative environmental impacts due to the large amounts of water and chemicals required, the use of solvents to assist dyeing is seen as a viable alternative for the textile industry to reduce effluent generation and avoid environmental challenges [10]. Solvent dyeing is one of the oldest approach to reduce the consumption of water and minimize generation of effluents in the textile dyeing sector. Solvent dyeing is a process that can be carried out in a nonaqueous phase, where a solvent is used as a dyeing medium instead of water, though a small amount of water may be required to aid the dyeing process [65–68].

Chlorinated hydrocarbon solvents such as trichloroethylene, perchloroethylene, and 1,1,1-trichloroethane were used as early nonaqueous dyeing solvents because they were relatively cheap, readily available, and nonflammable [69,70]. During the early period of the solvent dyeing process development, perchloroethylene was a preferred dyeing solvent due to wide commercial availability, convenient physical properties, acceptable stability to dyeing and finishing reagents, and more acceptable physiological properties than those of most of the possible alternatives [68,71,72]. Solvents for textile dyeing should ideally be low-cost, readily available, nontoxic, nonflammable, easily recoverable, noncorrosive, and inert to fabrics [69]. Perchloroethylene solvent dyeing of polyester fiber has been the focus of a lot of researches in this area during the 1970s and 1980s [65–75]. The optimism and enthusiasm following early development of nonaqueous solvent dyeing using perchloroethylene started waning soon after due to associated difficulties and environmental concerns [10,71]. This led to research efforts in identifying alternative solvent-based dyeing processes. The Dacsol process (J. & P. Coats Ltd), a totally nonaqueous bicomponent homogeneous system (a perchloroethylene and a silicone fluid), was developed principally for dyeing polyester fiber with existing disperse dyes enabling fastness levels to be maintained [66]. The Dacsol process included four key steps: dyeing, after-cleaning (removal of unfixed dye and residual solvents), drying, and solvent recovery. The dyeing technique was similar to traditional high-temperature aqueous dyeing, but dyebath preparation was emphasized, and a nonaqueous medium was required for subsequent cleaning processes in order to get the most out of the process. The efforts in finding suitable solvent systems for dyeing of different types of fabrics are still in progress. In the last few years, many reports appeared on finding novel solvent-based dyeing processes. Some representative reports are discussed below:

Xu et al. [76] developed a new hydrolysis-free solvent dyeing process for PLA fabrics using liquid paraffin as a nonaqueous dyeing medium and compared it with aqueous dyeing. Disperse Blue 56 and Disperse Blue 79 were chosen as the representative low and high energy dyes for anthraquinone and azo classes, respectively. At optimized dyeing time and temperature, high-quality dyed PLA fabrics were obtained having excellent color consistency and colorfastness comparable to that of aqueously dyed fabrics. Minimal strength loss of dyed fibers was achieved by postheat setting treatment. Wang et al. [77] employed a nonionic surfactant reverse-micellar method with a poly(ethylene glycol)-based surfactant to study the dyeability of cotton fabrics with reactive dye in nonaqueous alkane medium of heptane and octane (Figure 11.3).

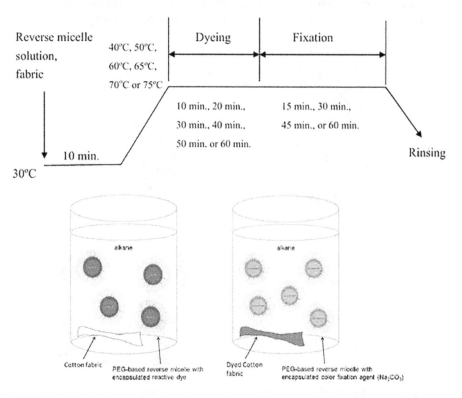

FIGURE 11.3 Dyeing cotton in alkane solvent using polyethylene glycol-based reverse micelle as a reactive dye carrier. (Reprinted with permission from Ref. [77].)

Under the optimized process conditions, the reactive dye was successfully encapsulated in the reverse micelle cavity. The use of octane and heptane does not affect surface of the cotton fibers. Cotton fabric swells in reverse micelle solution. Cotton fiber and reactive dye molecules show a strong reduction ionization effect in PEG-reverse micelle solution due to this better rate of adsorption of concentrated reactive dye observed in PEG-based reverse micelle solution than the water without electrolyte. Higher color strength was achieved using optimized concentration of dye and soda ash. The absorption capacity of dye in alkane media was found comparable than the water system. The alkane-assisted dyeing procedure has improved dyeability than traditional water-based dyeing without the use of electrolytes.

Gao et al. [78] prepared organic silicone nanomicelle microemulsions containing highly dispersed dyes for dyeing of PET fabrics at low liquor ratio. Nanomicelle dyeing had higher color yield, exhaustion, and levelling properties, as well as good fastness and softness properties compared to traditional dyeing. During the dyeing process, the dyes release slowly from the nanomicelles to PET fabrics and improve the levelling property (Figure 11.4). The organic silicon also served as a softening agent, adsorbing on the fabric's surface to provide a soft handle.

Chen et al. [79] demonstrated recyclable reactive dyeing by switching from water to a nonnucleophilic solvent system. They investigated the swelling of cotton by nonnucleophilic organic solvents systematically and concluded that by ensuring sufficient swelling, fully recyclable reactive dyeing could be realized in a mixture of

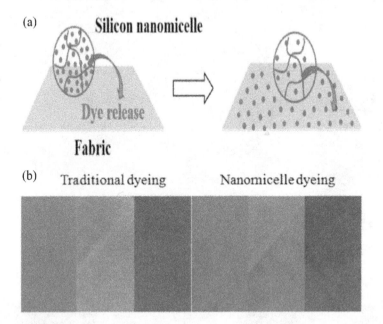

FIGURE 11.4 (a) Mechanism of dye releasing from the nanomicelles onto fabric during dyeing. (b) PET fabrics dyed by the nanomicelle dyeing and the traditional dyeing. (Adapted with permission from Ref. [78].)

nonnucleophilic solvents, maximizing material utilization and minimizing wastes for reactive dyeing of cotton. Zhao et al. [80] achieved recyclable reactive dyeing of cotton fabrics in a heterogeneous nonaqueous binary solvent system composed of ethyl octanoate and dimethyl sulfoxide as exhaustion medium and solvating medium, respectively. The solvents were immiscible at room temperature, but when heated and agitated, they formed a continuous phase, allowing for simple separation after dyeing. When compared to traditional aqueous dyeing, the solvent-based process used up to 40% less dye and no inorganic salts. A ten-cycle reuse sequence demonstrated excellent color consistency and fastness. At the end of each dyeing cycle, 98.5 v% EO and DMSO were recovered and reused. The method worked well with commercial monochlorotriazine dyes (Figure 11.5), resulting in consistent shade build-up and colorfastness.

Yuan et al. [81] employed ionic liquid (1-butyl3-methylimidazolium chloride) to improve the dyeability of wool. Initially, wool was treated with 1-butyl-3-methylimidazolium chloride, and then surface morphology, tensile strength, and wettability of ionic liquid treated wool were analyzed. Ionic liquid-treated wool showed improved wettability, dyeing behavior, rate of dyeing, and exhaustion. Washing and rubbing fastness of ionic liquid-treated wool showed comparable fastness rating; however, a slight strength loss was observed in ionic liquid-treated wool fabric. Andrade et al. [82] investigated an alternative method for dyeing cotton fabrics that used only two dyeing agents, a polyfunctional reactive dye Marinho Sidercron, containing a combination of triazines and vinyl sulfone groups, and a protic ionic liquid as an alternative solvent. The effectiveness of 13 protic ionic liquids as dyeing solvents was investigated in terms of conventional dyeing quality characteristics such as absorption color, tensile strength, and surface morphology of the cotton dyed. When compared to the aqueous procedure under the same operating conditions, the results achieved employing protic ionic liquids as dyeing solvents in the absence of

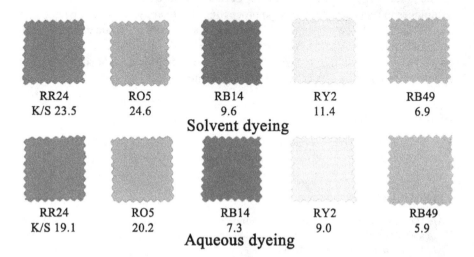

FIGURE 11.5 Photos of fabrics dyed with solvent dyeing and aqueous dyeing. (Reprinted with permission from Ref. [80].)

any auxiliary agent exhibited an outstanding effect. The PIL structure (ionic chains) appears to have an impact on dyeing quality. The organic residual chain in the ions has a strong influence on their thermodynamic properties, with steric hindrance being a critical factor for accommodation into a liquid net and, thus, access to active points of the cellulose fiber structure. Among many structural factors, such as anion chain length, bulk character of substitution, and cation degree of substitution, the latter appears to be the most important. PILs formed by monosubstituted amines outperformed those formed by disubstituted amines in terms of dyeing quality. Protic ionic liquids can be easily removed from the fabrics since they are infinitely soluble in water, a real advantage of the PILs when compared to aprotic ionic liquids which show low solubility in aqueous media. Since protic ionic liquids are infinitely soluble in water, they can be easily removed from fabrics, which is a major benefit over aprotic ionic liquids, which have minimal solubility in aqueous media. Eutectic solvents such as deep eutectic solvents (DES) and glycerin-based eutectic solvents (GES) have also been used as dyeing medium for polyester dyeing [83,84] and swelling agent during silk dyeing [85,86].

11.3 CONCLUSION AND FUTURE OUTLOOK

Today's world is shifting toward more sustainable solutions. Savings in terms of water, time, and temperature are extremely important in the modern textile industry. Dyeing of textile material is an essential value addition process in wet processing. Conventional dyeing processes require many chemicals such as dyestuff, acids, levelling agents, etc., which causes complex textile wastewater. Dyeing of textile materials requires a large amount of water. Two major challenges for the development of textile industry are large consumption of water and hazardous effluent. An eco-friendly approach needs to be developed to minimize consumption of water in textile processing sector. In the last few decades, researchers developed new dyeing techniques to replace water in textile processing. There are several new dyeing techniques that have low environmental impacts like supercritical carbon dioxide dyeing and nonaqueous solvent dyeing. For conservation of water, solvent dyeing could be an alternative approach to replace water in textile dyeing. In case of solvent dyeing, instead of water, solvent is used as a dyeing medium.

ACKNOWLEDGMENT

Mohammad Shahid acknowledges funding through UGC **DSKPDF** scheme (F.4-2/2006(BSR)/CH/18–19/0074), University Grants Commission (UGC), Govt. of India.

REFERENCES

1. Hasanbeigi, A. and L. Price, A technical review of emerging technologies for energy and water efficiency and pollution reduction in the textile industry. *Journal of Cleaner Production*, 2015. **95**: p. 30–44.
2. Volmajer Valh, J., et al., 4.20- Water in the Textile Industry, in *Treatise on Water Science*, P. Wilderer, Editor. 2011, Elsevier: Oxford. p. 685–706.

3. Bisschops, I. and H.J.E.T. Spanjers, Literature review on textile wastewater characterisation. *Environmental Technology*, 2003. **24**(11): p. 1399–1411.
4. Arputharaj, A., A.S.M. Raja, and S. Saxena, Developments in Sustainable Chemical Processing of Textiles, in *Green Fashion: Volume 1*, S.S. Muthu and M.A. Gardetti, Editors. 2016, Springer Singapore: Singapore. p. 217–252.
5. Raja, A.S.M., et al., 9- Water requirement and sustainability of textile processing industries, in *Water in Textiles and Fashion*, S.S. Muthu, Editor. 2019, Woodhead Publishing: Cambridge. p. 155–173.
6. Zheng, H., et al., An industrial scale multiple supercritical carbon dioxide apparatus and its eco-friendly dyeing production. *Journal of CO_2 Utilization*, 2016. **16**: p. 272–281.
7. Hussain, T. and A. Wahab, A critical review of the current water conservation practices in textile wet processing. *Journal of Cleaner Production*, 2018. **198**: p. 806–819.
8. Davies, N., The sustainability of waterless dyeing. *AATCC Review*, 2016. **16**(1): p. 36–41.
9. Vajnhandl, S. and J.V. Valh, The status of water reuse in European textile sector. *Journal of Environmental Management*, 2014. **141**: p. 29–35.
10. Tang, A.Y. and C.W.J.C.T. Kan, Non-aqueous dyeing of cotton fibre with reactive dyes: a review. *Coloration Technology*, 2020. **136**(3): p. 214–223.
11. Burkinshaw, S.M., *Physico-Chemical Aspects of Textile Coloration*. 2016: John Wiley & Sons: Hoboken.
12. Weibel, G.L. and C.K. Ober, An overview of supercritical CO_2 applications in microelectronics processing. *Microelectronic Engineering*, 2003. **65**(1): p. 145–152.
13. Reid, R.C., Prausnitz, J.M. and Poling, B.E. The properties of gases and liquids. U.S. Department of Energy Office of Scientific and Technical Information, United States, 1987.
14. Ahangari, H., et al., Supercritical fluid extraction of seed oils – A short review of current trends. *Trends in Food Science & Technology*, 2021. **111**: p. 249–260.
15. Nikolai, P., et al., Supercritical CO_2: properties and Technological Applications - A Review. *Journal of Thermal Science*, 2019. **28**(3): p. 394–430.
16. Santana, Á.L., D.T. Santos, and M.A.A. Meireles, Perspectives on small-scale integrated biorefineries using supercritical CO_2 as a green solvent. *Current Opinion in Green and Sustainable Chemistry*, 2019. **18**: p. 1–12.
17. Banchero, M., Recent advances in supercritical fluid dyeing. *Coloration Technology*, 2020. **136**(4): p. 317–335.
18. Knittel, D., W. Saus, and E. Schollmeyer, Water-free dyeing of textile accessories using supercritical carbon dioxide. *Indian Journal of Fibre & Textile Research*, 1997, **22**: p.184–189.
19. Banchero, M., Supercritical fluid dyeing of synthetic and natural textiles – a review. *Coloration Technology*, 2013. **129**(1): p. 2–17.
20. Bach, E., E. Cleve, and E. Schollmeyer, Past, present and future of supercritical fluid dyeing technology – an overview. 2002. **32**(1): p. 88–102.
21. Knittel, D., W. Saus, and E. Schollmeyer, Application of supercritical carbon dioxide in finishing processes. *The Journal of The Textile Institute*, 1993. **84**(4): p. 534–552.
22. Zhong, Z., S. Zheng, and Y. Mi, High-pressure DSC study of thermal transitions of a poly(ethylene terephthalate)/carbon dioxide system. *Polymer*, 1999. **40**(13): p. 3829–3834.
23. Montero, G.A., et al., Supercritical fluid technology in textile processing: an overview. *Industrial & Engineering Chemistry Research*, 2000. **39**(12): p. 4806–4812.
24. Saus, W., D. Knittel, and E. Schollmeyer, Dyeing of textiles in supercritical carbon dioxide. *Textile Research Journal*, 1993. **63**(3): p. 135–142.
25. Abou Elmaaty, T. and E. Abd El-Aziz, Supercritical carbon dioxide as a green media in textile dyeing: a review. *Textile Research Journal*, 2018. **88**(10): p. 1184–1212.

26. Xu, S., D. Shen, and P. Wu, Fabrication of water-repellent cellulose fiber coated with magnetic nanoparticles under supercritical carbon dioxide. *Journal of Nanoparticle Research*, 2013. **15**(4): p. 1577.
27. Xu, Y.-Y., et al., Water/oil repellent property of polyester fabrics after supercritical carbon dioxide finishing. *Thermal Science*, 2015. **19**(4); p. 1273–1277.
28. Zefirov, V.V., et al., Durable crosslinked omniphobic coatings on textiles via supercritical carbon dioxide deposition. *The Journal of Supercritical Fluids*, 2018. **133**: p. 30–37.
29. Zheng, H.-D., et al., Investigations on the effect of carriers on meta-aramid fabric dyeing properties in supercritical carbon dioxide. *RSC Advances*, 2017. **7**(6): p. 3470–3479.
30. Hendrix, W.A., Progress in supercritical CO_2 dyeing. *Journal of Industrial Textiles*, 2001. **31**(1): p. 43–56.
31. Guzel, B. and A. Akgerman, Solubility of disperse and mordant dyes in supercritical CO_2. *Journal of Chemical & Engineering Data*, 1999. **44**(1): p. 83–85.
32. Yamini, Y., et al., Solubilities of some disperse yellow dyes in supercritical CO_2. *The Journal of Chemical & Engineering Data*, 2010. **55**(9): p. 3896–3900.
33. Joung, S.N., K.-P.J.J.o.C. Yoo, and E. Data, Solubility of disperse anthraquinone and azo dyes in supercritical carbon dioxide at 313.15 to 393.15 K and from 10 to 25 MPa. *The Journal of Chemical & Engineering Data*, 1998. **43**(1): p. 9–12.
34. Bao, P., J.J.J.o.C. Dai, and E. Data, Relationships between the solubility of ci disperse red 60 and uptake on PET in supercritical CO_2. *The Journal of Chemical & Engineering Data*, 2005. **50**(3): p. 838–842.
35. Alwi, R.S. and C.J.C.P. Garlapati, A new semi empirical model for the solubility of dyestuffs in supercritical carbon dioxide. *Chemical Papers*, 2021. **75**(6): p. 2585–2595.
36. Shi, W., et al., Experimental determination and correlation of the solubility of a new reactive disperse orange SCF-AOL2 dye in pure supercritical carbon dioxide fluid. *Fluid Phase Equilibria*, 2018. **463**: p. 1–10.
37. Abate, M.T., et al., Colouration and bio-activation of polyester fabric with curcumin in supercritical CO_2: part I - Investigating colouration properties. *The Journal of Supercritical Fluids*, 2019. **152**: p. 104548.
38. Jaxel, J., et al., Facile synthesis of 1-butylamino- and 1,4-bis(butylamino)-2-alkyl-9,10-anthraquinone dyes for improved supercritical carbon dioxide dyeing. *Dyes and Pigments*, 2020. **173**: p. 107991.
39. Abou Elmaaty, T., et al., Green approach to dye PET and nylon 6 fabrics with novel pyrazole disperse dyes under supercritical carbon dioxide and its aqueous analogue. *Fibers and Polymers*, 2019. **20**(12): p. 2510–2521.
40. Abate, M.T., et al., Colouration and bio-activation of polyester fabric with curcumin in supercritical CO_2: Part II – Effect of dye concentration on the colour and functional properties. *The Journal of Supercritical Fluids*, 2020. **157**: p. 104703.
41. Hou, A. and J.J.C.t. Dai, Kinetics of dyeing of polyester with CI disperse blue 79 in supercritical carbon dioxide. *Coloration Technology*, 2005. **121**(1): p. 18–20.
42. Yiğit, İ., et al., An investigation of process parameters on colour during the dyeing of polyester in supercritical carbon dioxide media. *Coloration Technology*, 2021. https://doi.org/10.1111/cote.12553
43. Long, J.-J., Y.-Q. Ma, and J. P. Zhao, Investigations on the level dyeing of fabrics in supercritical carbon dioxide. *The Journal of Supercritical Fluids*, 2011. **57**(1): p. 80–86.
44. Prorokova, N., T.Y. Kumeeva, and L.J. Nikitin, Ethylene terephthalate oligomers in the processes of modification of polyester fabrics in supercritical carbon dioxide. *Russian Journal of Physical Chemistry B*, 2012. **6**(7): p. 827–833.
45. Kawahara, Y., et al., Oligomer deposition on the surface of PET fiber in supercritical carbon dioxide fluid. *Macromolecular Materials and Engineering*, 2006. **291**(1): p. 11–15.
46. Gong, D., et al., One-step supercritical CO_2 color matching of polyester with dye mixtures. *Journal of CO_2 Utilization*, 2021. **44**: p. 101396.

47. Wang, Y., et al., Waterless beam dyeing in supercritical CO_2: establishment of a clean and efficient color matching system. *Journal of CO_2 Utilization*, 2021. **43**: p. 101368.
48. Zheng, L., et al., Dyeing procedures of polyester fiber in supercritical carbon dioxide using a special dyeing frame. *Journal of Engineered Fibers and Fabrics*, 2015. **10**(4): p. 155892501501000414.
49. Bai, T., et al., Supercritical CO_2 dyeing for nylon, acrylic, polyester, and casein buttons and their optimum dyeing conditions by design of experiments. *Journal of CO_2 Utilization*, 2019. **33**: p. 253–261.
50. Penthala, R., et al., Synthesis of azo and anthraquinone dyes and dyeing of nylon-6, 6 in supercritical carbon dioxide. *Journal of CO_2 Utilization*, 2020. **38**: p. 49–58.
51. Liao, S., Y. Ho, and P.J.C.T. Chang, Dyeing of nylon 66 with a disperse-reactive dye using supercritical carbon dioxide as the transport medium. *Coloration Technology*, 2000. **116**(12): p. 403–407.
52. van der Kraan, M., et al., Dyeing of natural and synthetic textiles in supercritical carbon dioxide with disperse reactive dyes. *The Journal of Supercritical Fluids*, 2007. **40**(3): p. 470–476.
53. Abou Elmaaty, T., et al., Optimization of an eco-friendly dyeing process in both laboratory scale and pilot scale supercritical carbon dioxide unit for polypropylene fabrics with special new disperse dyes. *Journal of CO_2 Utilization*, 2019. **33**: p. 365–371.
54. Sawada, K., et al., Dyeing natural fibres in supercritical carbon dioxide using a nonionic surfactant reverse micellar system. *Coloration Technology*, 2002. **118**(5): p. 233–237.
55. Jun, J., et al., Effects of pressure and temperature on dyeing acrylic fibres with basic dyes in supercritical carbon dioxide. *Coloration Technology*, 2005. **121**(1): p. 25–28.
56. Gao, D., et al., Supercritical carbon dioxide dyeing for pet and cotton fabric with synthesized dyes by a modified apparatus. *ACS Sustainable Chemistry & Engineering*, 2015. **3**(4): p. 668–674.
57. Gao, D., et al., Synthesis and measurement of solubilities of reactive disperse dyes for dyeing cotton fabrics in supercritical carbon dioxide. *Industrial & Engineering Chemistry Research*, 2014. **53**(36): p. 13862–13870.
58. Jun, J., et al., Supercritical carbon dioxide containing a cationic perfluoropolyether surfactant for dyeing wool. *Coloration Technology*, 2005. **121**(6): p. 315–319.
59. Yan, K., et al., Development of a special SCFX-AnB3L dye and its application in ecological dyeing of silk with supercritical carbon dioxide. *Journal of CO_2 Utilization*, 2020. **35**: p. 67–78.
60. Elmaaty, T.A., et al., Water free dyeing of polypropylene fabric under supercritical carbon dioxide and comparison with its aqueous analogue. *The Journal of Supercritical Fluids*, 2018. **139**: p. 114–121.
61. Wen, H. and J.J. Dai, Dyeing of polylactide fibers in supercritical carbon dioxide. *Journal of Applied Polymer Science*, 2007. **105**(4): p. 1903–1907.
62. Zheng, H., J. Zhang, and L. Zheng, Optimization of an ecofriendly dyeing process in an industrialized supercritical carbon dioxide unit for acrylic fibers. *Textile Research Journal*, 2017. **87**(15): p. 1818–1827.
63. Gao, D., et al., Synthesis and measurement of solubilities of reactive disperse dyes for dyeing cotton fabrics in supercritical carbon dioxide. *Industrial & Engineering Chemistry Research*, 2014. **53**(36): p. 13862–13870.
64. Li, F., et al., Constructing of dyes suitable for eco-friendly dyeing wool fibers in supercritical carbon dioxide. *ACS Sustainable Chemistry & Engineering*, 2018. **6**(12): p. 16726–16733.
65. Roberts, G.A.F. and R.K. Solanki, Carrier dyeing of polyester fibre part III – The role of water in solvent dyeing. *Journal of the Society of Dyers and Colourists*, 1981. **97**(5): p. 220–223.

66. Lister, G.H., Water conservation—An alternative to solvent dyeing? *Journal of the Society of Dyers and Colourists*, 1972. **88**(1): p. 9–14.
67. Love, R.B. and A. Robson, The use of non-aqueous solvents in dyeing III - Dyeing of polyester and cotton-polyester fibre blends by the dacsol 2 process. *Journal of the Society of Dyers Colourists*, 1978. **94**(12): p. 514–520.
68. Love, R.B., The use of non-aqueous solvents in dyeing I - dyeing polyester fibre by the dacsol process. *Journal of the Society of Dyers Colourists*, 1978. **94**(10): p. 440–447.
69. Milićević, B., The use of non-aqueous solvents in coloration and textile processing I—literature survey. *Review of Progress in Coloration and Related Topics*, 1967. **1**(1): p. 49–52.
70. Gebert, K., The dyeing of polyester textile fabric in perchloroethylene by the exhaust process. *Journal of the Society of Dyers Colourists*, 1971. **87**(12): p. 509–513.
71. Furness, W. and J. Rayment, Some practical experiences in solvent dyeing. *Journal of the Society of Dyers and Colourists*, 1971. **87**(12): p. 514–520.
72. Milićević, B., Dyeing from solvents. *Journal of the Society of Dyers and Colourists*, 1971. **87**(12): p. 503–508.
73. Pkmkins, W.S. and D.M. Hall, A fundamental study of the sorption from trichloroethylene of three disperse dyes on polyester. *Textile Research Journal*, 1973. **43**(2): p. 115–120.
74. Love, R.B., The use of non-aqueous solvents in textile processing. *Review of Progress in Coloration and Related Topics*, 1975. **6**(1): p. 18–28.
75. Love, R.B., The use of non-aqueous solvents in dyeing II-solvent clearing of dyed polyester yarns. *Journal of the Society of Dyers Colourists*, 1978. **94**(11): p. 486–492.
76. Xu, S., et al., Sustainable and hydrolysis-free dyeing process for polylactic acid using nonaqueous medium. *ACS Sustainable Chemistry & Engineering*, 2015. **3**(6): p. 1039–1046.
77. Wang, Y., et al., Dyeing cotton in alkane solvent using polyethylene glycol-based reverse micelle as reactive dye carrier. *Cellulose*, 2016. **23**(1): p. 965–980.
78. Gao, A., et al., Silicone nanomicelle dyeing using the nanoemulsion containing highly dispersed dyes for polyester fabrics. *Journal of Cleaner Production*, 2018. **200**: p. 48–53.
79. Chen, L., et al., Comprehensive study on cellulose swelling for completely recyclable nonaqueous reactive dyeing. *Industrial & Engineering Chemistry Research*, 2015. **54**(9): p. 2439–2446.
80. Zhao, J., et al., A heterogeneous binary solvent system for recyclable reactive dyeing of cotton fabrics. *Cellulose*, 2018. **25**(12): p. 7381–7392.
81. Yuan, J., Q. Wang, and X. Fan, Dyeing behaviors of ionic liquid treated wool. *Journal of Applied Polymer Science*, 2010. **117**(4): p. 2278–2283.
82. S. Andrade, R., et al., Sustainable cotton dyeing in nonaqueous medium applying protic ionic liquids. *ACS Sustainable Chemistry & Engineering*, 2017. **5**(10): p. 8756–8765.
83. Pawar, A.B., Patankar, K.C., Madiwale, P. and Adivarekar, R. Application of chemically modified waste Allium cepa skin for one bath dyeing of polyester/wool blend fabric. *Pigment & Resin Technology*, 2019. 48(6): 493–501. https://doi.org/10.1108/PRT-11-2018-0118
84. Pawar, S.S. and Adivarekar, R. A novel approach for dyeing of polyester using non-aqueous deep eutectic solvent as a dyeing medium. *Pigment & Resin Technology*, 2020. 50(1): 1–9. https://doi.org/10.1108/PRT-09-2019-0085
85. Pawar, A., Biranje, S., Patankar, K. and Adivarekar, R.V. Statistical modelling for optimisation of dyeing of silk with semisynthetic azo dye made by chemical modification of areca nut. *Research Journal of Textile and Apparel*, 2020. 24(1): 20–37. https://doi.org/10.1108/RJTA-07-2019-0032
86. Pawar, S.S., Athalye, A. and Adivarekar, R.V. Solvent Assisted Dyeing of Silk Fabric Using Deep Eutectic Solvent as a Swelling Agent. *Fibers and Polymers*, 2021. **22**(2): 405–411.

12 Plasma Technology for Textile Coloration
Solutions for Green Dyeing of Textiles

Mina Shakeri
Tarbiat Modares University
Industrialization Center for Applied Nanotechnology (ICAN)

Azadeh Bashari
Amirkabir University of Technology
Industrialization Center for Applied Nanotechnology (ICAN)

CONTENTS

12.1 Introduction	203
12.2 Plasma Technology in Coloration	204
12.3 Different Types of Plasma	206
12.4 Effect of Plasma Treatment on Coloration of Textiles	206
12.5 Dyeing of Plasma-Treated Textiles	207
12.5.1 Cotton	207
12.5.2 Wool	208
12.5.3 Silk	211
12.5.4 Polyester	212
12.6 Hurdles of Using Plasma in the Textile Industry	214
12.7 Conclusion and Future Scope	214
Acknowledgment	215
References	215

12.1 INTRODUCTION

Textile industries play an important role in every society's economy. Earlier, clothing was only a way for coverage and protection of body; however, nowadays due to the rising demands about better life, people prefer multifunctional properties (Shahid and Mohammad 2013). Textile sector is a developed global competition due to the changing of human's needs, and therefore, there should always be utilization of new tools for higher performance of the garments (Gorjanc et al. 2013).

Environmental pollution has been an important issue in the textile industry. The textile industry uses several hundred million gal of water daily in producing and finishing of fabrics. Historically, green chemistry has focused on pollution prevention and reducing the environmental impact of wastewater released from textile wet processing activities. The presence of dye and finishing materials requires some water treatments before releasing the water in the environment. Therefore, any operation that can help with the cleaner environment will reduce pollution of dyeing of textiles and can be a substitute for the conventional methods (Reife and Freeman 1996).

Dyeing and its conventional methods for preparation and printing of textiles have a great environmental effect due to the materials and equipment that are used. Therefore, innovative and nonpolluting methods that can replace the conventional ones are drawing industries' attention (Yusuf, Shabbir, and Mohammad 2017). Scientists are investigating new methods that can replace old processes and techniques. Regarding environmental issues, not only the methods but environmentally friendly products are also introduced in dyeing and finishing industries (Mohammad 2014).

12.2 PLASMA TECHNOLOGY IN COLORATION

Recently, plasma treatment technology has drawn more and more attention in the textile industry. It is introduced as a promising and environmentally friendly method as an alternative to conventional methods of dyeing conventional methods. It is a different method from the wet process of dyeing due to being solvent-free. It is also a time-saving and cost-effective method. Plasma treatments of the surface won't change the properties of the textile; but its efficiency depends on several factors including the nature of the substrate and the treatment operating conditions (Jelil 2015).

Plasma is a partially ionized gas which is the fourth state of the matter. It is produced with applying an electrical field between two electrodes by inducing radiofrequency. Plasma has a complex mixture of electrons, neutral particles, ion, photons and radicals. In the ambient pressure, fabrics can be treated by plasma for in order to have more cohesion between the fabric and dyes and also finishing materials. It is categorized into different types such as hot and cold plasmas. Due to the sensitive structure of textiles and synthetic polymers, only cold plasma is suitable for textiles. It can affect the most top layer of the fabric and impart many functional groups on the surface without any changes in other properties. Plasma treatments of the surface of the fabric can increase the wettability, dyeability, and printability of the surface as well as its adhesion (Herbert 2007; Takke et al. 2009). Plasma treatment affects only the surface of a material which is an important feature. The thickness of this layer varies from 100 A° to several micrometers, according to different estimates (Knittel and Schollmeyer 2000; Jelil 2015) (Figures 12.1 and 12.2).

Plasma Technology for Textile Coloration

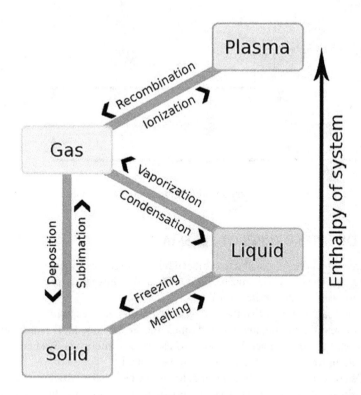

FIGURE 12.1 Different states of matter (Muthu and Gardetti 2020).

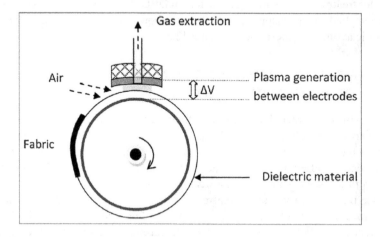

FIGURE 12.2 Schematic of corona plasma equipment in the textile industry (Labay et al. 2013).

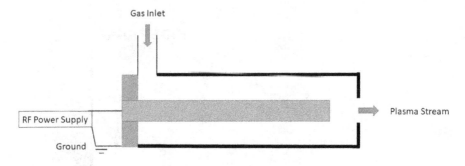

FIGURE 12.3 Schematic presentation of (Atmospheric pressure plasma jet) APPJ.

12.3 DIFFERENT TYPES OF PLASMA

There are many types of plasma treatment devices, but three of them including corona discharge, dielectric barrier discharge (DBD), and atmospheric-pressure plasma jet (APPJ) are used in the textile industry.

Corona treatment is one of the widely used plasma devices, which is not uniform enough to treat the textile, and it mat damage the surface due to its nonhomogeneity. In addition, the small space between two electrodes in corona discharge makes it inappropriate for bulky and thick things to be treated.

DBDs consist of an insulating layer at top of the electrodes which avoids gas spark or breakdown. It is highly electron-dense which can result in better changes in the treated surface structure.

APPJ is the latest development in atmospheric plasma which can use different gases for treatment and is more uniform than DBD. The key aspect of APPJ is that it can be applied to any shapes of the materials; however, it can only treat one side of the fabric at once (Herbert 2007) (Figure 12.3).

12.4 EFFECT OF PLASMA TREATMENT ON COLORATION OF TEXTILES

When textiles are treated with plasma, their surface will be bombarded with different particles (electrons, ions, radicals, and neutrals). In addition, emission of UV beam on the surface of the textiles will also affect their structure. There are many parameters that affect the performance of plasma treatment on textiles, namely textile substrate, the working gas, pressure, and input power.

Since the injected gas to the plasma device is an important matter in the treatment process, the resulting changes depend on the nature of the fabric. It is also important to mention the working gas of the plasma treatment device. Oxygen, air, argon, helium, carbon dioxide, nitrogen, hydrogen, tetrafluoromethane, water vapor, methane, and ammonia are some of the gases, which can be used individually or as a mixture for plasma treatment of fabrics. Every gas generates a unique plasma composition and results in different surface properties. Some of the gases such as helium, argon, nitrogen, and oxygen can activate the surface by means of ablation or etching. The amount of etching and roughening of the surface is governed by the impact energy of the ions which in turn is governed by the pressure and input power (Shahidi, Wiener, and Ghoranneviss 2013).

The plasma process effects on the surface of a textile substrate have a vast variety. Surface cleaning, surface etching, surface activation, and polymerization are some of the effects which plasma treatment can have on textiles. They can be helpful in some finishing operations and also the dyeing process.

12.5 DYEING OF PLASMA-TREATED TEXTILES

In this section, different textiles and the effect of plasma on them are being discussed.

12.5.1 COTTON

Cotton is the most used fiber that consists of several O-H groups, and because of that, it has no problem with dyeability with almost every method. When cotton is immersed in water, the ionization of O-H and C-H groups happens, and as a result, cotton fiber possesses a mild negative surface charge. Therefore, it can be deducted that the dyes that are appropriate for cotton should possess positive charge. This can make a problem when using natural dyes on cotton fiber. Natural dyes have also negative surface charge, which results in repulsion between cotton and the dye and unsuccessful dyeing of the cotton fabric. Table 12.1 shows some of the natural materials that were used as dyes with the help of plasma treatment

In one of the researches, Ebrahim et al. used plasma treatment on different cellulosic fabrics. In this study, N_2 plasma was used to produce NH_2 groups on the surface of the fabrics. Plasma treatment of the cellulosic fabrics results in surface activation and creation of active sites on the substrate. Figure 12.4 shows a high increase in nitrogen content of plasma treated cotton, linen, viscose, and lyocell cellulosic substrates with N_2-plasma. N_2, $-NH_2$, and $-OH$ groups are the most important content of the plasma which can modify fiber surface (Ibrahim, Eid, and Abdel-Aziz 2017).

As it is shown in Table 12.2, it is deducted that production of the NH_2 group on the surface of the cellulosic fabric has better color results when comparing both K/S values.

Another research on cotton fabric investigates the result of argon and air plasma on the coloration of cotton. Activated surfaces were cross-linked with two different amine compounds: ethylene diamine and triethylenetetramine. Pretreated cotton was dyed with acid dye, and the effects of pretreatment on the color strength, as well as the washing, rubbing, and the light fastness of the dyeing were investigated (Karahan et al. 2008).

The color strength (K/S) values of the dyed fabrics are shown in Figure 12.5. Dyeing results can evaluate the grafting value of the samples. The best result of the grafting conditions shows that the plasma process of 20s is the suitable time for having reaction between plasma-modified and grafted surface with the acid dye. It could be seen that using more chemical compound did not improve color strength. This was because of the quantity of the groups on the surface, for attaching more chemical compound (Karahan et al. 2008).

TABLE 12.1
Effect of Different Plasma Treatments on Dyeing of Cotton with Various Natural Dyes

	Plasma Parameters					
Types of Gas	Pressure	Power (W)	Duration (s)	Natural Dye	Effect	Reference
O_2	Low	800	4	Onion skin and onion Pulp	Deeper shades, antibacterial activity	(Chen and Chang 2007)
O_2	Low	100	180	Madder and Weld	Improved color strength and fastness	(Haji 2019)
O_2	Low	800	5	17 Taiwanese medicinal plants	Improved dyeability and antibacterial activity up to 14 days of incubation	(Chen and Chang 2007)
O_2	Low	100	120	Berberis Vulgaris	Improved dyeability	(Haji 2013)
Water Vapor	Low	500	10	Fallopia japonica	Improved dye uptake and antibacterial activity	(Gorjanc et al. 2016)
Air	Atmospheric	3–15	180–600	Cochineal	Improved printing and fastness properties	(Ahmed et al. 2017)
O_2 + Chitosan	Low	150	150–300	Pomegranate rinds	Increased color strength and fastness	(Haji 2017)
O_2 + Amminia	Low	100	1–300	Curcumin and green tea	Increased color yield for ammonia plasma-treated sample, decreased color yield for oxygen plasma treatment	(Gorjanc, Mozetič, and Vesel 2018)

Source: Haji and Naebe 2020.

12.5.2 WOOL

Wool is one of the mostly used textile fibers because of its properties such as high absorption ability, light weight, elasticity, as well as its resiliency with some technical problems such as wettability (which affects the dyeability). It consists of three important parts namely cuticle, cortex, and medulla. It also has another layer, named epicuticle, which is the outer layer of the wool fiber and has hydrophobic properties, which makes it hard for water and dyes to penetrate it (Haji and Naebe 2020). Wool surface consists of a high number of disulfide cysteine functional groups and fatty acids. Therefore, finding a way to improve wettability of wool and increasing its surface roughness can result in better dyeing of the fiber surface and increased color depth of the samples (Ahmed and El-Shishtawy 2010).

One of the recommended methods to improve wettability, wicking, and surface energy of the wool fiber is to apply DBD treatment on the surface of the fiber. In one

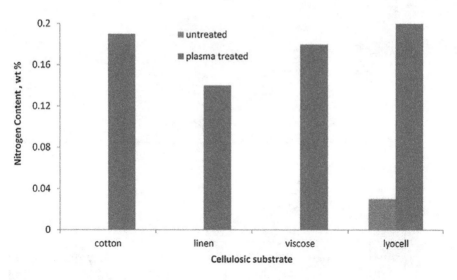

FIGURE 12.4 Effect of N_2-plasma treatment on the nitrogen content, wt.% of the treated substrate (Ibrahim, Eid, and Abdel-Aziz 2017).

TABLE 12.2
K/S Values for Cellulosic Fabrics

Cellulosic Substrate	Plasma Gas	K/S Acidic Dyeing
Cotton	N_2	1.05
	O_2	-
Linen	N_2	0.85
	O_2	-
Viscose	N_2	1.20
	O_2	-
Lyocell	N_2	1.46
	O_2	-

Source: Ibrahim, Eid, and Abdel-Aziz 2017.

of the researches, physical and chemical properties of wool fabric were investigated, after being treated by DBD plasma with oxygen and nitrogen working gases (Navik et al. 2018). Oxygen plasma treatment increased the dye diffusion rate and decreased the required time for the coloration of fibers. It also uses less energy than conventional methods (Sun and Stylios 2004) (Figure 12.6).

As it can be seen in Figure 12.3, the surface of the treated wool has changed after plasma treatment, and the roughness that is made on the surface will result in better penetration of the dye to the structure of the fiber (Navik et al. 2018).

It can be deducted that plasma treatment duration time was the most important factor on the treated surface. The DBD-treated wool specimens obtained improved wettability, wick ability, surface energy, and mechanical strength. It is also understood that with more treatment time, the more C=O and C–O and C–N groups

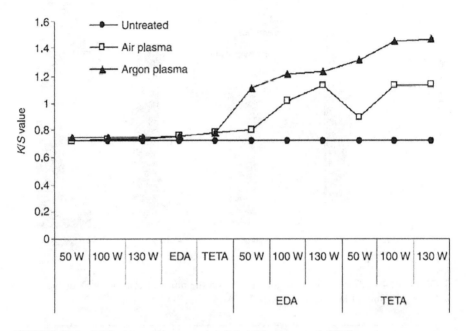

FIGURE 12.5 K/S values of the treated samples (Karahan et al. 2008).

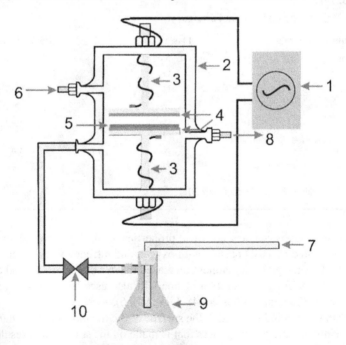

FIGURE 12.6 Atmospheric pressure DBD plasma generator; 1—high voltage power supply, 2—reactor body, 3—Al2O3 ceramic electrode, 4—barrier film (polyimide), 5—wool substrate, 6 and 7—gas inlet, 8—gas outlet, 9—solvent box (ethanol or acetone), 10—gas flow control valve (Navik et al. 2018).

FIGURE 12.7 SEM micrograph of (a) the control wool fiber and (b) the plasma-treated wool for 450 s (Navik et al. 2018).

appear on the surface of the wool. Due to the damage to the hydrophobic layer on the surface of the wool and the –S–S–, several pores and channels appeared which result in the quality of the dying process. It is also declared that no decrease in the tensile of the fibers was seen after being treated with oxygen plasma treatment (Figure 12.7).

As far as there are different hydrophobic and hydrophilic dyes, Naebe et al. showed that the effect of plasma treatment of the textiles largely depends on the hydrophobicity of the used dye. It is claimed that the plasma-treated fabric adsorbed hydrophilic dyes faster, showing the greater importance of the electrostatic effect. Thus, plasma treatment will be more effective on the dyeing rate of the dyes with higher water solubility compared to the dyes with lower water solubility (Naebe et al. 2011)

12.5.3 Silk

Similar to wool, silk has a protein nature and is famous for its luxurious handle, comfort, softness, and water absorbency. There is limited report on plasma pretreatment of silk fabric because of its delicacy.

In one of the studies, the researchers investigate the plasma treatment on silk fabric. It was showed that DBD air plasma of silk can increase the amide and OH groups on the surface. It was also shown that pretreatment of silk with air/O_2 plasma increases C=O, C–N, and C–O groups. In several studies, it is declared that plasma treatment of the silk improves color strength as well as color yield. However, there is no evidence whether pretreatment of plasma on silk fabric can weaken its mechanical properties (Haji and Naebe 2020).

In one of the studies, the effect of plasma treatment using helium and nitrogen gas mixture was evaluated on silk fabric. The plasma-treated samples showed better K/S values than the untreated samples, and it also could decrease the amount of discharged dyes. As it is shown in Figure 12.8, in all of the temperatures that were investigated, the K/S value in plasma-treated samples is higher than that of the untreated samples (Teli et al. 2015).

Figure 12.9 shows the distribution of functional groups such as CH– (red color), O– (green color), and NH– (blue color) in the raw and treated silk fabric. It can be

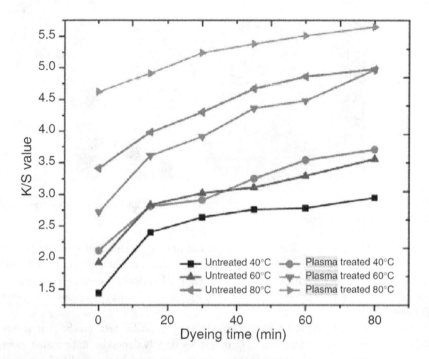

FIGURE 12.8 K/S values of raw and plasma-treated silk fabrics (Teli et al. 2015).

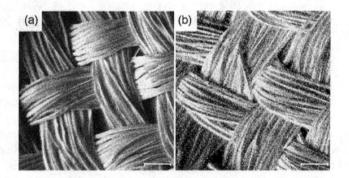

FIGURE 12.9 Distribution of CH– (red color), O– (green color), and NH– (blue color) in the (a) untreated and (b) plasma-treated samples (Teli et al. 2015).

clearly observed that a high number of functional groups are created on the surface of the treated fabric due to the plasma treatment (Teli et al. 2015).

12.5.4 Polyester

Polyester is mainly dyed with disperse dyes by a pad-dry-bake process at high temperature and pressure. This process can pollute the environment due to use of phenol-based carriers. In one of the researches, it has been reported that low-temperature

Plasma Technology for Textile Coloration 213

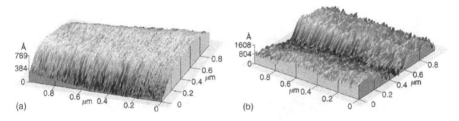

FIGURE 12.10 The 3-D views of noncontact mode AFM images of PET fabrics after dyeing: (a) untreated surface; (b) surface air plasma-treated for 60 s. Operating treatment conditions: P = 0.1 mbar; WRF = 120 W; d = 10 cm (Raffaele-Addamo et al. 2006).

argon and air plasma can increase the color depth of polyester fabric. Also, the hydrophilic groups made plasma treatment on the surface, may help in increasing water swelling pace and the affinity of PET fibers for dyes containing polar groups, which is related to the increase of K/S values of dyed PET specimens. (Raffaele-Addamo et al. 2006) (Figure 12.10).

PET fabrics that are exposed to plasma treatment generally exhibit more depth in coloration compared to the untreated fabric. This happens when the exposed side is being examined. The percent increase in color depth on such side showed that the depth of the coloration largely depends on the adopted plasma treatment conditions (Raffaele-Addamo et al. 2006).

In one of the studies, the coloration of polyester was investigated using air plasma pretreatment and padding the fabric with chitosan. The coloration was evaluated by measuring the color yield (K/S) and the durability properties of the colored fabrics. In this study, two methods of coloration namely exhaustion and padding were compared. The results show that the exhaustion method leads to better color strength and durability. On the other hand, in the padding method, lower water consumption can be achieved and with the aid of plasma technology, functional groups for wettability and adhesion increase can be obtained (Agnhage, Perwuelz, and Behary 2016) (Figure 12.11).

Figure 12.12 indicated that surface treatment of PET enhanced the coloration process by the padding method. These K/S values were measured after the first wash of the samples. As it is shown in the figure, the untreated PET surface had the lowest affinity to the coloring substances in the madder dye among all. However, pretreatment of the samples resulted in increase of color yield on surface-modified PET. Chitosan treatment alone increased the K/S value much more than air plasma treatment alone. The K/S reached its highest value when a prior plasma treatment followed by chitosan treatment was used. It can be deducted that surface activation by

FIGURE 12.11 Schematic flowchart of the padding process studied. White boxes indicate surface modification processes/pretreatments (Agnhage, Perwuelz, and Behary 2016).

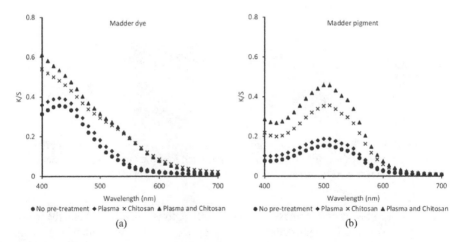

FIGURE 12.12 Effect of surface modification/pretreatment(s) on the color yield (K/S), (a) dye and (b) pigment (Agnhage, Perwuelz, and Behary 2016).

plasma and then functionalization with chitosan using the padding method is one of the suitable ways for coloration of polyester fabric (Agnhage, Perwuelz, and Behary 2016).

12.6 HURDLES OF USING PLASMA IN THE TEXTILE INDUSTRY

Although plasma treatment has several applications in the textile industry and dyeing process, it is still limited due to some reasons.

Plasma treatment of the textiles is highly affected by the environmental conditions. Being dependent to the system makes it very difficult for precise control of functional groups on the surface of the samples. Therefore, there should be strong restrictions for designing the plasma device in order to reproduce the same conditions in all samples.

Time is another important factor for plasma-treated textile substrates. Since the changes which are made on the surface of the treated fabrics are not stable over a long time, it is recommended that the substrates will be dyed as soon as possible after the plasma pretreatment so that the significant amount of surface energy that is obtained as a result of plasma treatment can be technically useful (Jelil 2015).

In some of the studies, atmospheric-pressure plasma treatments work at lower paces in comparison to target line speed, which is a serious obstacle that would limit commercializing of the in-line plasma devices in the future. Thus, for a continuous dyeing process on textile industry, there is a need for technically and economically feasible plasma sources.

12.7 CONCLUSION AND FUTURE SCOPE

Plasma treatment of the textile can be used for different types of fabrics using plasma devices. There are several important factors that should be under control to increase the ability of plasma equipment to reproduce samples. These factors including working gas, treatment time, power, pressure, and gas flow rate can also make limits for plasma treatment of the textile substrates. By exact selection of the plasma parameters, the wettability, surface roughness, functionality, and dyeability of the fibers can be tuned. As far as its's a green, environmentally friendly, and cost-effective method, it can be predicted that with the precise selection of the plasma parameters, it has the potential to be an alternative for conventional dyeing methods in the textile industry.

ACKNOWLEDGMENT

This chapter is dedicated to the sublime and generous soul of our great master, Mr. Firoozmehr Mazaheri, who taught us the principles of dyeing and printing and passed away on 2020-11-30.

REFERENCES

Agnhage, Tove, Anne Perwuelz, and Nemeshwaree Behary. 2016. "Eco-innovative coloration and surface modification of woven polyester fabric using bio-based materials and plasma technology." *Industrial Crops and Products* 86:334–341.

Ahmed, HM, KA Ahmed, HM Mashaly, and AA El-Halwagy. 2017. "Treatment of cotton fabric with dielectric barrier discharge (DBD) plasma and printing with cochineal natural dye." *Indian Journal of Science and Technology* 10 (10):1–10.

Ahmed, Nahed SE and Reda M El-Shishtawy. 2010. "The use of new technologies in coloration of textile fibers." *Journal of Materials Science* 45 (5):1143–1153.

Chen, Chonyu, and Wen-Ya Chang, 2007. "Antimicrobial activity of cotton fabric pretreated by microwave plasma and dyed with onion skin and onion pulp extractions." *Indian Journal of Fibre and Textile Research* 32: 122–125.

Chen, Chonyu, and Wen-Ya Chang. 2007. *Antimicrobial activity of cotton fabric pretreated by microwave plasma and dyed with onion skin and onion pulp extractions.*

Gorjanc, Marija, Marija Gorenšek, Petar Jovančić, and Miran Mozetič. 2013. Eco-Friendly Textile Dyeing and Finishing, Melih Gunay, Vol 1:5-31.

Gorjanc, Marija, Miran Mozetič, and Alenka Vesel. 2018. "Natural dyeing and UV protection of plasma treated cotton." *The European Physical Journal D* 72 (3):1–6.

Gorjanc, Marija, Aleksandar Savić, Ljiljana Topalić-Trivunović, Miran Mozetič, Alenka Vesel, and Dragana Grujić. 2016. "Dyeing of plasma treated cotton and bamboo rayon with Fallopia japonica extract." *Cellulose* 23 (3):2221–2228.

Haji, A. 2019. "Dyeing of cotton fabric with natural dyes improved by mordants and plasma treatment." *Progress in Color, Colorants and Coatings* 12 (3):191–201.

Haji, Aminoddin. 2013. "Eco-friendly dyeing and antibacterial treatment of cotton." *Cellulose Chemistry and Technology* 47 (3–4):303–308.

Haji, Aminodin. 2017. "Improved natural dyeing of cotton by plasma treatment and chitosan coating; optimization by response surface methodology." *Cellulose Chemistry and Technology* 51 (9–10):975–982.

Haji, Aminoddin and Maryam Naebe. 2020. "Cleaner dyeing of textiles using plasma treatment and natural dyes: a review." *Journal of Cleaner Production* 265:121866.

Herbert, T. 2007. "Atmospheric-pressure cold plasma processing technology." *Plasma Technologies for Textiles*:79 (1).

Ibrahim, Nabil A, Basma M Eid, and Mohamed S Abdel-Aziz. 2017. "Effect of plasma superficial treatments on antibacterial functionalization and coloration of cellulosic fabrics." *Applied Surface Science* 392:1126–1133.

Jelil, R Abd. 2015. "A review of low-temperature plasma treatment of textile materials." *Journal of Materials Science* 50 (18):5913–5943.

Karahan, HA, E Özdoğan, A Demir, H Ayhan, and N Seventekin. 2008. "Effects of atmospheric plasma treatment on the dyeability of cotton fabrics by acid dyes." *Coloration Technology* 124 (2).106–110.

Knittel, D, and E Schollmeyer. 2000. "Technologies for a new century. Surface modification of fibres." *Journal of the Textile Institute* 91 (3):151–165.

Labay, Cedric, C Canal, C Rodríguez, G Caballero, and JM Canal. 2013. "Plasma surface functionalization and dyeing kinetics of Pan-Pmma copolymers." *Applied Surface Science* 283:269–275.

Mohammad, Faqeer. 2014. "Emerging green technologies and environment friendly products for sustainable textiles." In *Roadmap to sustainable textiles and clothing*, Subramanian Senthilkannan Muthu, 63–82. Springer, Singapore.

Muthu, Subramanian Senthilkannan, and Miguel Angel Gardetti. 2020. *Sustainability in the textile and apparel industries: production process sustainability*: Springer, Cham.

Naebe, Maryam, Ron Denning, Mickey Huson, Peter G Cookson, and Xungai Wang. 2011. "Ageing effect of plasma-treated wool." *Journal of the Textile Institute* 102 (12):1086–1093.

Navik, Rahul, Sameera Shafi, Md Miskatul Alam, Md Amjad Farooq, Lin Lina, and Cai Yingjie. 2018. "Influence of dielectric barrier discharge treatment on mechanical and dyeing properties of wool." *Plasma Science and Technology* 20 (6):065504.

Raffaele-Addamo, Antonino, Elena Selli, Ruggero Barni, Claudia Riccardi, Francesco Orsini, Giulio Poletti, Laura Meda, Maria Rosaria Massafra, and Bruno Marcandalli. 2006. "Cold plasma-induced modification of the dyeing properties of poly (ethylene terephthalate) fibers." *Applied Surface Science* 252 (6):2265–2275.

Reife, Abraham and Harold S Freeman. 1996. *Environmental chemistry of dyes and pigments*: John Wiley & Sons New York.

Shahid, Mohammad and Faqeer Mohammad. 2013. "Recent advancements in natural dye applications: a review." *Journal of Cleaner Production* 53:310–331.

Shahidi, Sheila, Jakub Wiener, and Mahmood Ghoranneviss. 2013. "Surface modification methods for improving the dyeability of textile fabrics." *Eco-Friendly Textile Dyeing and Finishing*: (1):34–50.

Sun, D and GK Stylios. 2004. "Effect of low temperature plasma treatment on the scouring and dyeing of natural fabrics." *Textile Research Journal* 74 (9):751–756.

Takke, V, N Behary, A Perwuelz, and C Campagne. 2009. "Studies on the atmospheric air–plasma treatment of PET (polyethylene terephtalate) woven fabrics: effect of process parameters and of aging." *Journal of Applied Polymer Science* 114 (1):348–357.

Teli, MD, Kartick K Samanta, Pintu Pandit, S Basak, and SK Chattopadhyay. 2015. "Low-temperature dyeing of silk fabric using atmospheric pressure helium/nitrogen plasma." *Fibers and Polymers* 16 (11):2375–2383.

Yusuf, Mohd, Mohd Shabbir, and Faqeer Mohammad. 2017. "Natural colorants: historical, processing and sustainable prospects." *Natural Products and Bioprospecting* 7 (1):123–145.

13 Plasma Technology in Processing of Textile Materials

D. Saravanan
Kumaraguru College of Technology

M. Gopalakrishnan
Bannari Amman Institute of Technology

CONTENTS

13.1 Introduction 217
    13.1.1 Types and Characteristics of Plasma 218
13.2 Influence of Substrate Structures on Plasma Processing 221
13.3 Yarn and Fabric Manufacturing 221
13.4 Fabric Preparation 222
13.5 Dyeing and Printing 223
13.6 Functional Finishing 225
13.7 Effects of Plasma Treatment on Properties 227
13.8 Conclusion 229
References 229

13.1 INTRODUCTION

Among various processes involved in the manufacturing of textile materials, chemical processing often remains as the centre of focus on account of various advantages and value additions imparted to the final products. However, most of the unit operations involved in chemical (wet) processing result in significant amounts of pollution loads in terms of toxic gases and liquid and solid effluents and also necessitate more energy to accomplish the processes. Novel methods and techniques are attempted in coloration and finishing of textile substrates, frequently, to identify environmentally friendly ones. Water-free dyeing, processes assisted by foam techniques and low wet pick-up have been attempted to mitigate the hazardous materials and pollutants generated at processing stages. Plasma treatments can be potentially carried out on different textile substrates from preparation to finishing.

    Plasma is considered as the fourth state of matter, after solid, liquid, and gas, and contains ionized gas comprising ions, electrons, atoms, and molecules (Saravanan and Nalankilli, 2008). Sir William Crookes identified plasma effect as early as in the year 1879 and subsequently Irving Langmuir in the year 1929. A gaseous substance requires the external potential of 1–30 eV per particle to alter the state to plasma, which is higher than the potential required for solid-to-liquid or liquid-to-gas transitions. Higher levels of collisions in the gaseous substances, at higher temperatures, result in the conversion of neutral molecules or atoms into ionized states with equal densities of positively charged and negatively charged particles. The presence of free electrons makes the plasma electrically conducting and respond to electric and magnetic fields as an efficient source of radiation.

13.1.1 Types and Characteristics of Plasma

The plasma modifies the surface of the fabrics by the bombardment with high-energy electrons and ions. Various milestones achieved in the plasma technology are shown in Table 13.1.

Gaseous atmosphere used in the plasma processes varies depending upon the effects or the type of finish required on the substrates; for example, absorbency/repellency, temporary/durable, and accordingly reactive/inert gases, water vapors, and their combinations are employed during the process (Bhat et al., 2003; Kutlu and Cireli, 2004; Jhala, 2005). Combinations of two or more gases are simultaneously used for better process performance and wider characteristics in the substrates after processing. Different types of plasma used in the textile processing for altering the properties of the substrates are given below (Table 13.2).

Classification of plasmas into various groups depends on process conditions – temperature, pressure levels, type of electric current, and nature of gas used in the process (Kutlu and Cireli, 2004; Jhala, 2005; Shanmugasundaram and Rengasamy, 2006). Plasma surface treatments can be affected by either active or passive plasma exposure (Roth, 2001). Active-plasma (abnormal glow discharge treatment) treatments are carried out in an atmosphere where an electrically conducting work piece, cathode of DC abnormal glow discharge, is used to cover the substrates completely

TABLE 13.1

Milestone in Plasma Research

Year	Milestone in Plasma Research
1879	Sir William Crookes identified ionized gas
1920s	Research started in plasma
1929	Irving Langmuir coined the word "plasma"
1960s	Focus on low pressure plasma technology
1970s	Initial reviews on plasma applications in textiles
	Attempts to use intermediate pressure plasma on wool, cellulosics
1980s	Vacuum plasma for processing of spun bonded webs.
	Intermediate pressure glow discharge plasma using DC/RF plasma/coplanar magnetron plasma results limited success due to batch processing and expensive vacuum system
1990s	Treatments using atmospheric plasma with continuous treatment facility
2000s	Attempts to establish prototype and commercial scale of operations

TABLE 13.2
Plasma – Types and Nature

Plasma	Requirements	Nature
Glow Discharge	Direct Current, microwave, radio frequency waves	Plasma develops at low pressure
Vacuum Glow Discharge	Microwave	
OAUGDP	Radio frequency waves	Uniform active species with high concentration
Corona Discharge	Atmospheric pressure conditions LF or pulsed HV	Nonhomogeneous lightning-type-discharge
Dielectric Barrier Discharge	Application of pulsed voltage over an electrode pair, in which one is covered with a dielectric material	Uniform lightning-type-discharge

OAUGDP, one atmospheric unit glow discharge plasma.

for treatment, while in the passive exposure occurs when the active species bombard the surface of the substrates without the support of real currents.

The two broad categories of plasmas, on the basis of temperature (thermal plasma), include hot plasma and cold plasma. Thermal plasma is characterized by equal temperature of electrons/ions/atoms and small field strength/pressure, e.g., high intensity arcs, shock waves, and nuclear fusion, whereas in the nonthermal plasmas, the field strength or pressure is high. This necessitates the careful selection of plasma to carry out surface modification on various substrates. Radicals/ionic species generated inside the plasma region needs either direct or indirect access to the energy sufficient enough to approach and bombard the fiber surface. Accordingly, substrates can be placed and treated between electrodes or closer to plasma zone, and the path of active species between the points of generation and reaction is influenced by the distance between single fibers and active species.

In hot plasma (aka equilibrium plasma), the active species, ions and electrons, exist in thermal equilibrium and are processed at high/atmospheric pressures. The potential applications of hot plasma are found in typical processes that involve deposition/spraying type of reactions, welding/cutting, production of ultra-fine powders, decomposing toxic/hazardous substances including gases, liquids, and solid substances. Nonequilibrium cold plasma is one, where only a small fraction of atoms in a gas are ionized, and the electrons reach a very high temperature. Cold plasma reactions are carried out at low-pressure conditions, where electrons, ions, and UV radiation break covalent bonds of gaseous molecules and polymers and also result in etching of polymer surfaces including fibers. Polymerization, vapor deposition, surface-level treatments, and sputtering are some of the applications carried out using cold plasma. Cold plasma (also called technological plasma) is grouped into (i) glow-discharge, (ii) corona-discharge, and (iii) RF-discharge plasma. Inductively or capacitively coupled electric fields, with a power source, are used to accelerate ionization of gases.

Low-pressure plasma requires a closed vacuum system that makes the system less amenable for continuous treatment of textile materials. However, the atmospheric cold plasma is now a well-established system for modifying the surface of textile materials by a surface reaction without altering the internal composition. In the case of pressure maintained inside the plasma reactor, the applied pressure value must be matched to the characteristic structure of the (textile) materials meant for the treatment. In the low-pressure regions (P < 1 mbar), the mean free path in the gas phase exceeds the typical distances in textile materials, and the very low pressure causes a relatively low radical concentration per unit volume. In the pressure range P > 100 mbar, especially at atmospheric pressure, the mean free path in the gas phase is much lower than textile distances. Most of the collisions happen among the gas particles and reduce the lifetime of the radicals and may not reach the target sites in the substrates, and the temperature of the electrodes needs to be maintained at appropriate levels. At higher pressure levels, the gas transportation mechanism shifts from diffusion to convection or combination of diffusion and convection.

TABLE 13.3
Effect of Ionization on Nature of Plasma

Degree of Ionization	Nature of Plasma
< 1	Neutral particles are present in the plasma
≈ 1	Strongly ionized
≈ 0	Weakly ionized

TABLE 13.4
Plasma Applications and Their Effects on Substrates

Objective/Parameter of Interest	Nature of Process	Effects Obtained
Surface Energy	Modifies chemical nature of surface by bombardment of active species	Absorbency (wetting, wicking, washing), Easy to process (printing dyeing and functional finishes)
Cohesion	Enhances surface-to-surface cohesion	Crosslinking of polymers, modification of handle value of fabrics
Adhesive Properties	Alter surface energy/adhesion by combination of mechanical, chemical, electrostatic contributions	Painting of surfaces without VOCs, medical applications
Electrical Characteristics	Alter surface resistivity, deposition of electrostatic charges	Functional finishing - Antistatic, charging/discharging, filtration
Surface Finish	Addition of removal of substances from the surface of fibers/fabrics inducing chemical reactions	Binding, etching, scratch-resistance and optical properties
Bulk Properties	Modification of surface properties and/or cohesion at microscopic levels	Mechanical properties – tenacity, compression and handle value
Inactivation of Microorganisms	Deactivation/cleaning	Sterilization, disinfection

The amount of originally present gaseous atoms or molecules that are decomposed into charge carriers decides the efficiency of the process, and the following table shows degrees of ionization and relevant effects on plasma (Table 13.3).

Table 13.4 shows certain applications of plasma and the related effects obtained on different substrates (Roth, 2001; Kutlu and Cireli, 2004; Shanmugasundaram and Rengasamy, 2006).

13.2 INFLUENCE OF SUBSTRATE STRUCTURES ON PLASMA PROCESSING

The presence of different functional groups and difference in the physical structures lead to different end results in the textile fibers during the plasma treatments. However, plasma treatment as a preprocessing step/plasma-assisted processing facilitates both dry and wet processes of the textile substrates. Further, morphology of fibers, structure, and nature of yarns and fabrics play crucial roles in responding to the plasma treatments besides the process environment. The presence of natural impurities and added impurities present in the yarns and fabrics influences the plasma processing, thereby making positive or unwanted impacts depending upon the reaction aimed at. Mean free path of gas particles and statistical distances like inter-fiber and inter-yarn impact the plasma processing (Poll et al., 2001; Rashidi et al., 2004). The ion bombardment actions on fibers can reveal the details of the surface and internal structures of fibers and make the molecules of surface layers activated. Considering the activation states of interactions between plasma particles and substrates, a three body-collision is considered, which results in nonactivated particles as the final state, which includes accommodation coefficient of energy, i.e., probability of energy transfer in the case of collisions between particles from gas phase and surface sites and active particle penetration into the textile structures (Poll et al., 2001; Saravanan and Nalankilli, 2008).

13.3 YARN AND FABRIC MANUFACTURING

The influence of plasma on mechanical processing of staple fibers has been analyzed as early as 1970s as an attempt to enhance the efficiency at various stages, using cotton fibers, in terms of spinnability, tenacity, optimum twist levels, spinning performance (reduced end-breaks), and preparatory processes in fabric formation (Abbot and Robinson, 1977).

Corona treatment of cotton slivers, using carbon dioxide gas atmosphere and 450 W power, results in the reduction of breakages in the predraft (creel) zones in the ring spinning and end-breakages. Such improved results are obtained mainly due to improvement in the cohesion values that in turn leads to enhanced tenacity of the yarn from 116 to 142 mN/Tex with the count-strength-product from 1970 to 2330, without affecting elongation and unevenness values of the yarn. Indirectly, improvement in the tenacity of the yarn helps to reduce the twist levels in the roving (16%) and yarn (11%) and enhance the softness of the yarn and thereby handle the value of the fabrics in addition to reducing the energy consumed in the yarn manufacturing stage.

Spinnability of wool fibers after glow discharge plasma has been analyzed in the past by many researchers (Ralkowski, 1997; Kang and Sarmadi, 2004). Predominantly, the influence of plasma treatment is limited to the surface of the fibers and effects similar to etching, ablation of surface results in higher frictional coefficient, cohesion among the fibers, and hence, spinnability, yarn strength, and felting resistance (reduction in directional friction effect). Because of surface ablations, fatty matters present in the cuticle of raw fibers reduced by 1/3 after the treatment. Top (sliver) cohesion factor increased 50%– 100% (1.5–2.0) though significant changes in the neither strength nor elongation could be observed in the treated samples. Improved cohesion among the fibers results in enhanced performance in spinning with lower end-breaks by 10%– 25%. Fiber-to-fiber friction among the wool fibers increases by >50% after plasma treatment in the tops (Anon, 1997). In the case of fabrics, tearing strength reduces in warp and weft directions, after the plasma treatment, due to decreased sliding actions of the warp and weft yarns (Kan et al., 2004). In the case of tensile strength, breaking load and elongation-at-break of the plasma-treated fabrics increased because of higher inter-yarn and inter-fiber frictions due to higher surface roughness. The elasticity of the scales improved due to cleavage of disulfide linkages.

The presence of relatively inert scale-structure present in the surface of wool fibers leads to various problems including felting and resistance to the diffusion of dyes. Chemical compositions of the surface of wool fibers change to different extents, after plasma treatment, based on the process and environmental conditions that exist during the plasma generation (Kan and Yuen, 2006). Sharp and round edges of the scales present in the untreated fibers become nonuniform after low-temperature plasma treatment due to surface ablations. Low-temperature plasma under oxygen atmosphere results in formation of –OH, C=O, and –COOH functionalities and is often considered to be an effective alternative for wet chlorination process carried out for higher wettability and shrink resistance (Kang and Sarmadi, 2004). On the other hand, nitrogen-plasma results in formation of amine groups in the wool fibers that are capable of attaching dye molecules efficiently as a receptor site, thereby increasing the dye absorption. Plasma treatments at low temperatures using O_2/N and mixture of N+H result in the cleavage of di-sulfide link leading to formation of Bunte salt and cysteic acid moieties. Conversion of –S–S– to cysteic acid is almost 90%, which is relatively higher than those observed in UV or ozone oxidation of the wool fibers.

13.4 FABRIC PREPARATION

Plasma-assisted desizing of PVA-sized viscose rayon, starch-sized cotton yarns, PVA and polyacrylic acid-sized polyester fabrics, and scouring of cotton and wool fibers using cold plasma have been attempted by many researchers in the past (Cai et al., 2002, 2003; Bhat et al., 2003; Sun and Stylios, 2004; Bae et al., 2006). Current methods of removal of hydrophobic impurities, starch, involve the use of surfactants, alkalis, and oxidants at high temperatures, which result in very high effluent loads. Desizing using rot steeping of 100% cotton fabrics results in a weight loss of about 6.0% after 24 hours while the weight loss in the case of plasma-treated fabrics increases with time of treatment and the power (supply) applied. At higher power inputs (15 W), the weight loss reached the maximum levels within 5 minutes, and subsequent washing reduces the wetting time significantly. This mechanism of the process includes predominantly ablation and etching of substances.

PVA is used as the secondary sizing agent along with starch for cotton yarns due to its better film forming abilities. Desizing of polyvinyl alcohol-treated fabrics requires very high temperature levels, 90°C–95°C, with risk of redeposition over the fabrics because of its gel-type nature. Capacitive-coupled plasma reactors with frequency levels between 1 and 12 kHz, at a pressure of 3.4 kPa, using air-helium plasma has been used to desize the cotton fabrics treated with 8% polyvinyl alcohol, and analyzed weight loss, desizing efficiency, loss in tensile strength, and surface appearance of the fabrics after the treatment. Desizing efficiency was observed up to 97% in a duration of 10 min, while conventional oxidative treatment resulted in the desizing efficiency of 76%. In the case of PVA-sized viscose rayon fabrics, air-O_2-He plasma results in better weight loss values (up to 2.8%) compared to air-He plasma (2.3%) though air-O_2-He and air-He combinations are able to degrade polyvinyl alcohol to smaller molecules like carbon dioxide and water. Weight loss increases to 9% with air-oxygen-helium plasma-treated samples after cold wash, 7.5% for air-helium plasma treatment, and against 4.7% in the control samples (Cai et al., 2002).

Bae et al. (2006) evaluated plasma-assisted processing as an alternative method for desizing the polyester fabrics, containing sizing agents including polyvinyl alcohol, polyacrylic acid, and their esters and compared the results with the conventional method. The study proved higher efficiency of oxygen plasma desizing compared to the conventional methods in the cases of polyacrylic acid and their esters, with relatively lower levels in the case of PVA sized-polyester fabrics.

Conventionally, scouring of natural fibers requires high amounts of alkali and temperature conditions. Nonpolymerizing oxygen plasma is used in a study to scour cotton and wool fibers and dye them subsequently. Contact angle value, a measure of wettability, decreases considerably after oxygen plasma treatment compared to untreated samples with very low residual wax contents in both cotton and wool samples.

In general, wetting time >200 seconds is considered to be an unwettable surface. Plasma treatment enhances the wettability and brings down the wetting time to < 1 second (Kan, 1998a). On treatment with plasma, cystine groups present on the surface are converted into cysteic acid, thereby making the substrates more hydrophilic and reducing the wetting time and improving the dyeability. Formation of hydrophilic sulfonate and carboxylate groups and partial elimination of covalently-bound fatty acids from the outermost surface enhance the wettability of the substrates (Jhala, 2005).

13.5 DYEING AND PRINTING

Modification of substrates and creation of functional groups in the textile materials suitable for dyeing and finishing treatments have been attempted in the past by many research groups (Carneiro et al., 2001; Rashidi et al., 2004; Virk et al., 2004; Bozzi et al., 2005; De Geyter, Morent and Leys, 2006). Plasma as the pretreatment in processing of raw cotton fibers resulted in the final exhaustion levels similar to that of bleached fabrics. Carboxylic acid groups formed during the treatment within the cuticle layers result in increase of surface acidity, which might be nullified with the addition of ionic or nonionic wetting agents. Plasma pretreated fabrics are found darker in color with higher dye uptake levels than the untreated samples.

Detailed analyses on wool fibers, yarns, and fabrics were carried out using glow discharge plasma to analyze the effects including changes in surface morphology, chemical compositions, and physical properties (Ralkowski, 1997; Saravanan and Nalankilli, 2008). Combination of plasma-treated and regular scouring process resulted in the lower residual wax content at 8% in the wool fibers, and the contact angle decreased considerably after oxygen plasma treatment compared to untreated samples. Oxidation of functional groups in the surface of fibers, reduction of covalently bound hydrophobic content (methyleicosanoic acid), and increase in the content of oxidized sulfur species cause improvements in the dyeing and shrink-proofing of plasma-treated fabrics (Sun and Stylios, 2004).

Plasma treatments are attempted in reducing pollution levels of chrome dyeing, used for dark shades (Kan, 1998b). Oxygen, nitrogen, and mixture (25%:75%) of hydrogen and nitrogen plasma treatments result in better half-time-of-dyeing and improvements in the final exhaustion (%) levels (Table 13.5). Samples of fabrics processed with the combination of hydrogen and nitrogen plasma resulted in the fastest rate of dyeing compared to the nitrogen, oxygen, and plasma-treated fabrics and untreated fabrics. Exhaustion rate of hexavalent-chromium was observed faster in the samples treated with plasma than untreated fabrics. The decrease in the chromium concentration in the dye bath remains constant after certain duration of treatment, i.e., 40–50 minutes. Fixations of trivalent chromium also improved for low-temperature plasma-treated fabrics.

In the case of dyeing, regular dyeing process and poor wettability of untreated fibers result in poor dye exhaustion initially but then proceeds similar to that of treated fabrics. Nevertheless, total time required for exhaustion was low in the case of plasma-treated samples. Modification of fiber surface often changes the dyeing isotherms and diffusion of dyes into the fibers. Both untreated and plasma-treated samples showed similar depth of dyeing though overall improvements were observed in the case of plasma-treated samples including dry-rubbing fastness, wash fastness, and perspiration fastness values (Kan et al., 2004).

In printing, the corona treatment helps to improve the binding among the fiber-binder-pigment system which is responsible for increase in wet/dry rubbing fastness (Carneiro et al., 2001).

TABLE 13.5
Influence of Plasma on Dyeing Time and Exhaustion

Material	$T_{1/2}$	Final Exhaustion (%)
Untreated	22.53	97.16
Oxygen Plasma Treated	21.06	98.73
Nitrogen Plasma Treated	17.34	98.67
$H+N_2$ Plasma Treated	12.52	98.91

13.6 FUNCTIONAL FINISHING

Plasma treatment results in the effects that are decaying in nature with respect to time or a durable ones (Roth, 2001) depending on nature of gases used. Hydrophilization of the substrate increases the penetration depth and velocity of penetrating front, and it was evident in the plasma treatment carried out on cotton using RF plasma at 20 kHz using plasma power of 0.64 W/cm^2 at a pressure of 0.6–8 mbar (Poll et al., 2001).

Shrinkage in a fabric takes place due to (i) relaxation, (ii) consolidation, and (iii) felting (Kan et al., 2004, Sun and Stylios, 2004). Shrinkage resistance can be imparted on wool fabrics using low-temperature plasma treatments and etching effects, which decrease the natural shrinking tendency (Kang and Sarmadi, 2004; Kan et al., 2004; Jhala, 2005). Felting dimensional change is an irreversible dimensional change, which occurs during agitations in laundering and dimensional changes obtained in warp direction relatively higher than weft. Felting dimensional change reduces from 9.6% to 1.1% for the treated fabrics in warp direction and 12.3% to 1.5% in the area shrinkage. Combinations of oxygen low temperature plasma and silicone polymer treatment improve the dimensional stability higher than silicone treatment alone (Kang and Sarmadi, 2004). Reduced surface tension helps to add polymer onto the surface, and Hercosett/PMS treatment subsequent to plasma treatment results in high shrinkage resistance (3.8%–6.4%) compared to only-plasma treatment (27.0%–14.0%) (Ralkowski, 1997).

Plasma treatments can be used to coat different kinds of gases without changing bulk properties of the fibers. Plasma treatments on wool fibers enhance shrink resistance and dye absorption to higher levels than the conventional treatments using DCCA, Dylan though it is lower than oxidation treatments like chlorination treatment. Yellowing effects during steaming operation of wool fibers reduce with plasma treatments, carried out in setting and finishing processes. Both abrasion resistance and breaking force of the fabric increase along with reduced time of half dyeing.

The effect of plasma treatment on fabrics made of various fibers and the relevant results are detailed in the following Table 13.6.

Ratio of F/C, F/O (fluorine content) of the finishing agents in the textile fabrics influences the water repellency effects. Surface resistivity of the samples, after treatment, reduces due to formation of the polar groups on the surface of cotton fibers and formation of voids and cracks, which also influences electrical conductivity. Attempts have been made to compare repellency effects obtained by conventionally finished fabrics and fabrics treated with plasma of CF_4 and C_3F_6 gases, to assess the barrier property actions of blood or body fluids on medical/surgical gowns (Virk et al., 2004), where plasma-treated fabrics showed better repellency and zone inhibition against microbes.

Grafting of comonomers has been carried out using plasma treatment to obtain durable flame-retardant finish (Tsafack and Grutzmacher, 2006). Acrylate monomers containing phosphorous like diethyl(acryloyloxyethyl)phosphate, diethyle-(methacryloyloxyethyl) phosphate, diethyl(acryloyloxymethyl)phosphate, and dimethyl(acrylooyloxymethyl) phosphorate are grafted using air plasma. Phosphoramidate monomers resulted in LOI values up to 29.5 exhibiting higher retardancy (Yusuf, Mohd and Faqeer, 2017; Yusuf, 2018).

TABLE 13.6
Effect of Plasma Finishing on Various Fabrics

Fiber	Process	Process Conditions	Effects	Results
Cotton	Absorbency	RF at a pressure 0.6–8 mbar, air-oxygen RF, at 9 Pa pressure and 120 W, air-oxygen	Weight reduction, formation of C=O, COOH, enhanced wicking	Etching
	Hydrophobicity	RF, 300W, 75 mTorr for CF_4 160W, 150 mTorr for C_3F_6	Lower hydrophobicity for CF_4 than C_3F_6	Plasma induced polymerization
Linen	Wicking	RF, pressure 15 Pa with power 200W, argon-oxygen	Wicking rate decreases	No significant effects on prolonged exposure
Flax	Surface Modification	RF 13.56 MHz, pressure 15 Pa and 200W, argon, oxygen	Etching and revelation of fibrillar structure	Formation of micropores
Wool	Hydrophilicity	RF, power 100 W and pressure 100 Pa with water vapor	Elimination of fatty layer, formation of hydrophilic groups	Removal of epicuticle
	Shrink Resistance	Glow discharge, low temperature at 10 Pa pressure, oxygen	Decreased felting, higher shrink resistance and dye rate	Creation of micropores and cleft
BOPP*	Increased surface energy and wettability	RF, pressure (1–10 Pa), oxygen glow discharge	Formation of –OH, -C=O. Increase in surface energy (24 to 71 mJ/m^2)	Surface roughness, loss of hydrogen and low reflectance
Nylon	Water Repellency	RF, power 100 W and 4 Pa, CF_4	Absorption of F on surface	Improved glossiness

* BOPP, biaxially oriented polypropylene film.

Functional groups that anchor TiO_2 on the textile surface are formed in the plasma treatment using RF/MW plasma and UV radiation and that also results in the formation of TiO_2 crystallites with 5–7 nm size, from the precursor (Bozzi et al., 2005). This, in turn, results in the self-cleaning cotton textiles for various applications.

Improvement in wettability, hydrophobicity, and biocompatibility of silk sutures and biological support materials was observed in the low-temperature plasma treatment (Demura et al., 1992; Selli et al., 2001; Gu, Yang and Zhu, 2002; Chen et al., 2004; Chaivan et al., 2005). Radio frequency plasma treatment using SiF_6 gas on silk material improves the hydrophobicity and exhibits ageing effects at a frequency of 13.56 MHz, at a pressure of 1–7 mTorr, and the power supply in the range of 25–75 Watts (Selli et al., 2001; Chaivan et al., 2005). Attachment of fluorine atoms onto the polymer surface and increase in crystallinity result in water repellency and higher surface oxygen level (%) as a result of radical scavenging and formation of peroxides, after plasma treatment.

Antithrombogenicity of SO_2 gas/two-step SO_2–NH_3 gas plasma-treated films increases on account of surface sulfonation/nitration (Gu et al., 2002). Treatments with low-temperature plasma using oxygen and tetrafluoro methane gases on silk fibroin result in successful immobilization of alkaline phosphatase and enhanced activity levels (Demura et al., 1992). Improved loading capacities of silk sutures treated with low temperature oxygen plasma are observed due to formation of surface flutes and etching effects due to plasma result in reduction of suture weights (Chen et al., 2004).

13.7 EFFECTS OF PLASMA TREATMENT ON PROPERTIES

Glossiness index of the plasma-treated samples increases and then decreases subsequently in the warp direction (Kan, 1998a). At different angles, glossiness values remained unchanged and decreased toward the viewing angle 85°. However, glossiness distribution curves of both treated and untreated fabrics remain unchanged since the plasma does not change overall construction of the fabrics. Electrostatic charges present in the surface of plasma-treated fabrics elapse faster compared to untreated fabrics with 243–249 seconds and 263–283 seconds, respectively (Kan, 1998a).

The effect of plasma reactions on the textile substrates depends on both gases used in the process and nature of materials treated (Bajaj, 1998; Carneiro et al., 2001). Free radicals produced in the plasma environment often undergo various reactions depending on the gases present in the system and the other process variables. The presence of carboxylic groups in the cellulose/cotton fibers decides the pH values of the fibers, which are evident in the plasma-treated samples. Aqueous extracts of raw cotton fabrics and corona-treated fabrics showed significant difference at 6.60 and 5.0, respectively (Carneiro et al., 2001) with a number of passages in the corona reactor that result in the formation of carboxylic groups. Acidification, formation of carboxylic acid groups, increases with increase in purity levels of cotton fibers, and two-fold increase in pH was observed in the case of bleached fibers.

Plasma treatments cause surface ablation/etching of fiber surfaces, resulting in different kinds of surface roughness, cracks, and/or fissures. Even very high power supply with prolonged treatment results in no significant change in tenacity and elongation at break values for both PET and cotton fabrics. The sonic velocity of plasma-treated fabrics increased due to etching of primary wall and exposing the ordered secondary layers of cellulose in cotton fibers. But the treatment of PET fabrics results in reduction of sonic velocity due to disruption in the surface orientation, which decreases with the power of treatment. Because of these reasons, decrease in moisture content was observed with cotton, while it increased with PET. The voids and cracks developed in PET also aid the penetration of moisture. Moisture content of the hydrophilized polyester fabrics increases to 2%.

Reduction in the tensile strength has been reported in the plasma-treated cotton fabrics, and interestingly, such samples showed lower weight loss values in the subsequent cellulase enzyme treatment. On the other hand, combination of oxygen plasma-protease treatment resulted in higher rates of weight loss values. Since the plasma treatments are carried out under dry state of substrates, the fibers do not swell, and changes in the surfaces are restricted. Directional frictional effects in the wool fibers are reduced even though fiber-to-fiber friction, surface roughening, and harsh handle increase due to plasma treatment.

An attempt has been made to study the effect of nonpolymerizing gases like oxygen, argon, and tetrafluoromethane on Nylon 6 woven fabrics using glow discharge generator with a discharge power of 100 W and 4 Pa pressure (Yip el al., 2002). Prolonged treatment results in ripple-like patterns oriented perpendicular to the fiber axis. Oxygen plasma gives more distinct effects than argon plasma due to its inertness. In the case of nylon 6, 6 fabrics subjected to atmospheric pressure, helium and helium-oxygen plasma for selected exposure intervals led to a wide range of results; no surface changes take place with helium plasma while aligned grooves and bubbling in the surfaces of helium-oxygen-treated fibers are observed (Canup et al., 2009).

With a view to improving the processability of wool fibers, low-temperature plasma treatments are attempted on wool fibers and fabrics, and changes in physical and chemical properties have been reported (Anon, 1997; Ralkowski, 1997; Kan, 1998a,b; Osenberg, 1999; Molina, 2003; Kan, Chan and Yuen, 2004; Kang and Sarmadi, 2004; Sun and Stylios, 2004; Jhala, 2005; Kan and Yuen, 2006; Yusuf, 2018). Wool fibers consist of cuticular cells that are located in the outermost part of the fiber, surrounding the cortical cells. The cuticular cells consist of endocuticle, exocuticle, and exterior hydrophobic epicuticle (thin membranes). Modification in the endocuticle and interscale cell membrane facilitates the diffusion of water vapor and liquid water (Kan, 1998a). UV radiations can interact with the substrates depending upon absorption coefficient of substrates, capable of breaking the C–C and C–H bonds and leading to the formation of free radicals (Molina, 2003). TEM analysis of plasma-treated samples showed modification of cuticle, layer 'A', to varied degrees due to groove formation resulting due to the sputtering effect (Kan, 1998b). Wool fibers show stable free radicals attributed to carbon- or oxygen-centered radicals. Electron spin resonance suggests the new nitrogen radicals which is stable under UV/Vis irradiation.

Plasma treatment on knitted fabrics and various woven fabrics shows very interesting results (Ralkowski, 1997). On account of surface roughening, some of the knitted fabrics showed harsh handle that would adversely affect the innerwear fabrics. The effects of plasma treatment on low stress mechanical properties of nylon fabrics have been analyzed with reference to the control fabrics (Yip et al., 2002). Plasma treatment results change in different properties to various extents. The removal of uneven surfaces at the beginning that results in smoothness becomes ripples for both oxygen and argon plasmas, after the prolonged treatment. The frictional effects of the plasma-treated fabrics reduce slightly at the beginning and then increase because of the etching effects (ripples) observed in the samples. Decrease in tensile energy, bending rigidity, and shear stiffness has been observed at the beginning, but all the properties improve significantly after prolonged treatment due to the formation of rough surfaces and more contact points and also the cohesion. The compressional resilience and tensile resilience decrease with treatment due to the action of cohesive forces. Air permeability is altered according to the change in fabric thickness and alteration in the fabric surface. Surface roughness voids and spaces created in the plasma treatment increase the air trapped in the fabrics and yarns, and the trapped air acts as a good insulation medium and prevents heat loss of fabrics.

13.8 CONCLUSION

Processing of textile materials involves operations starting from fiber to garments, huge amounts of water, chemicals, formation of byproducts, and discharge of these products into the environments in different forms. The recent focus on sustainability of various processes emphasizes protecting the environment with as minimal damage as possible with different measures. Plasma technology and plasma-assisted processing of textile materials eliminates stress on the processing needs in terms of various resources, time, and energy, thereby inherently including sustainability into every processor stage involved in the textile manufacture. A wide range of techniques are available for a manufacturer if one decides to adopt plasma techniques in the process line. However, maintaining a uniform plasma field and access to the bulk of the substrate needs to be probed further to ensure large-scale commercial success.

REFERENCES

Abbot, G.M. and Robinson, G.A. 1977. The corona treatment of cotton – yarn and fabric properties. *Textile Research Journal*, 47, 3: 199–202.
Anon. 1997. Europlasma Technical Paper Techtextil Symposium 1997 Session 5.1 Lecture 516. http://acms.lodestar.be/europlasma/bestanden/4.8.EP%20TP%20RolltoRoll%20Treatment.pdf (accessed 21 Jan, 2019).
Bae, P.H., Hwang, Y.J., Jo, H.J., Kim, H.J., Lee, Y., Park, Y.K., Kim, J.G. and Jung, J. 2006. Size removal on polyester fabrics by plasma source ion implantation device. *Chemosphere*, 63, 6: 1041–0147.
Bajaj, P. 1998. Eco-friendly finishes, in *Environmental Issues – Technology Option for Textile Industry*. Eds. Chavan R.B., Radhakrishnan J. 93–96. IIT Delhi Publications, New Delhi.
Bhat, N.V., Benjamin, Y.N., Gore, A.V. and Upadhyay, D.J. 2003. Plasma Processing of Textile Fibres. *44th Joint Technological Conference (SITRA, NITRA, ATIRA, BTRA)*, March, 60–66.
Bozzi, A., Yuuranova, T., Guasaquillo, L., Laub, D. and Kiwi, J. 2005. Self-Cleaning of modified cotton textiles by tio$_2$ at low temepratures under day light irradiation. *Journal of Photochemistry and Photobiology A: Chemistry*, 174, 2: 156–164.
Cai, Z., Hwang, Y.J., Par, Y.C., Zhang, C., McCord, M. and Qiu, Y. 2002. Preliminary investigation of atmospheric pressure plasma aided desizing for cotton fabrics. *AATCC Review*, 2, 12: 18–21.
Cai, Z., Qiu, Y., Hwang, Y.J., Zhang, C. and McCord, M. 2003. The use of atmospheric plasma treatment in desizing pva on viscose fabrics. *Journal of Industrial Textiles*, 32, 3: 223–232.
Canup, L., McCord, M., Hauser, P., Qiu, Y., Cuomo, J., Hankins, O. and Bourham, M.A. 2012. Modification of nylon fabrics with atmospheric pressure plasmas. www.ntcresearch.org/pdf-rpts/Bref0602/C99-NS09-02.pdf (accessed 23 Jan, 2012).
Carneiro, N., Souto, A.P., Silva, E., Marimba, A., Tena, B., Ferreira, H. and Magalhaes, V. 2001. Dyeability of corona treated fabrics. *Coloration Technology*, 117, 5: 298–302.
Chaivan, P., Pasaja, N., Boonyawan, D. and Suanpoot Vilaithong, T. 2005. Low temperature plasma treatment for hydrophobicity improvement of silk. *Surface and Coatings Technology*, 193, 1–3: 356–360.
Chen, Y.Y., Lin, H., Ren, Y., Wang, H.W. and Zhu, L. 2004. Study on bombyx mori silk treated with oxygen plasma. *Journal of Zhejiang University of Science*, 5, 8: 918–922.
De Geyter, N., Morent, R. and Leys, C. 2006. Surface modification of polyester nonwoven with dielectric barrier discharge in air at medium pressure. *Surface Coating Technology*, 201, 6: 2460–2466.

Demura, M., Takekawa, T., Asakura, T. and Nishikawa, A. 1992. Characterization of low temperature plasma treated silk fibroin fabrics by esca and the use of the fabrics as an enzyme immobilization support. *Biomaterials*, 13, 5: 276–180.

Gu, J., Yang, X. and Zhu, H. 2002. Surface sulfonation of silk film by plasma treatment and in vitro anti-thrombogenicity study. *Materials Science and Engineering: C*, 20, 1–2: 199–202.

Jhala, P.B. 2005. *Plasma textile technology.* NID Publications, India, 3–32.

Kan, C.W. and Yuen, C.W.M. 2006. Surface characterisation of low temperature plasma treated wool fibre. *Journal of Materials Processing Technology*, 178, 1–3: 52–60.

Kan, C.W., Chan, K. and Yuen, C.W.M. 2004. Low temperature plasma treatment of wool fabric for its industrial use. *Indian Journal of Fibre and Textile Research*, 29, 4: 385–390.

Kan, C.W., Chen, K., Yuen, C.W.M. and Miao, M.H. 1998a. The effect of low temperature plasma on chrome dyeing of wool fibres. *Journal of Materials Processing Technology*, 82, 1–3: 122–126.

Kan, C.W., Chan, K., Yuen, C.W.M. and Miao, M.H. 1998b. Surface properties of low temperature plasma treated wool fabrics. *Journal of Materials Processing Technology*, 83, 1–3: 180–184.

Kang, J.Y. and Sarmadi, M. 2004. Textile plasma treatment review – natural polymer based textiles. *AATCC Review*, 4, 10: 28–32.

Kutlu, B. and Cireli, A. 2004. Plasma technology in textile processing. *3rd Indo-Czech Textile Research Conference*, 14–16, June Liberec, No. Turkey 10 through http://www.ft.vslib.cz/indoczech-conference/conference-proceedings/index.html (accessed 21 Jan 2019).

Molina, R., Erra, P., Julia, L. and Bertran, E. 2003. Free radical formation in wool fabrics by low temperature plasma. *Textile Research Journal*, 73, 11: 955–959.

Osenberg, F., Theirich, D., Decker, A. and Engemann, J. 1999. Process control of a plasma treatment of wool by plasma diagnostics. *Surface and Coatings Technology*, 116–119, 9: 808–811.

Poll, H.U. and Scwaditz, U. 2001. Penetration of plasma effects into textile structures. *Surface Coatings and Technology,* 142–144: 489–493.

Ralkowski, W. 1997. Plasma treatment of wool today – Part 1 – Fibre properties, spinning and shrink-proofing. *Journal of Society of Dyers and Colorists*, 113, 9: 250–255.

Rashidi, A., Moussavipourgharbi, H. and Mirjabili, M. 2004. Effect of low temperature plasma treatment on surface modification of cotton and polyester fabrics. *Indian Journal of Fibre and Textile Research*, 24, 1: 74–78.

Roth, J.R. 2001. *Industrial Plasma Engineering – Volume 2*, Institute of Physics Publishing, Philadelphia.

Saravanan, D. and Nalankilli, G. 2008. Atmospheric plasma induced modifications on natural fibres. *Indian Textile Journal*, 118, 11: 29–36.

Sarmadi, M. 2013. Advantages and disadvantages of plasma treatment of textile materials. *21st International Symposium on Plasma Chemistry*, 4–9 August, Queensland, Australia, 1–4.

Selli, E., Claudia Riccardi, C., Massafra, M. R. and Marcandalli, B. 2001. Surface modifications of silk by cold SF_6 plasma treatment. *Macromolecular Chemistry and Physics*, 202, 9: 1672–1678.

Shanmugasundaram, O.L. and Rengasamy, K.S. 2006. Application of plasma in textile industry. *Textile Asia*, 5: 44–47.

Subbulakshmi, M.S. and Kasturiya, N. 1998. Effect of plasma on fabrics. *Indian Textile Journal*, 10: 12–16.

Sun, D. and Stylios, G.K. 2004. Effect of low temperature plasma treatment on the scouring and dyeing of natural fabrics. *Textile Research Journal*, 74, 9: 751–756.

Tsafack, M.J. and Grutzmacher, J.L. 2006. Flame retardancy of cotton textiles by plasma induced graft polymerisation. *Surface and Coating Technology*, 6, 4: 2599–2610.

Virk, R.K., Ramasamy, G.N., Bourham, N. and Bures, B.L. 2004. Plasma and antimicrobial treatment of nonwoven fabrics for surgical gowns. *Textile Research Journal*, 74, 12: 1073–1079.

Yip, J., Chan, K., Sin, K.M. and Lau, K.S. 2002. Low temperature plasma treated nylon fabrics. *Journal of Materials Processing Technology*, 123, 1: 5–12.

Yusuf, M. 2018. A review on flame retardant textile finishing: current and future trends. *Current Smart Materials*, 3, 2: 99–108.

Yusuf, M., Mohd, S. and Faqeer, M. 2017. Natural colorants: historical, processing and sustainable prospects. *Natural Products and Bioprospecting*, 7, 1: 123–145.

14 State of the Art Colorization Techniques in Textile Industry

Aleem Ali
Glocal University

Farah Jamal Ansari
University Polytechnic

Anushka Dhiman
Delhi University India

CONTENTS

14.1 Introduction 234
14.2 Nanotechnology 234
   14.2.1 Nanosized Pigment Particle Applications 235
   14.2.2 Nanocomposites 235
14.3 Electrochemical Coloration 237
   14.3.1 Indirect Electrochemical Reduction 237
   14.3.2 Direct Electrochemical Reduction 238
14.4 Supercritical Carbon Dioxide Coloration 238
   14.4.1 Theoretical Background 239
   14.4.2 Coloration Approaches 239
14.5 Plasma Technology 240
   14.5.1 Coloration of Plasma-Treated Polyester Fibers 240
   14.5.2 Coloration of Plasma-Treated Wool Fibers 241
   14.5.3 Coloration of Plasma-Treated Cotton Fibers 241
14.6 Ultrasonics 241
   14.6.1 Ultrasonic-Assisted Coloration of Natural Fibers 242
   14.6.2 Ultrasonic-Assisted Coloration of Synthetic Fibers 243
14.7 Microwave 243
14.8 Conclusion 244
References 244

DOI: 10.1201/9781003140467-14

14.1 INTRODUCTION

Color affects every moment of our lives strongly in many forms: the clothes we wear, the furnishings in our homes, etc. It has always played an important role in the formation of different cultures of human being all over the world. From prehistoric times, humans had used natural colors obtained from various places such as minerals, plants, and insects/animals for their daily use articles as well as walls and the surrounding of their own houses (Yusuf et al. 2017, 2018; Tripathi et al. 2015). The unique character of their works was the result of using different mixtures of dyes and mordants, as varnishes and lacquers responsible for the cohesion of the pigments and protection of the layers destroyed by environmental effects (Yusuf et al. 2016, 2017; McDonald 1997). In the present scenario, colors obtained from the natural origin are nowadays witnessing very rapid growth for their use in sustainable apparel production as well as in other advanced application disciplines such as food sector and dye-sensitized solar cells (Yusuf et al. 2016).

Coloration is the application of color onto fiber, yarn, or cloth, typically with natural colors (Yusuf et al. 2013, 2016, 2017; Hosseinnezhad et al. 2021) or synthetic organic dyes (Yusuf et al. 2017; Hosseinnezhad et al. 2021). Dye and required chemicals are applied to the textile in this process to achieve a uniform and a fast color property appropriate for the end user. According to the requirement of the end user of the textile, different fastness criteria can apply. For example, swimsuits that must not bleed in sunlight, as well as car fabrics that must not fade after extended sun exposure (Nie et al. 2008; Rajoriya et al. 2018; Yusuf et al. 2017; Hosseinnezhad et al. 2021). Dyes are used in various applications of textiles in many ways, including continuous pad applications or batch dyeing. Knitted fabric is dyed in batch equipment using exhaust methods, while woven fabric is dyed continuously. Per year, aged products and processes are phased out in favor of the technological invention of new ones. The demand for safer, more cost-effective, and value-added textile commodities is driving this trend (Yusuf et al. 2017; Hosseinnezhad et al. 2021; Mani and Bharagava 2018).

The latest ongoing inventions in the coloration of textile fiber are discussed here based on physical chemical technologies such as electrochemistry, microwave, carbon dioxide coloration, ultrasonic, plasma as well as in nanotechnology.

14.2 NANOTECHNOLOGY

Nanotechnology has been around for over 40 years. Nanotechnology is characterized as activities that take place at the atomic and molecular level. Nanotechnology is characterized as activities that take place at the atomic and molecular levels and have real-world applications (Joshi and Bhattacharyya 2011; Yusuf et al. 2017). Nanoparticles within the span of 1–100 nm are typically used in commercial items. Nanotechnology is gaining popularity around the world as a technology with huge potential for the different kind of products.

The textile commercial has a lot of promise with nanotechnology. This is primarily due to the fact that traditional methods for imparting various properties to fabrics often do not last for long and lose their functionality after being washed or used.

State of the Art Colorization Techniques

Furthermore, a nanoparticle coating on fabrics has no effect on their comfort. As a consequence, the use of nanotechnologies in the cloth industry is increasing.

A subsidiary of Burlington Industries, Nano-Tech, United States, conducted the first research on nanotechnology in textiles. Since then, many countries have contributed to the nanotechnology. Coating is a common method for applying nanoparticles to textiles (Joshi and Bhattacharyya 2011; Joshi 2008). Coating fabrics can be done in a variety of ways, like showring, cleaning, rinsing, and padding. Padding is the maximum utilized in this sort. The nanoparticles are adhered to the fabrics using a padder set to the appropriate pressure and pace and then dried and cured. Water repellent, wrinkle resistant, soil resistant, antistatic, antibacterial, UV defence, flame retardation, and dye absorption enhancement are few of the properties added to the textile by nanotechnology (Yusuf 2017, 2018;; Sayed and Dabhi 2014; Saleem and Zaidi 2020). Water abhorrence, land obstruction, wrinkle resistance, antibacteria, antistatic, UV defence, flame retardation, dyeability improvement, and more are some of the properties imparted to textiles by nanotechnology. Properties related to textile coloration are outlined here, as there are several possible applications of nanotechnology in the textile industry (Figure 14.1).

14.2.1 Nanosized Pigment Particle Applications

The advancement of nanotechnologies has ignited interest in nanosized pigment particle applications in textile manufacturing. Exhaust dyeing of cationized cotton with nanoscale pigment dispersion has shown that the dyeing is better in soft handle, and shade obtained is also better with less pigmentation as obtained by pigment dispersion (Nie et al. 2020; Fang et al. 2005; Song et el. 2020).

Langhals is the method having the possibility of obtaining nanodispersion of lipophilic perylene bisimide pigment. The UV/Vis spectra obtained with the pigment nanodispersion is very much same similar to that of the homogeneous lipophilic solution, for example, chloroform. This shows that the dispersant completely covered the nanosize chromophore. Similar dispersions were shown for chloroform dispersions.

14.2.2 Nanocomposites

Nanocomposites are materials made by combining nanoparticles with a macroscopic sample material. This is a branch of the rapidly developing field of nanotechnology. The properties of the resulting nanocomposite may be greatly improved after nanoparticles are added to the matrix material (Nie et al. 2020; Fang et al. 2005; Song et al. 2020).

Polypropylene (PP) is less expensive in comparison of polyester and nylon, but it is difficult to dye. Copolymerization, polyblending, and grafting are a few typical processes that can increase the dyeability of PP; however, all of these processes are expensive, which boosts the total cost of the PP fiber (Ahmed and El-Shishtawy 2010). Corrosive and scatter colors can be utilized to color this new polypropylene. The ionic fascination between the contrarily charged corrosive color and the emphatically charged quaternary ammonium salts in the nanoclay makes the nanoPP corrosive dyeable. Van der Waals powers, just as potentially hydrogen holding, assume

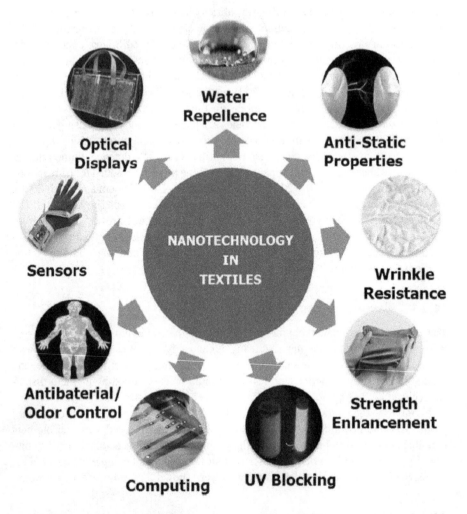

FIGURE 14.1 Representation of various utilizations/applications of nanotechnology-based textiles. Reproduced from Saleem and Zaidi (2020) under the Creative Commons Attribution (CC BY) license.

a part in the corrosive tinge of nanoPP. In the tinge utilizing scattered color, additionally, the fascination between the nanoclay and scatter color is because of van der Waals powers and hydrogen holding. The corrosive dyeability of nanoclay/PP nanocomposites utilizing three compound constructions, in particular anthraquinone, premetallized, and monoazo, with four significant tones, i.e., C.I Acid Blue 80, C.I Acid Orange 74, C.I Acid Red 266, and C.I Acid Yellow 135 was additionally observed (Razafimahefa et al. 2008).

Moreover, it was likewise found that scattered colors color nanoclay/polyamide 6 (PA6) nanocomposite yarn quicker than unfilled PA6 yarn, while corrosive colors and 2:1 metal complex colors do the inverse (Razafimahefa et al. 2005). In all of these cases, the nanoclay holds fast to the amino locales, halting the corrosive or metal complex colors from clinging to them. Polyester (PET) nanocomposite fiber was made utilizing polyester and silica (SiO_2) nanoparticles determined to build PET's worth and further developing dyeability with scatter color. When contrasted with unadulterated PET filaments, PET/SiO_2 nanocomposite strands showed a more noteworthy level of weight reduction during soluble hydrolysis (Adak et al. 2020). More and harder superfine constructions, like breaks, cavities, and pits, were added, taking into account further tinge in explicit applications.

14.3 ELECTROCHEMICAL COLORATION

Vat and sulfur dyes account for a significant portion of the dyestuff demand in the coloration of cellulose fibers. The situation appears to be stable in the near future, owing to the fact that vat dyes produce colored fibers best all-around quickness, especially to chlorine bleaching. Likewise, sulfur dyes are especially necessary for the manufacturing of low-cost products with a common fastness need. The dyes are light and washing-resistant, but not chlorine-resistant (Park et al. 1999; Tarábek et al. 2006).

The diminishing specialists utilized in the hue interaction for vat and sulfur colors are not recyclable, bringing about dangerous waste. Different electrochemical lessening techniques have been recognized in ongoing examinations to work on the biocompatibility of the vatting system much further, including circuitous electrochemical decrease utilizing a redox middle person and direct electrochemical decrease of indigo. These methodologies have major natural advantages since they lessen synthetic utilization and gushing burden. In this vein, the European exploration drive EUREKA extended the Lillehammer Award in June 2006 for DyStar Textilfarben GmbH's electrochemical coloring project, which was completed as a team with the Institute for Textile Chemistry at the University of Innsbruck and Textile Physics (TID) in Dornbirn, Austria, and Getzner Textil AG. An electric flow is combined with a recyclable go between in this exclusive innovation. It replaces the nonregenerative decreasing specialists at present used to add tank and sulfur colors to materials, which debilitate dyebath reusing and dirty yield profluent. Therefore, the honor winning undertaking marks can be a turning point in the mechanical use of electrochemical coloring.

14.3.1 INDIRECT ELECTROCHEMICAL REDUCTION

The electrons move from one cathode layer to the other layer of the dispersed dye pigment microcrystals is the rate-limiting stage in electrochemical reduction. This is particularly true if electrons must be directly transferred between solid surfaces. To improve the rate of electron transfer, an unintended electrochemical mitigation process with the help of a soluble redox mediator was developed.

Regenerative triethanolamine or gluconic acid as ligands is used as mediators in this process. These mediators, on the other hand, are costly and not completely safe from a toxicological standpoint. Furthermore, the mediator must be isolated from the soluble leuco dye by ultrafiltration after the reduction and before the coloration step, significantly increasing the costs of this vatting process.

14.3.2 DIRECT ELECTROCHEMICAL REDUCTION

A novel electrochemical vatting measure for vat and sulfur colors has been recognized as an option in contrast to the intervened electrochemical decrease. This decrease doesn't require the presence of a redox go between endlessly. To start the decrease, a limited quantity of leuco color, which goes about as an electron-transport between the cathode and the outside of the color shade, should be delivered first. The leuco color, which goes about as an electron-transport between the terminal and the color shade's surface, should initially be delivered in a modest quantity to begin the decrease, which then at that point forges ahead its own.

In terms of economics and the environment, electrocatalytic hydrogenation is promising and appealing. Adsorbed hydrogen works with adsorbed organic substrates at the electrode surface in a process known as electrochemical hydrogenation. The hydrogenation stage competes with the hydrogen evolution reaction for efficiency, and this competitiveness determines the electrocatalytic hydrogenation efficiency. Electrocatalytic hydrogenation has a number of benefits over traditional catalytic hydrogenation (e.g., elevated temperatures and pressures can be avoided, and the explosion risk is minimized). To summarize, the latest research concerning direct electrochemical reduction on graphite granules appears to be the most desirable method in terms of stability, availability, and costs, and the results are clearly a promising basis for further development. The mediator method is about to hit the market, and the message is clear: electrochemistry in the textile industry is coming our way.

14.4 SUPERCRITICAL CARBON DIOXIDE COLORATION

Supercritical carbon dioxide (scCO$_2$) has of late arisen as an elective innovation for making an all the more earth feasible tinge strategy in the material business. In a few modern material applications, for example, yarn readiness, shading, and completing, scCO$_2$ hue innovation can settle numerous natural and specialized issues (Hou et al. 2004). As a result of their thermo-physical and transport properties, scCO$_2$ might be an exceptional mechanism for shipping synthetic compounds into or out of a polymeric substrate. Supercritical liquids have viscosities and diffusivities that are equivalent to those of gases, just as densities that are like those of fluids. Besides, under specific conditions, carbon dioxide is nontoxic, nonflammable, harmless to the ecosystem, and synthetically dormant.

At DTNW in Krefeld, Germany, the dissolving power of scCO$_2$ for disperse dyes and its use as a transport medium for coloration of polyester were investigated from all theoretical angles. From 1995 to 1997, our laboratory worked with the DTNW in Krefeld, Germany, on research projects (Karthikeyan et al. 2021). According to these findings, the existence of intramolecular hydrogen bonds and the dye molecule's hydrophobicity are both positive factors for better solubility in supercritical carbon dioxide, as shown by increased dye uptake. Recently, Abate et al. (2020) using a water-free supercritical CO$_2$ system investigated the sustainable approach to the fabrication of multifunctional polyester fabric with bioactive natural colorant curcumin. In this study, the dyed fabrics showed desirable UV-protection, antioxidant, and antimicrobial properties with acceptable color fastness to washing and rubbing characteristics.

14.4.1 THEORETICAL BACKGROUND

Very few fluids are generated by the pressure and temperature changes on gases and liquids. A rise in temperature above the triple point pushes liquid into the vapor phase, while a rise in pressure pushes vapor back into liquid. At a pressure of 73.8 bar and a temperature of 31.1°C, carbon dioxide reaches its critical stage.

14.4.2 COLORATION APPROACHES

Van der Kraan made the accompanying discoveries in a new postulation on the Process and Equipment Creation for Textile Dyeing in Supercritical Carbon Dioxide. To test polyester bar coloring in $scCO_2$ at 300 bar and 120°C, a specialized scale 100 L coloring machine was planned and created. Another sort of pressing factor vessel was utilized, comprising a steel liner enclosed by carbon filaments to assimilate outspread powers and a burden construction to ingest hub powers. Since the measure of steam used to warm the vessel is not exactly for a completely steel vessel, this plan lessens both the venture and working expenses.

The interaction time is additionally diminished in light of the fact that the carbon fiber vessel needs less warming because of the low warmth limit of the carbon strands. A low-pressure radial siphon was intended for use in $scCO_2$ and situated inside the coloring vessel to circle the CO_2 with the broke down color through the material. A 1000 L supercritical coloring machine was additionally made for treating 300 kg polyester while reusing all color and 96% of the CO_2. Accordingly, the supercritical interaction presently has a 50% lower measure cost. Cotton was changed to make it less hydrophilic and in this way more reasonable for coloring with scatter colors in $scCO_2$. Utilizing scatter responsive colors dependent on bromoacrylate and dichlorotriazine, it was additionally conceivable to shading cotton in $scCO_2$ without modification. Scatter responsive colors dependent on fluorotriazine were observed to be more receptive with cotton than chlorotriazine within the sight of methanol as a cosolvent in $scCO_2$ with an obsession level of 85, as per new examination. Moreover, hazier shades can be accomplished with fluorotriazines since the color focus can be raised up to 10% owf without causing harm to the cotton filaments.

Water addition has been analyzed in supercritical carbon dioxide ($scCO_2$) to perceive what it means for the dyeability of polyester, nylon, silk, and fleece with scattered receptive colors. Scatter colors with a responsive vinyl sulfone or dichlorotriazine bunch have been observed to be fitting for tinge of polyester, nylon, silk, fleece, or mixes of these filaments in $scCO_2$. Polyester color take-up is unaffected by the presence of water. The hue of amino-containing materials increases with the measure of water in the $scCO_2$ and the materials. Water had an advantageous impact either on account of its capacity to grow strands or as a result of its impact on the color fiber framework's reactivity. Profound tones were gotten with all colors for polyester, nylon, silk, and fleece at the immersion stage, with obsession rates going from 75 to 94 (Periyasamy and Militky 2020).

Nonetheless, endeavors were made to color regular strands with water-solvent colors utilizing the opposite micelle technique, where an ionic color that has been solubilized in a water-pool passes into the fiber alongside a limited quantity of water following contact. Due to the electrostatic repugnance between the color and the outside of the cotton fiber, good outcomes were gotten for proteinic strands yet not for cotton filaments.

14.5 PLASMA TECHNOLOGY

Plasma is described as to some extent ionized gas containing particles, electrons, and nonpartisan particles that are made when an electromagnetic field associates with a gas at a specific pressing factor. Present-day plasma-substance strategies are ecologically manageable and much better than synthetic change moves toward the utilization of destructive reagents like acids, hydroxides, antacid earth metals, and their mixtures. Plasma innovations are progressively supplanting wet substance applications in the pretreatment and completing of material textures in this setting (Abdelghaffar et al. 2020; Haji et al. 2020; Periyasamy and Militky 2020).

Low-temperature plasma treatment is quite possibly the most encouraging and progressed polymer alteration strategy, permitting surface properties to be changed across a wide reach and the application scope of polymeric materials to be enormously extended. The hydrophilicity of the treated fiber is expanded because of this surface change. Plasma treatment enjoys the benefit of just influencing the outside of the substance being dealt with, just as an extremely slight close surface layer whose thickness goes from 100 A to a few micrometers, contingent upon gauges. Under these conditions, the heft of the polymer stays unblemished, keeping up with the first mater's mechanical, physicochemical, and electrophysical properties (Abdelghaffar et al. 2020; Haji et al. 2020). The plasma's UV photons have sufficient energy to cut off synthetic bonds (e.g., C–C, C–H) and produce revolutionaries that can move along the chain and recombine. The action of the plasma causes initiation (extremist development), unsaturation (twofold bond arrangement), chain scission, and cross-connecting, contingent upon plasma conditions and the idea of the polymer.

14.5.1 Coloration of Plasma-Treated Polyester Fibers

Polyester textiles are usually dyed with disperse dyes using a pad-dry-bake (thermosol) process or at high temperatures and pressures. Otherwise, phenol-based carriers are expected to swell the fiber during coloration at atmospheric pressure, posing a possible environmental hazard. As a result, instead of using a chemical process to treat PET fibers, low-temperature plasma was used. After treating PET fabrics with both argon and air plasma, an increase in color depth was found after coloration. This was due to an increase in surface roughness and surface area caused by plasma. Furthermore, the introduction of hydrophilic groups, which can be caused by both reactive and chemically inert plasmas, can increase the water swelling capacity and affinity of PET fibers for dyes with polar groups, resulting in an increase in K/S values of dyed PET specimens. Nanostructured surfaces are appealing because they have a large surface area, and ultrathin coatings can thus have high functionality (Abdelghaffar et al. 2020; Haji et al. 2020).

Plasma polymerization of acetylene/ammonia onto PET fiber was carried out in a regime that included both deposition and etching, resulting in a nanoporous, crosslinked network of usable functional groups. The coated polyester fiber showed high color strength values per film thickness and was acid dyeable. Furthermore, the plasma-deposited and dyed polyester fabrics exhibited excellent rubbing and washing fastness, suggesting coating-functional permanency. The coating's high abrasion resistance proved that it was permanently adhered to the substrate. The dyeability of water-soluble acid dyes was found to be substantially improved due to the increased surface area and hydrophilicity of the PET fiber.

14.5.2 Coloration of Plasma-Treated Wool Fibers

In contrast to other regular strands, this has a mind boggling surface construction, making it perhaps the most fundamental material filaments. It does, notwithstanding, have some specialized issues, like wettability (which influences the dyeability). The presence of a high number of disulfide cystine crosslinkages (–S–S–) in the A-layer of the exocuticle, just as unsaturated fats on the fiber's surface, makes the fleece fiber surface hydrophobic. The dispersion in fleece fiber is though not really set in stone by the surface morphology (Abdelghaffar et al. 2020; Haji et al. 2020). Accordingly, rather than utilizing a substance cycle to treat fleece, low-temperature plasma (LTP) was utilized. LTP treatments have been displayed to improve the hue conduct of fleece strands in an assortment of shading frameworks. This might be because of plasma-initiated cystine oxidation, which decreases the quantity of crosslinkages in the fiber surface, considering transcellular just as intercellular color dissemination.

14.5.3 Coloration of Plasma-Treated Cotton Fibers

Cotton fibers are generally dyed with reactive dyes because they offer a wide color range and are easy to apply, particularly in exhaust coloration. Cotton fiber, on the other hand, has just a mild tolerance for reactive dyes. Attempts to resolve this restriction have been made in the past. Enhancing dye-fiber contact with cationized cotton is one of the most feasible methods. As compared to untreated cotton fiber, plasma-treated cotton fiber in the presence of amine compounds demonstrated better dyeability with reactive dyes.

14.6 ULTRASONICS

The phenomenon known as cavitation in a fluid medium, which is the development and breakdown of tiny air bubbles, can work on a wide scope of actual and chemical process. The abrupt and explosive breakdown of these air bubbles will bring about "problem areas," which are spaces of high temperature, strain, shock waves, and outrageous shear power fit for breaking substance bonds. As a major wet process that devours a great deal of energy and water and emanates a ton of gushing into the climate, a few endeavors have been made to research this strategy in material tinge (Vouters et al. 2004). Cavitation marvels are for the most part credited to upgrades in ultrasound-helped shading measures, and thus, other mechanical and synthetic impacts are delivered, for example,

- Dispersion (separating of totals with high relative atomic mass);
- Degassing (ejection of broke up or entangled air from fiber vessels);
- Diffusion (speeding up the pace of dispersion of color inside the fiber);
- Intense disturbance of the fluid;
- Destruction of the dispersion layer at color/fiber interfaces;
- Generation of free revolutionaries; and
- Dilation of polymeric nebulous districts.

Many employees have noticed an increase in coloration rates, which may be due to the combined effects of the above. Furthermore, the Marco Company of Korea produced an ultrasonic retrofit module for jet-dyeing machines that includes a generator, transducer, and electronic wire during the previous decade. Ultrasounds, an industrial solution to optimize prices, environmental demands, and efficiency for textile finishing, are another recorded attempt to manufacture production machines.

14.6.1 Ultrasonic-Assisted Coloration of Natural Fibers

Thakore's noteworthy work in 1988 exhibited that cotton texture could be ultrasonically colored with direct colors. The utilization of ultrasonic coloring enormously decreased coloring time, coloring temperature, color focus, and electrolytes in the colors shower, as per the discoveries. The utilization of C.I. Receptive Red 120 and C.I. Receptive Black 5 in ultrasonic-helped hue of cellulosic textures was examined. The hues were looked at as far as rate weariness, aggregate sum of color moved to the washing shower after hue, speed properties, and shading esteems utilizing conventional and ultrasonic strategies. Ultrasound upgraded color obsession and expanded rate depletion for both receptive colors and yet had no impact on the colored materials' quickness properties, as indicated by the outcomes.

The immediate colors Solophenyl Blue FGL 220 and Solophenyl Scarlet BNL 200 were utilized to color cotton. For exhaustion, the two colors required a significant measure of salt. Hue with direct colors at lower salt levels moves toward similar last weakness as tinge at higher salt levels without ultrasound, ultrasound has the best impact on hue at low temperatures, and it can limit the measure of salt and energy required when contrasted with an ordinary cycle. Both customary and sonicator techniques were utilized to examine the hue of cotton texture utilizing Eclipta as a characteristic color. The presentation of sonicator hue is 7%–9% higher than that of conventional tinge. Additionally, ultrasonic color take-up of cationized cotton texture with lac regular color was productive, with the upgraded impact after balance shading being about 66.5% more prominent than the traditional warming.

Colorization of silk utilizing cationic, corrosive, and metal-complex colors at low temperatures with the guide of a low-recurrence ultrasound of 26 kHz was researched, and the color take-up outcomes were contrasted with those acquired by customary cycles. In contrast with conventional shading, silk tinge within the sight of ultrasound builds color assimilation for all gatherings of colors at lower hue temperatures and a more limited hue time. Moreover, there was no proof of cavitation-instigated fiber injury. Lac as a characteristic color has been utilized to smudge fleece textures utilizing both conventional and ultrasonic strategies (Anwar et al. 2021; Helmy

State of the Art Colorization Techniques

2020). Moreover, the extractability of lac color from normal sources utilizing power ultrasonic was contrasted with customary warming. Ultrasonics were observed to be fruitful in the color extraction and color retention of lac colored fleece strands, with an upgraded impact of 41% and 47% over customary warming, individually.

14.6.2 Ultrasonic-Assisted Coloration of Synthetic Fibers

The impact of low-recurrence ultrasound on the colorization of polyester strands with C.I. Scatter Orange 25 and C.I. Scatter Blue 79 colors was considered, and unswollen PBT and PET strands were colored with and without low-recurrence ultrasound under different time and temperature conditions. The discoveries of this examination demonstrate that ultrasound further developed color particle dissemination through strands; however, the degrees of hue are not as high as in ordinary business hue measures. Likewise, utilizing C.I. Scatter Red 60, which has an emphatically translucent design, and C.I. Scatter Blue 56, which has a poor glasslike structure, an examination of the impact of ultrasound on the shading conduct of PET filaments was conducted (Periyasamy and Militky 2020). The discoveries uncovered that ultrasound significantly affects the decrease in molecule size of C.I. Scatter Red 60; however, it's important that ultrasound has no impact on color assimilation or shading rate for C.I. Scatter Blue 56. At various temperatures and under various responsive conditions, like ultrasound strength, medium pH, and introductory color fixation, nylon-6 fiber is inclined to low-recurrence ultrasound-helped hue with different gatherings of colors (Ahmed and El-Shishtawy 2010). At a low ultrasound field (27 and 38.5 kHz), hues with scatter, corrosive, corrosive severe, and responsive colors were explored, and expanding the hue rate diminished the enactment energy for all cases. Kamel likewise discovered that during the tinge dispersion measure, color take-up is expanded. The de-accumulation of the color particles prompts the improved impact, which lead to all the more likely color dispersion and expected help for color fiber bond obsession, as indicated by these test discoveries.

14.7 MICROWAVE

Microwave-assisted organic reactions are well known as environmentally friendly methods for speeding up a wide range of chemical reactions. In particular, in reactions that are run for a long time at high temperatures under traditional conditions, the reaction time and energy input are expected to be greatly reduced. Figure 14.2 depicts the comparison of conventional/traditional heating – surface heating (slow), with microwave heating – volumetric (fast) (Shahid et al. 2016). Many scholars have recognized this fact in textile coloration. In this regard, it has been stated that for dichlorotriazine reactive dyes, a short exposure period of 30–50 seconds yields good results. The effect of batching time on coloration of cotton with monochlorotriazine reactive dyes using microwave irradiation and traditional heating was investigated using the pad-batch process. The results show that using a microwave for a short time (2 minutes) was better than using traditional heating for a 12hour batching time.

Due to flax fiber's poor dyeability, a new method based on microwave treatment with urea to boost dyeability with reactive dyes was recently published. The dyeability

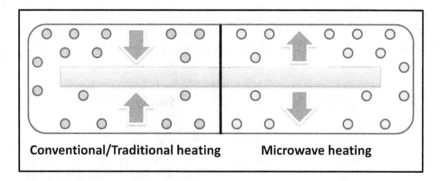

FIGURE 14.2 Conventional/Traditional *vs.* Microwave heating.

of the treated flax fibers was found to be substantially improved. Increased dye absorption on the fiber and increased reaction probability between the dye and the fiber were discovered to be the causes of improved dyeability of flax fiber.

The use of microwave irradiation to color polyester fiber was investigated. There was a significant rise in dye absorption as well as a speeding up of the coloration rate. The influence of microwave irradiation on the degree of aqueous sodium hydroxide hydrolysis of PET fiber and its coloration with disperse dyes was investigated. When the effects of microwave irradiation and traditional heating methods were compared, it was discovered that microwave irradiation resulted in a faster rate of hydrolysis. The treated fabric was then dyed in a dyebath heated by microwave irradiation. Increased dye absorption was observed as the weight of the hydrolyzed polyester fabric was reduced.

14.8 CONCLUSION

The innovations are very good and environment suitable, but there are still many hurdles which are to be overcome. The textile industry is a manufacturing industry; therefore, there is very tough competition in garment prices. Still, a lot of optimization is required in the technologies discussed for reducing the cost of production and increasing the commercial viability for meeting the customer demand. As raw-material costs have increased, manufacturers are unable to produce a garment without increasing the prices and failing to meet the paying capacity of the client.

REFERENCES

Abate, M.T., Zhou, Y., Guan, J., Chen, G., Ferri, A. and Nierstrasz, V. 2020. "Colouration and bio-activation of polyester fabric with curcumin in supercritical CO_2: part II–effect of dye concentration on the colour and functional properties." *The Journal of Supercritical Fluids*, 157, 104703. https://doi.org/10.1016/j.supflu.2019.104703

Abdelghaffar, F., Abdelghaffar, R.A., Rashed, U.M. and Ahmed, H.M. 2020. "Highly effective surface modification using plasma technologies toward green coloration of polyester fabrics." *Environmental Science and Pollution Research*, 27, 28949–28961.

Adak, B., Joshi, M. and Ali, S.W. 2020. "Dyeability of polymer nanocomposite fibers." In Joshi, M. (ed.), *Nanotechnology in Textiles*, 145–178. Boca Raton, FL: Jenny Stanford Publishing.

Ahmed, N.S. and El-Shishtawy, R.M. 2010. "The use of new technologies in coloration of textile fibers." *Journal of Materials Science*, 45, 5: 1143–1153.

Anwar, M., Shazia, S., Haq, I.U., Saleem, M., AbdEl-Salam, N.M., Ibrahim, K.A., Mohamed, H.F. and Khan, Y. 2021. "Ultrasonic bioconversion of silver ions into nanoparticles with azadirachta indica extract and coating over plasma-functionalized cotton fabric." *Chemistry Select* 6, 8: 1920–1928.

Fang, K., Wang, C., Zhang, X. and Xu, Y. 2005. Dyeing of cationised cotton using nanoscale pigment dispersions. *Coloration Technology*, 121, 6: 325–328.

Haji, A., Ashraf, S., Nasiriboroumand, M. and Lievens, C. 2020. Environmentally friendly surface treatment of wool fiber with plasma and chitosan for improved coloration with cochineal and safflower natural dyes. *Fibers and Polymers*, 21, 4: 743–750.

Helmy, H.M. 2020. "Extraction approaches of natural dyes for textile coloration." *Journal of Textiles, Coloration and Polymer Science* 17, 2: 65–76.

Hosseinnezhad, M., Gharanjig, K., Jafari, R., Imani, H. and Razani, N. 2021. "Cleaner colorant extraction and environmentally wool dyeing using oak as eco-friendly mordant." *Environmental Science and Pollution Research* 28, 6: 7249–7260.

Hou, A., Xie, K. and Dai, J. 2004. "Effect of supercritical carbon dioxide dyeing conditions on the chemical and morphological changes of poly (ethylene terephthalate) fibers." *Journal of Applied Polymer Science* 92, 3: 2008–2012.

Joshi, M. 2008. "The impact of nanotechnology on polyesters, polyamides and other textiles." In Deopura, B. L., Alagirusamy, R., Joshi, M. & Gupta, B. (eds.), *Polyesters and Polyamides*, 354–415. Cambridge: Woodhead Publishing.

Joshi, M. and Bhattacharyya, A. 2011. "Nanotechnology–a new route to high-performance functional textiles." *Textile Progress* 43, 3: 155–233.

Karthikeyan, A., Kasparek, E., Kietzig, A.-M., Girard-Lauriault, P.-L. and Coulombe, S. 2021. "Synthesis and characterization of MWCNT-covered stainless steel mesh with Janus-type wetting properties." *Nanotechnology* 32, 14: 145719.

Mani, S. and Bharagava, R.N. 2018. "Textile industry wastewater: environmental and health hazards and treatment approaches." In Bharagava, R. N. (ed.), *Recent Advances in Environmental Management*, 47–69. Boca Raton, FL: CRC Press.

McDonald, R. 1997. *Colour physics for industry*. Bradford: Society of Dyers and Colourists

Nie, Q.Q., Zhang, B.S., Li, H.C., Wang, W.J., Han, W. and Si, X.F. 2008. "Research on application for color matching in textile dyeing based on BP neural network adapted by Bayesian regularization algorithm, Qingdao University." *Journal of Qingdao University (Engineering & Technology Edition)* 4: 1–15.

Park, Y.H., Shin, H.C., Lee, Y., Son, Y. and Baik, D.H. 1999. "Electrochemical preparation of polypyrrole copolymer films from PSPMS precursor." *Macromolecules* 32, 14: 4615–4618.

Periyasamy, A.P. and Militky, J. 2020. "Sustainability in textile dyeing: recent developments." In Muthu, S.S. & Gardetti, M. A. (eds.), *Sustainability in the textile and apparel industries: Production Process Sustainability*, pp. 37–79, Switzerland: Springer Nature.

Rajoriya, S., Bargole, S., George, S. and Saharan, V.K. 2018. "Treatment of textile dyeing industry effluent using hydrodynamic cavitation in combination with advanced oxidation reagents." *Journal of Hazardous Materials* 344: 1109–1115.

Razafimahefa, L., Chlebicki, S., Vroman, I. and Devaux, E. 2005. Effect of nanoclay on the dyeing ability of PA6 nanocomposite fibers. *Dyes and Pigments*, 66, 1: 55–60.

Razafimahefa, L., Chlebicki, S., Vroman, I. and Devaux, E. 2008. Effect of nanoclays on the dyeability of polypropylene nanocomposite fibres. *Coloration Technology*, 124, 2: 86–91.

Saleem, H. and Zaidi, S.J. 2020. Sustainable use of nanomaterials in textiles and their environmental impact. *Materials*, 13, 22: 5134.

Sayed, U. and Dabhi, P. 2014. "Finishing of textiles with fluorocarbons." In Williams, J. (ed.), *Waterproof and water repellent textiles and clothing*, 139–152. Cambridge: Woodhead Publishing.

Shahid, M.O.H.A.M.M.A.D., Yusuf, M.O.H.D. and Mohammad, F. 2016. Plant phenolics: a review on modern extraction techniques. *Recent Progress in Medicinal Plants*, 41, 265–287.

Song, Y., Fang, K., Bukhari, M.N., Ren, Y., Zhang, K. and Tang, Z. 2020. Green and efficient inkjet printing of cotton fabrics using reactive dye@ copolymer nanospheres. *ACS Applied Materials & Interfaces*, 12, 40: 45281–45295.

Tarábek, J., Jähne, E., Rapta, P., Ferse, D., Adler, H-J. and Dunsch, L. 2006. "New acetophenone-functionalized thiophene monomer for conducting films on electrodes in chemical ion-sensorics: the synthesis and spectroelectrochemical study." *Russian Journal of Electrochemistry* 42, 11: 1169–1176.

Tripathi, G., Yadav, M.K., Upadhyay, P. and Mishra, S. 2015 "Natural dyes with future aspects in dyeing of Textiles: a research article." *International Journal of PharmTech Research* 8, 1: 96–100.

Vouters, M., Rumeau, P., Tierce, P. and Costes, S. 2004. "Ultrasounds: an industrial solution to optimise costs, environmental requests and quality for textile finishing." *Ultrasonics Sonochemistry* 11, 1: 33–38.

Yusuf, M. 2018. A review on flame retardant textile finishing: current and future trends. *Current Smart Materials*, 3, 2; 99–108.

Yusuf, M. (ed.) 2018. *Handbook of renewable materials for coloration and finishing*. Hoboken, NJ: John Wiley & Sons.

Yusuf, M., Shahid, M., Khan, S.A., Khan, M.I., Islam, S.-U., Mohammad, F. and Khan, M.A. 2013. "Eco-dyeing of wool using aqueous extract of the roots of Indian madder (Rubia cordifolia) as natural dye." *Journal of Natural Fibers* 10, 1: 14–28.

Yusuf, M., Khan, M.A. and Mohammad, F. 2016. "Investigations of the colourimetric and fastness properties of wool dyed with colorants extracted from Indian madder using reflectance spectroscopy." *Optik* 127, 15: 6087–6093.

Yusuf, M., Shabbir, M. and Mohammad, F. 2017. "Natural colorants: Historical, processing and sustainable prospects." *Natural Products and Bioprospecting* 7, 1: 123–145.

Index

A

acid dyes 114
air-dyeing 7, 25–26
airflow dyeing 175–176
anthraquinone dyes 43–44
antimicrobial properties 102–103
antioxidant finishes 105
antistatic finishes 104

B

basic dyes 114
bio-mordants 49–51

C

carbonaceous nanomaterials 123–124
carotenoid dyes 44
cellulosic materials 14–16
chrome mordant dyes 38–39
condensed tannins 48–49
conventional dyeing 172–174
cotton 207–208
cotton materials 14–16

D

digital printing 9
dihydropyran dyes 45
direct dyes 113
direct electrochemical reduction 61–68
discoloration processes 122–127
disperse dyes 114
dye degradation 125–127
dyeing quality 83–84

E

ecological aspects 52–53
electrocatalytic hydrogenation 79–83
electrochemical coloration 237–238
electrochemical dyeing 5, 60–68, 176–179
electrochemical techniques 16–17
enzymes mordant 51–52

F

flame retardant 166–167
flavonoid dyes 44–45

foam dyeing 8–9
functional finishing 225–227

G

gamma radiation 158
gamma ray irradiation technology 159–160

H

hydrolysable tannins 48

I

indigoid dyes 44

M

magnetic nanomaterials 125
mediator-enhanced electrochemical reduction 69
metal-based mordants 47–49
metal complex dyes 39–41
metallic nanomaterials 124–125
microbial colorants 8
microwave 243–244
microwave-assisted dyeing 5–6, 184–185
milestone in plasma research 218

N

nanobubble dyeing 144–148
nanodisperse dye 116, 137–144
nanofiltration 148–151
nanopigments 134–137
nanotechnology 7, 45, 115–116, 234–237
α-naphthoquinone dyes 44
natural dyes 114–115
natural mordants 27–29
nonaqueous dyeing 191–198
nylon materials 21–22

O

oil-mordants 49

P

plasma-assisted dyeing 24–25
plasma dyeing 179–180, 204–206, 240–241

247

plasma processing 221
plasma technology 4, 203–215
polyester 212–214
polyester materials 18–21
polyethylene terephthalate 90
polypropylene 90–91
protein materials 17

R

reactive dyes 112–113
resources of natural dye 42–47

S

silk 211–212
spin-dyeing 4
sulfur dyes 113
supercritical carbon dioxide 180–182, 238–240
supercritical carbon dioxide dyeing 22–24
supercritical fluid dyeing 4–5
sustainable technologies 2
synthetic dyes 112–114

T

tannin 48–49
traditional mordant dyeing practices 38–40

U

ultrasonic dyeing 6–7, 182–184
ultrasonics 241–243
ultrasound-assisted dyeing 26–27
UV-protective finishes 104–105

V

vat dyes 113

W

waterless dyeing 174
water repellency 165–166
wool 208–211

CPSIA information can be obtained
at www.ICGtesting.com
Printed in the USA
BVHW091744190422
634676BV00002B/34